7th International Conference on Key Engineering Materials (ICKEM 2017)

IOP Conference Series: Materials Science and Engineering
Volume 201

Penang, Malaysia
11 – 13 March 2017

ISBN: 978-1-5108-4175-8
ISSN: 1757-8981

Printed from e-media with permission by:

Curran Associates, Inc.
57 Morehouse Lane
Red Hook, NY 12571

Some format issues inherent in the e-media version may also appear in this print version.

Printed by Curran Associates, Inc. (2017)

For permission requests, please contact the Institute of Physics
at the address below.

Institute of Physics
Dirac House, Temple Back
Bristol BS1 6BE UK

Phone: 44 1 17 929 7481
Fax: 44 1 17 920 0979

techtracking@iop.org

Additional copies of this publication are available from:

Curran Associates, Inc.
57 Morehouse Lane
Red Hook, NY 12571 USA
Phone: 845-758-0400
Fax: 845-758-2633
Email: curran@proceedings.com
Web: www.proceedings.com

Table of contents

Volume 201

7th International Conference on Key Engineering Materials (ICKEM 2017)
11–13 March 2017, Penang, Malaysia

Accepted papers received: 10 May 2017
Published online: 2 June 2017

Preface

Design and test of composite structure

Performance optimization and evaluation of building materials

Simulation of mechanical properties of materials

Analysis of chemical properties of materials

Material performance analysis and test

PREFACE

It is our great pleasure to welcome you to The 7th International Conference on Key Engineering Materials which will take place in Penang, Malaysia on March 11-13, 2017. ICKEM2017 is dedicated to issues related to Key Engineering Materials.

The major goal and feature of the conference is to bring academic scientists, engineers, industry researchers together to exchange and share their experiences and research results, and discuss the practical challenges encountered and the solutions adopted. Professors from Britain, France and Japan are invited to deliver keynote speeches regarding latest information in their respective expertise areas. It will be a golden opportunity for the students, researchers and engineers to interact with the experts and specialists to get their advice or consultation on technical matters, sales and marketing strategies.

These proceedings present a selection from papers submitted to the conference from universities, research institutes and industries. All of the papers were subjected to peer-review by conference committee members and international reviewers. The papers selected depended on their quality and their relevancy to the conference. The volume tends to present to the readers the recent advances in the field of key engineering materials and various related areas, such as performance analysis and testing of composites, synthesis and preparation of metal materials, thin film and energy materials, application of building materials, biochemical materials, etc.

We would like to thank all the authors who have contributed to this volume and also to the organizing committee, reviewers, speakers, chairpersons, sponsors and all the conference participants for their support to ICKEM 2017.

Prof. Alexander M. Korsunsky MA DPhil CPhys Minst
Trinity College, Oxford, UK
April 20, 2017

Peer review statement

All papers published in this volume of *IOP Conference Series: Materials Science and Engineering* have been peer reviewed through processes administered by the proceedings Editors. Reviews were conducted by expert referees to the professional and scientific standards expected of a proceedings journal published by IOP Publishing.

Study of stinging nettle (urtica dioica l.) Fibers reinforced green composite materials : a review

I G P Agus Suryawan[1,6,7], N P G Suardana[2], I N Suprapta Winaya[3], I W Budiarsa Suyasa[4], T G Tirta Nindhia[5]

[1,2,3,5]Department of Mechanical Engineering, Udayana University, Kampus Bukit Jimbaran, Bali Indonesia
[4] Department of Chemistry, Udayana University, KampusBukit Jimbaran, Bali Indonesia
[6] Doktoral Student of engineering science, Udayana University, Denpasar, Bali Indonesia.
[7] Email : agus88@unud.ac.id & suryaagus77@gmail.com

Abstract. Stinging Nettle (Urtica dioica L., latin) is a wild plant that grows in Indonesia, Asia, and Europe. Nettle in Bali, Indonesia is called as *Lateng, Jelatang*. Nettle plant has a very strong fiber and high fixed carbon. Nettle plants are covered with fine hairs, especially in the leaves and stems. When it is touched, it will release chemicals, sting and trigger inflammation that causes redness, itching, bumps and irritation to the skin. Nettle plants grow in the wild, regarded as a weed in the agricultural industry, easy to grow and snatch food from the parent plant. The main objective of this paper is to review of the potential nettle fibers and then explain about the potential of local nettle plant in Indonesia. Nettle is a plant group at the end of bast. Its plant fibers taken from the bark, as reinforcement in composite materials. Nettle fibers have three main advantages such as strong, lightweight and low environmental impact.

1. Introduction

Natural fiber reinforced polymer composites is a term used in composites journals as a term in producing developed materials from polymer that is reinforced by natural fiber. Natural fiber has a great impact as a potential substitute for synthetic conventional fiber such as aramid and glass fiber during the last decade. Because of the mechanic characteristics of natural-polymer fiber, namely: good in isolation, low density, non-abrasive, easily obtained from renewable materials, cheap in price and can be recycled, it has attracted the composite industry for automotive application, structure and non-structure. Glass fiber that is difficult to be decomposed triggers serious health and environment problems. They cannot be recycled easily by heat because they melt in very high temperature and still produce residues that may spoil the environment and are relatively abrasive in nature, as what are mentioned in the result of some studies [1-5]. The main focus of this study is to find out the potency of natural fiber of stinging nettle as the replacement of glass fibers as reinforced composite fibers.

1.1. Classification of Natural fibers

Natural fibers from plantation can be divided into six categories, namely: bast, leaf, seed, fruit, grasses/reeds, & wood fibers. Table 1, shows the hierarchy of the fibers variation and their family.

Table 1. Natural Fibers/Plant Fibers[6].

Classification	Name Plant
Bast	Jute, Flax, Isora, Mesta, Kenaf, Ramie, Toina, Totora, Urena, Banana, Roselle, Rattan, Spanish-broom, Nettle
Leaf	Sisal, Henequene, Manila, Curaua, Pineapple, Palm, Yucca, Piassava, Cabuja, Opunita, Screw-pine, Agaves, Abaca
Seed	Catton, Calotropis, Poplar, Kapok
Fruit	Coconut, Luffa, Coir
Grasses/Reed	Bamboo, Bagasse, Wheat, Oat, Pape, Rye, Rice, Esparto, Barley, Corn
Wood	Hard wood, Soft wood

1.2. The Usefulness of Stinging Nettle

Stinging nettle is classified into shrubs with 30–45 species. Stinging nettle grows in fertile soil and it grows until 40 up to 120 cm in height. Stinging nettle as natural biomass, the applications that have been developed are in livestock, medicine, cosmetics and fibers.

Table 2. Potential uses of stinging nettle[7].

Field of aplication	**Use**	**Part of the plant**
Textile/fiber	Tissuses and fabrics, ropes and fishing nets, silky fabric, biocomposites, paper and cloth, paper, natural dye (for yams, eggs, etc.)	Fiber tissues of stems. Root and leaf extracts for dyes
Medicine	Anaemia, gout, rheumatism, eczema, hypoglycaemia,diuretic, hypotension, benign prostatic hyperplasia, arthritis,cardiovascular problems, allergic rhinitis, antiviral, antifungal, and antioxidant, antimicrobial, antiulcer analgesic [8].	Leaves, roots, aqueous, seeds and alcoholic extracts
Cosmetics	Soap, skin lotion and shampoo	
Food	Salad, pier, decocted tea, soups and natural yellow colorant for egg yolk [9].	Leaves, young plant
Forage crop	Cattle, poultry, horses, and pigs for enhancing yolk yellowness	Whole plant
Animal housing	Bedding, lactating dairy cows [10].	Stem, shivers as fiber by-product and seeds
Bioenergy	Biochar	

2. Method

Stinging nettles have been being cultivated by some countries have been investigated in term of mechanic performance in which taking place in France, Tuscani, Netherland and India.

2.1. Material

Stinging nettles which were used in this study were harvested in France, Brittany Region [11]. Those stinging nettles were cultivated in 1-2 years, the study was done in Prato ($43^0$53'N, $11^0$06'E) Tuscany region [12]. Stinging nettles fiber are taken from Brennels BV, Kraggenburg, The Netherlands, fibers from Urtika dioica L. clone B13, cultivated in the Netherlands, harvested in August-September 2007[13].

2.2. Sample preparation

Stinging nettle stems were cut and dried for two days before they were decayed in the water for seven days. Those stinging nettle stemswere dried in room temperature for weeks. The fibers were extracted manually [11]. The stems of cultivated stinging nettle were tested on its ends, middle, and [12]. Fibers extraction were done in 4-6 weeks, done mechanically in the same July 2008 for decortications rami and flax [13].

Several works have been done by many researchers on stinging nettles composite study as shown in table 2 for potential uses of stinging nettles, table 3 for the tensile strength of the composite reinforcement fibers, table 4 for the diameter, length, tensile strength and elongation of fiber according to the position on the stem nettle, table 5 for the chemical composition, morphology, and mechanical properties of fiber extraction results, table 6 for influence of different fiber of the loads on the mechanical characteristics of compression molded on PLA (poly lactic acid) composites without adhesion promocers.

3. Result and Discusion of Mechanical Properties

Research results Bodros, as shown in Tabel 3, stinging nettle has the highest ultimite stress among the nature fibers that is 1594 (± 640) MPa. It means that if the stinging nettle fiber is used as composite material it will produce composite with a great strength.

Table 3. Ultimite stress of some composites [11]

Name	Young's modulus (Gpa)	Ultimite stress (MPa)	Strain to failure (%)	Density (g/cm^3)	Average diameter (μm)
Stinging nettle	59-115	2274-914	2.92-1.3	0.72	24.3-15.5
Flax ariane	73-43	1825-853	3.31-3.23	1.53	23.6-12.0
Flax Agatha	96-46	1800-962	2.9-1.3	1.53	15.6-14.4
Hemp	21.6-16.6	310-230	0.9-0.7	1.48	36.1-26.3
Ramie	24.5	560	2.5	1.51	34

Specific weight of stinging nettle that is 0,72 gram/cm^3 categorized as light fiber [12]. With light average specific weight, if stinging nettle is used as composite reinforcement there is a potency that it will produce a light and strong material.

Table 4. Diameter, length, tensile strength and elongation fiberbased on the position of stinging nettle stem [12]

		Bottom	**Middle**	**Top**
Diameter (μm)	Mean	47	32	19
	Range	31-63	21-42	10-26
Length (mm)	Mean	43	50	58
	Range	27-60	39-63	40-73
Tensile Strength (cN tex^{-1})	Mean	24	62.1	58.7
	Range	12-40	38-98	24-98
Elongation (%)	Mean	2.6	2.3	2.5
	Range	1.5-3	1.3-3.5	1-6

Stinging nettle fiber that is taken from the middle of the stem has better tensile strength and elongation as shown in Table 4. Most of natural fibers contain lignin-cellulose, but they also contain other components such as hemicellulose, pectin, hardwood, ash, silica, oil, wax & other water solutions. So many things to learn in order to understand the individual concentration of each component if natural fiber composite is produced. Cellulose is semi-crystalline polysaccharide, while hemicellulose is a highly branched amorphous polymer. In order to produce adhesion strength between fiber and matric minimum ash and wax elements are needed, wax reduce adhesion. Hydroxyl group from cellulose within the natural fiber describes natural hydrophilic, that reduces the bond between faces and makes composite absorbs water easily.

Table 5. Chemical composition, morphology, and mechanic characteristics of extracted fiber [13]

Treatment	CR	D	WR	D+WR	WR+D	MR	D+MR	ET	ET+CA
Cellulose (%)	81	65	78	83	85	78-84	75-85	80-82	81-83
Hemicellulose (%)	6	5	9	13	6	9-10	5-7	11-12	11-12
Lignin (%)	2	3	3	2	4	2-5	3-4	2-3	2-3
Diameter (μm)	23-37	23-47	37-41	40-46	29-43	24-31	16-40	30-40	25-35
Length (mm)	38-62	25-58	41-49	38-58	35-55	41-55	33-60	42-52	42-51
Tensile strength (cN tex^{-1})	38-81	70-182	8-94	41-83	23-71	33-65	7-98	21-72	32-76
Elongation (%)	4-7	2-3	2-4	1-3	1-2	2-4	0-2	3-6	3-6

Methods which were used to take the fiber include: chemical retting (CR), decortication (D), water retting (WR), microbiological retting (MR), enzymatical treatment (ET), chelating agent (CA).

Table 6. Influence of different fiber loads on the mechanical characteristics of compression on moulded PLA (Poly lactic acid) composites without adhesion promocers [14].

Fiber	Fiber load in wt-%	Note	Tensile strength in MPa	Tensile modulus in GPa	Elongation at break	Flexural modulus in GPa	Impact strength in kJ/m^2
Nettle	20	Press pressure 5.6 MPa, maintained for 20 min at 175 ^{0}C, fibres were oriented predominantly in length direction	45	4.8	1.2	4.2	14
	30		59	5.6	1.5	5.2	11
	40		40	4.8	1.3	4.6	6
Hemp	34	Fiber length 5-15 mm; random fiber orientation	41.1	5.65			
	44		44.6	7.4			
	55		43.7	7.0			
Flax	30	Enzyme retted fibers; random fiber orientation	53	8.3	1.0		
	40		40	7.3	0.9		
Jute	34	Water cleaned fibre. Fibre length 5-10 mm random fibre orientation	39.5				18.5
	44		42.0				23.5
	55		43.0				32.0

Research on composite tensile strength was done by combining PLA with some natural fibers (nettle, hemp, flax and jute). It is shown that the highest tensile strength of nettle is 59 MPa, with weight fraction 30%.

Natural fiber that was given chemicals to erase lignin and enrich the adhesion strength between fiber and matric can be seen in literature [17-21]. Meanwhile the textbook that explains about composite materials can be seen in literature [22-24]. The development of studies and cultivation of nettle in some countries are discussed in literature [25-27].

Figure 1. Stinging nettle plants

Figure 2. Stems of nettle

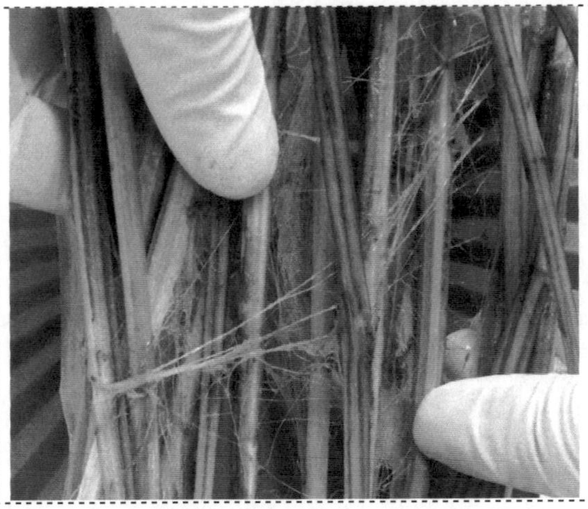

Figure 3. Stems that have been marinated in the water and the fibers are out

Figure 4. Nettle stinging fibers

IOP Conf. Series: Materials Science and Engineering **201** (2017) 012001 doi:10.1088/1757-899X/201/1/012001

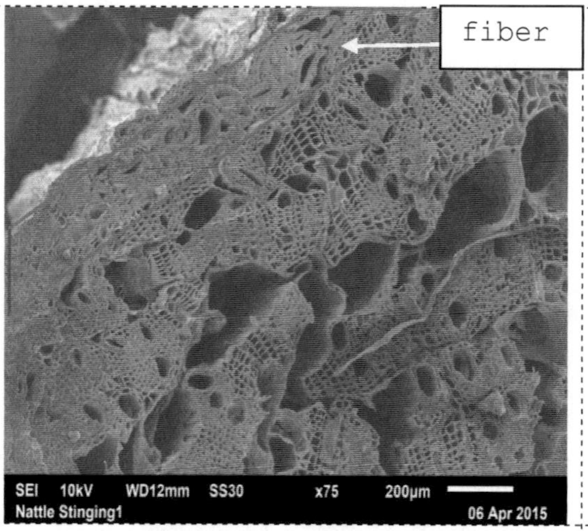

Figure 5. SEM stems crosswise

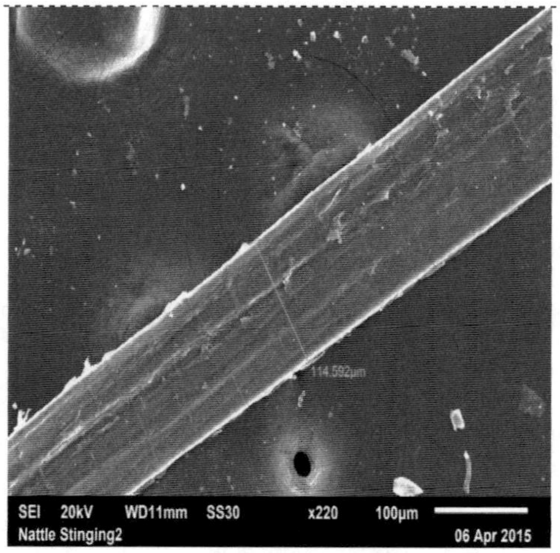

Figure 6. SEM stinging nettle fiber

Figure 1 until 6 is nettle plants which are available in Indonesia that have been SEM tested on its stems and fiber, we will apply chemical treatment with local nettle to increase the strength and toughness of the composite materials.

4. Conclusion

The potency of stinging nettle to be used as reinforcement of composite materials is so great, it can be seen from the result of the study that was conducted by the researcher. The fibers were taken from the stems of the nettle. Some treatments with chemicals on nettle fiber are needed to be done, form example on flax, hemp and ramie, and to reinforce the fiber and the bound between fiber and matric.

References

[1] Romanzinia D, Junior H L O, Amico S C and Zattera A J 2012Preparation and Characterization of Ramie-Glass Fiber Reinforced Polymer Matrix Hybrid Composites*Materials Research* **15**415–20

[2] Jie Z, Hua Z and Jianchun Z 2014Effect of Alkali Treatment on the Quality of Hemp Fiber *Engineered Fibers and Fabrics***9** Issue 2 pp 19-24

[3] Christian S J and Billington S L 2011 Mechanical response of PHB- and cellulose acetate natural fiber-reinforced composites for construction applications*Composites: Part B***42** 1920–28

[4] Ashrafi M, Vaziri A and Nayeb-Hashemi H 2011 Effect of processing variables and fiber reinforcement on the mechanical properties of wood plastic composites*Reinforced Plastics and Composites***30** 1939–45

[5] Goda K, Sreekala M S, Gomes A, Kaji T and Ohgi J 2006Improvement of plant based natural fibers for toughening green composites-Effect of load application during mercerization of ramie fibers*Composites: Part A***37** 2213–20

[6] Mohanty A K, Misra M and Drzal L T 2005 *Natural Fibers Biopolymers and Biocomposites* Taylor & Francis United States of America

[7] Virgilio N D 2015The potential of stinging nettle (Urtica dioica L.) as a crop withmultiple uses*Industrial Crops and Products***68** 42-49

[8] Gülçin I, Küfrevioglu O I, Oktay M and Büyükokuroglu M E 2004 Antioxidant, antimicrobial, antiulcer and analgesic activities of nettle (Urtica dioica L.) *Ethnopharmacology***90**205–15

[9] Loetscher Y, Kreuzer M andMessikommer R E 2013 Utility of nettle (Urtica dioica) in layer diets as a natural yellow colorant for egg yolk *Animal Feed Science and Technology***186** pp 158– 68

[10] Humphries D J and Reynold C K 2014The effect of adding stinging nettle (Urtica dioica) haylage to a total mixed ration on performance and rumen function of lactating dairy cow*s* *Animal Feed Science and Technology***189** 72–81

[11] Bodros E & Baley C 2008Study of the tensile properties of stinging nettle fibres (Urtica dioica) *Materials Letters***62** 2143–45

[12] Bacci L, Baronti S, Predieri S and Virgilio N D 2009Fiber yield and quality of fiber nettle (Urtica dioica L.) cultivated in Italy *Industrial Crops and Products***29** 480–84

[13] Bacci L, Lonardo S D, Albanese L,Mastromei G and Perito B 2010 Effect of different extraction methods onfiber quality of nettle (Urtica dioica L.) *Textile Research Journal***81** 827–37

[14] Fischer H, Werwein E, and Graupner N 2012 Nettle fibre (Urtica dioica L.) reinforced poly(lactic acid): A first approach *Composite Materials***46** 3077–87

[15] Akgul M 2013 Suitability of stinging nettle (Urtica dioica L.) stalks for medium densityfiberboards production *Composites: Part B***45** 925–29

[16] Pauksztа D, Mankowski J, Kołodziej J and Szostak M 2013 Polypropylene (PP) Composites Reinforced with Stinging Nettle (Utrica dioica L.) Fiber *Journal of Natural Fibers***10** 147–158

[17] Suardana N P G, Min-Seuck-Ku and Jae-Kyoo-Lim 2011 Effects of Diammonium Phosphate on The Flammability and Mechanical Properties of Bio-Composites *Materials and Design***32** 1990-99

[18] Sydenstricker T H D, Mochnaz S andAmico S C 2003 Pull-out and other evaluations in sisal-reinforced polyesterbiocomposites *Polymer Testing***22**375–380

[19] Aziz S H andAnsell M P 2004 The effect of alkalization and fibre alignment on the mechanical and thermal properties of kenaf and hemp bast fibre composites: Part 1 – polyester resin matrix *Composites Science and Technology***64**1219–30

[20] Sgriccia N, Hawley M C andMisra M 2008 Characterization of natural fiber surfaces and natural fiber composites *Composites: Part A* 39 1632–37

[21] Mohan T P and Kanny K 2012 Chemical treatment of sisal fiber using alkali and clay method *Composites: Part A***43** 1989–98

[22] Daniel I M & Ishai O 1994*Engineering Mechanics of Composite Materials*, Oxford University Press, New York.

[23] Chawla K K1987*Composite Materials Science and Engineering*, Springer Verlag, New York.

[24] Cheremisinaf-Mcholas P 2010 *Handbook of Ceramics and Composite* Vol II Mechanical p ropersties and Specialey, Marcel Dekker p

[25] Klimesova J 1995 *Population dynamics of Phalaris arundinacea L. and Urtica dioica L. in a floodplain during a dry period* (*Wetlands Ecology and Management* vol 3) no 2 pp 79-85

[26] Boufford D E 1992 *Urticaceae Nettle Family* (*The Arizona-Nevada Academy of Science* vol 26) issue 1 pp 42-49

[27] Damme E J M V, Broekaert W F and Peumans W J 1988*The Urtica dioica Agglutinin Is a Complex Mixture of Isolectins(Plant Physiol* vol 86) pp598 – 601

Acknowledgments

The authors thank the Ministry of Research, Tech., and Higher Education of the Republic of Indonesia and LPPM (Lembaga Penelitian dan Pengabdian Masyarakat) University of Udayana for supporting this research and paper through The Grant.

IOP Conf. Series: Materials Science and Engineering **201** (2017) 012002 doi:10.1088/1757-899X/201/1/012002

A study on biocomposite from local balinese areca catechu l. husk fibers as reinforced material

C I P K Kencanawati[1,2,4], N P G Suardana[2], I K G Sugita[2] and I W B Suyasa[3]

[1]Doctoral Student of Engineering Science of Udayana University, Bali-Indonesia 80632
[2]Mechanical Engineering of Udayana University, Bali - Indonesia 80632
[3]Chemistry Department of Udayana University, Bali – Indonesia 80632

[4]cok_putrikusuma@yahoo.com

Abstract. Untapped areca catechu l. husk fibers optimally can cause pollution to the environment. Therefore it is necessary to learn the characteristics of local balinese areca catechu l. husk fibers, such as physical, chemical, morphological, and mechanical. AHF testing the tensile strength with a single pull fiber test in accordance with ASTM D 3379 in the amount of 146-152 MPa. While the observation of the physical properties, of local balinese areca catechu l. husk fibers have a diameter and length variations of each 250-540 μm and 9.24 to 55.20 mm, with an aspect ratio of between 31.43 to 102.22, density ranges between 0:48 - 0.74 kg / cm^3, absorption lower water (90-150%) when compared to AHF grows in other areas. From this study it appears that local Bali AHF can be used as reinforcement in composite replacement for synthetic fibers.

1. Introduction

Areca fibers can be used as a composite reinforcement [1], which contains lignocellulosic plant tissue structures that are environment friendly and biodegradable [2], support karbondiokasida cycle in the atmosphere [3]. Applications areca as reinforcement fibers biocomposites has not been widely used. Natural fibers have been used are kenaf, sisal, hemp, jute and others [4] in the automotive, packaging [5], construction, interior and furniture [6]. Aplication of natural fibers as reinforcement composites have several advantages: lower density, renewable source, low cost and low investment, reduce wear gear, healthy working conditions, good acoustic and thermal insulation [7]. But natural fibers have some disadvantages low wetability, incompatibility with some hydrophobic polymer matrix, a low melting point and high water absorption [2]. However, this weakness can be overcome with some chemical treatment [8]. Areca trees bear fruit depends on soil conditions [9], the average in bali areca trees bear fruit after 5-8 years old, elliptic fruit shape between 3.5 - 7 cm. Stringy fibrous fruit wall hard covers endosperm [10] ,husk weight approximately 60-80% of the total weight. Fibers length 3.5 - 6 cm, diameter 0.3 mm-0.5 mm. In general, the chemical composition of areca fibers composed of cellulose (hemicellulose), lignin, pectin and protopectin [11]. Areca tree plantations developed on a large scale to take its seeds as raw material for cosmetics, healthcare, and industrial paints [12, 13]. AHF are finding use in fabrication of value added products such as cushion, hard boards, non-woven fabrics and housing insulation materials [14]. Areca based on the level of maturity there are three types of raw areca nut, ripe and dried [14], so that the waste from the skin also there are 3 types of leather waste of raw areca nut, ripe and dried. Raw areca green, soft leather and hard-shelled seeds, areca ripe golden yellow and the skin is chewy to contain a lot of liquid when compared to the raw and dried fruit. The dried areca obtained when it reaches full maturity and apart from fruit bunches. Dried areca show a brownish color with coarse fibers. Fibrous part of areca there are two types of fibers is thin and coarse fibers. Crude fibers surface roughness of AHF can be taken into consideration in choosing a good fibers as reinforcement in composites. This study investigates the physical, chemical, morphological, mechanical and thermal characteristics of areca leather fibers (Areca Husk Fibers / AHF). AHF is extracted from the fibrous bark near the shell of the fruit.

Some studies related to the development of AHF as a composite reinforcement has been done such as short fibers characterization of areca [10], AHF research on the tensile strength of polypropylene composites [2], studies of bending strength and impact AHF [15], AHF performance in epoxy composite [16] and others. Although it has been studied about the AHF for use as reinforcement in composite materials, AHF research has not found that the agricultural commodities local Bali. Several different fundamental between local balinese areca with others is elliptic shape and fruit size is smaller when compared with other areca, in addition to the soil and climate conditions will also affect the characteristics of plants including fruit and fibers skin [17].

2. Research Methodology

AHF obtained from local areca plantations in Bongaya village, Karangasem, Bali. Areca immersed in water at room temperature for 5 days to loosen the fibers from the husk. Furthermore, AHF is separated manually from the shell by hand. After stripping is washed with distilled water before being dried in an oven at 70 ° C for 24 hours. AHF has done is stored in an enclosed space to protect the fibers from atmospheric moisture. Fifteen samples of AHF is measured and the average fibers diameter and length was recorded. Density AHF obtained using a mathematical equation that is dividing the mass by volume.Water absorption is determined after soaking the samples in water at room temperature for 24 hours. Study on morphology of the AHF surface and cross section using a scanning electron microscopy (SEM). AHF chemical composition testing such as cellulose, hemicellulose, lignin and wax were determined using standard test procedures [18]. AHF for tensile test specimens were randomly selected from fibers of the skin. Fibers were observed using an optical microscope to measure the diameter of the fibers. Fibers affixed to a piece of cardboard, with a gauge length of 25 mm and tested according to standard test method for single fiber tensile test (ASTM D 3379-75). Cardboard cut in the mid-section before measuring the tensile test [9]. Specimens were tested with a crosshead speed of 1 mm / min with the replication of 15 specimens, with a load capacity of 5 kN.

3. Results and discussion

Physical properties of AHF

AHF there are 2 types of coarse fibers and fine fibers, so that the fibers diameter varies from 250-540 μm as well as fibers length 11.24 - 55.20 mm more varied dimensions when compared AHF investigated by Binoj and Yusriah. Fibers length and diameter of the fibers is an important parameter for determining the strength of natural fibers [19]. By dividing the length of the fibers diameter, the aspect ratio of natural fibers can be calculated, asek ratio AHF 31.43 - 102.22. Natural Fibers with a high aspect ratio is more favorable because it has a high tensile strength [4]. In addition, the aspect ratio of the fibers have a significant effect on the flexible nature and behavior of natural fibers rupture [20]. Local balinese areca fibers density ranges between 0:48 - 0.74 kg / cm3. water-absorption of fibers is the fiber's ability to absorb water vapor in a given time. This greatly affects the properties of the composite characteristics that use natural fibers as reinforcement, particularly in the use of composites in the outdoors and life time composites. Water absorption can decrease the bond interface between the fibers with the polymer so that it will affect the mechanical properties. Water absorption of local balinese AHF range 90-150%, lower than AHF investigated by Yusriah and Hasan. Moisture content is one of important parameters that should be Considered in Choosing natural fibers as reinforcement in polymer composites. Moisture content of natural fibers determines the dimensional stability, electrical resistivity, tensile strength, porosity and swelling behavior of natural fibers reinforced composites [21]. Composites reinforced with high moisture content of fibers are often prone to decay as compared to the composites reinforced with low moisture content fibers [40]. This is likely due to the ability of the fibers to hold up water molecules within the composites, which act as degradation agent in the composites. SEM of AHF figure 2. The size of lumen in natural fibers is proportional to the diameter of their respective fibers, where the lumen size increases with the increase in fibers diameter [22]. cross section of a fibers showing the arrangement of individual cells. The bundle is slightly flattened from being cut.

Table 1. Physical properties AHF

Materials	Diameter (μm)	Length (mm)	Aspec ratio	Density (gr/cm^3)	Water absorption (%)	Moisture (%)	Color fibers
Local balinese AHF	578 ± 9	36.2 ± 12.6	63.3 ± 22.9	0.63 ± 0.13	120.7 ± 30.0	10.4 ± 0.2	Golden brown
AHF [9]	396 - 476	10 - 60	-	0.7 – 0.8	-	7.32	-
AHF [14, 23]	401 – 470	49 – 57	-	0.19 – 0.38	116 - 231	3.7 - 60	-

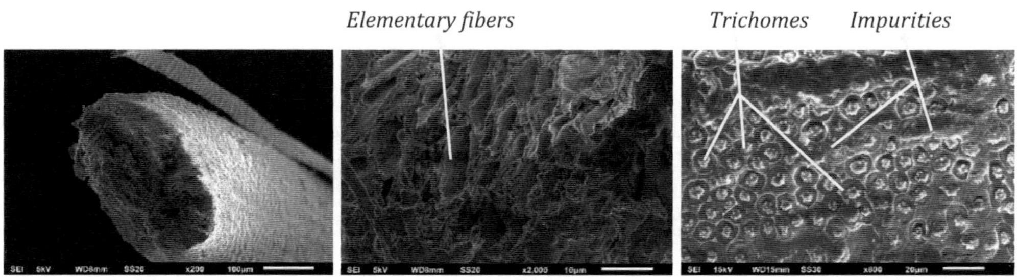

(a) (b) (c)

Figure 1. (a) SEM of AHF (b) cross section of AHF (c) surface of AHF

The chemical composition of the fiber can be different from the same species. The chemical composition of the fiber affects the mechanical strength, moisture, morphology and characteristics of the bond, so it will take effect when the fiber is used as a reinforcement in composite [9]. local balinese AHF chemical composition when compared with other regions, local balinese AHF sellulose content of 52.34 - 55.89%. Lower hemicellulose content is desirable, Because content will degrade its higher thermal stability and increases the moisture absorption capacity [24]. Lignin improves the rigidity and stiffness of the fiber by bonding together the cell wall structures. It also helps in bonding with polymers when used as a reinforcement in polymer matrix composites [25]. lignin is the main element of areca, which gives rigidity to the AHF. Lignin is also responsible for the color change of the fiber [4].

Table 2. Chemical properties AHF

Materials	Cellulose (%)	Hemicellulose (%)	Lignin (%)	Ash (%)	Wax(%)	Other
Local balinese AHF	54.15 ± 1.78	16.61 ± 1.58	21.02 ± 1.36	2.02 ± 0.42	0.77 ± 0.21	4.44 ± 1.49
AHF [9]	57.35 – 58.21	13 – 15.42	23.17 – 24.16	-	0.12	-
AHF [23]	53.20	32.98	7.20	1.05	0.64	3.12

Table 2 shows the tensile strength, Young's modulus, and elongation at break of AHF study ever conducted. There are many factors affecting the tensile properties of natural fibers such as species and varieties of plants, the level of maturity of plants, soil and weathering conditions and also the efficiency of the retting process [22]. Tensile strength of local Balinese AHF quite high when compared with other regions and the lowest elengationnya 4.9 - 10.9%.

Table 3. Tensile strength and young's modulus AHF

Mechanical Properties	Local balinese AHF	AHF [9]	AHF [14, 19]
Tensile strength (MPa)	148.32 ± 2.84	147 – 322	123.93 – 166.03
Break elongation (%)	7.15 ± 3.29	10.23 – 13.15	22.56 – 23.76
Young's modulus (GPa)	2.33 ± 0.83	1.12 – 3.16	1.26 – 2.57

Conclusion

The results of this study shows that the local balinese AHF can be used as a reinforcement in composite replacement for synthetic fibers. With varying aspect ratio and a lower water absorption thus allowing used according to the application and either used outdoors as well as a longer shelf life. In addition local balinese AHF has a lignin content is high enough so that the strength of the fiber and composite will increase.

References

[1] E. Jayamani, P. Ezhumalai, S. Hamdan, and M. R. Rahman 2014 Investigation on Sound Absorption Coefficients of Betel Nut Fiber Reinforced Polymer Matrix Composites presented at Applied Mechanics and Materials.

[2] E. Jayamani, S. Hamdan, S. K. Heng, and M. Rahman 2014 Sound Absorption Property of Agricultural Lignocellulsic Residue Fiber Reinforced Polymer Matrix Composites presented at Applied Mechanics and Materials.

[3] W. L. Lai, M. Mariatti, and M. Jani S 2008 The properties of woven kenaf and betel palm (Areca catechu) reinforced unsaturated polyester composites *Polymer-Plastics Technology and Engineering*, **47**, 1193-99.

[4] S. Bavan and G. C. M. Kumar 2010 Potential use of natural fiber composite materials in India *Journal Of Reinforced Plastics And Composites*, **29**, 3600-13.

[5] H. S. Tan, Y. Z. Yu, L. L. Liu, and L. X. Xing 2010 Effect of alkali treatment of coir fiber on its morphology and performance of the fiber/LLDPE bio-composites presented at Advanced materials research.

[6] E. Jayamani, S. Hamdan, M. R. Rahman, and M. K. B. Bakri 2015 Dielectric Properties of Lignocellulosic Fibers Reinforced Polymer Composites: Effect of Fiber Loading and Alkaline Treatment *Materials Today: Proceedings* **2** 2757-66.

[7] B. F. Yousif, N. S. Gill, S. T. W. Lau, and A. Devadas 2008 The potential of using betelnut fibres for tribo-polyester composites considering three different orientations presented at ASME 2008 International Mechanical Engineering Congress and Exposition.

[8] M. S. Salit 2014 Tropical Natural Fibre Composites: Properties, *Manufacture and Applications: Springer*.

[9] J. S. Binoj, R. E. Raj, V. S. Sreenivasan, and G. R. Thusnavis 2016 Morphological, Physical, Mechanical, Chemical and Thermal Characterization of Sustainable Indian Areca Fruit Husk Fibers (Areca Catechu L.) as Potential Alternate for Hazardous Synthetic Fibers *Journal of Bionic Engineering* **13**, 156-165.

[10] G. C. M. Kumar 2008 A study of short areca fiber reinforced PF composites Proceedings of the World Congress on Engineering **2**.

[11] R. P. Swamy, G. C. M. Kumar, Y. Vrushabhendrappa, and V. Joseph 2004 Study of areca-reinforced phenol formaldehyde composites *Journal Of Reinforced Plastics And Composites*

23 1373-82.

[12] D. Sampathkumar, R. Punyamurthy, B. Bennehalli, R. P. Ranganagowda, and S. C. Venkateshappa 2014 Natural Areca Fiber: Surface Modification And Spectral Studies J*ournal of Advances in Chemistry* **10**.

[13] C. V. Srinivasa, A. Arifulla, N. Goutham, T. Santhosh, H. J. Jaeethendra, R. B. Ravikumar, S. G. Anil, D. G. S. Kumar, and J. Ashish 2011 Static bending and impact behaviour of areca fibers composites *Materials & Design* **32** 2469-75.

[14] L. Yusriah, S. M. Sapuan, E. S. Zainudin, and M. Mariatti 2012 Exploring the potential of betel nut husk fiber as reinforcement in polymer composites: Effect of fiber maturity *Procedia Chemistry* **4** 87-94.

[15] C. V. Srinivasa and K. N. Bharath 2011 Impact and hardness properties of areca fiber-epoxy reinforced composites *J Mater Environ Sci* **2** 351-56.

[16] G. R. Reddy, A. M. Kumar, and K. V. P. Chakradhar 2011 Fabrication and Performance of Hybrid Betel Nut (Areca Catechu) Short Fiber/Sansevieria Cylindrica(Agaveceae) Epoxy Composites *International journal of materials and biomaterials applications* 1 6-13.

[17] J. E. Russell 2002 Soil conditions and plants growth: Daya Books.

[18] O. Faruk, A. K. Bledzki, H.-P. Fink, and M. Sain 2012 Biocomposites reinforced with natural fibers: 2000-2010 *Progress in polymer scienc*e **37** 1552-96.

[19] L. Yusriah, S. M. Sapuan, E. S. Zainudin, and M. Mariatti 2014 Characterization of physical, mechanical, thermal and morphological properties of agro-waste betel nut (Areca catechu) husk fibre *Journal of Cleaner Production* **72** 174-80.

[20] D. Puglia, J. Biagiotti, and J. M. Kenny 2005 A review on natural fibre-based composites Part II: Application of natural reinforcements in composite materials for automotive industry J*ournal of Natural Fibers* **1** 23-65.

[21] K. Sukumaran, K. G. Satyanarayana, S. G. K. Pillai, and K. K. Ravikumar 2001 Structure, physical and mechanical properties of plant fibers of Kerala *Metals Materials and Processes* **13** 121-36.

[22] C. Baley 2002 Analysis of the flax fibres tensile behaviour and analysis of the tensile stiffness increase C*omposites Part A: Applied science and manufacturing* **33** 939-48.

[23] M. M. Hassan, M. H. Wagner, H. U. Zaman, and M. A. Khan 2010 Physico-mechanical performance of hybrid betel nut (Areca catechu) short fiber/seaweed polypropylene composite *Journal of Natural Fibers* **7** 165-177.

[24] Z. N. Azwa, B. F. Yousif, A. C. Manalo, and W. Karunasena 2011 A review on the degradability of polymeric composites based on natural fibres *Materials & Design* **47** 424-42.

[25] M. Jawaid and H. P. S. A. Khalil 2011 Cellulosic/synthetic fibre reinforced polymer hybrid composites: A review *Carbohydrate polymers* **86** 1-18.

Mechanical properties and water sorption of geothermal silica/thermoset composites

D E O Santiago[1, 2] and B B Pajarito[1]

[1]Polymer Research Laboratory, Department of Chemical Engineering, University of the Philippines, Diliman, Quezon City 1101 Philippines
[2]Department of Chemical Engineering, University of the Philippines, Los Baños, College, Laguna 4031 Philippines

deniseester_santiago@yahoo.com, bryan.pajarito@gmail.com

Abstract. The effect of thermoset resin type and geothermal silica loading on the mechanical properties and water sorption of geothermal silica/thermoset composites are determined. Thermoset resin type has significant effect on the flexural strength, hardness, water uptake and weight loss of the composites. Composites based from isophthalic polyester resin has higher flexural properties and hardness and lower water uptake and weight loss over other types. Geothermal silica loading has significant effect on the hardness and water uptake of composites. The highest hardness and water uptake were observed at 5%w/w geothermal silica loading.

1. Introduction

Thermoset polymer resins are used in structural applications such as aircraft, sporting goods, marine and infrastructure [1]. To achieve the needed strength in structures, thermosets are reinforced with fillers such as glass flakes, bentonite, ceramic whiskers or silica [2-11]. Amorphous silica is commonly used as filler to thermoset polymers like epoxy and polyester because of the improvement in mechanical properties and barrier characteristics of thermosets [2, 4-6, 8-10]. The effect of silica on the mechanical properties of polymer composites is dependent on particle size and shape, aspect ratio, modulus and strength of silica, adhesion to resin, loading, viscosity, and dispersion of silica [4-6]. In general, the mechanical properties of the thermosets increase as the filler loading increases [4, 5]. However, agglomeration of silica could decrease the mechanical properties due to the presence of voids between the filler and the matrix [5]. Aside from mechanical behavior, water sorption of thermosets can also be affected by silica due to its hydrophilic nature [8].

Some studies which tested silica as filler used synthetically produced silica [4-6, 8-10]. An alternative source of silica that can be used as filler to thermosets is silica harvested from geothermal brine (geosil). Harvesting of geosil can reduce the problem of scaling in geothermal power plants [12]. Conversion of geosil into a useful product can minimize the blockage of pipes which can reduce the maintenance and operating costs, improve the energy utilization of geothermal resources and provide additional income to geothermal plants [12].

In this study, the harvested geosil from geothermal brine was used as reinforcement to isophthalic polyester (iso-polyester), epoxy, and orthophthalic polyester (ortho-polyester). The effect of type of thermoset resin and geosil loading on flexural properties, hardness, and water sorption of geosil/thermoset composites was determined.

2. Experimental

Thermoset resins used were iso-polyester (R31-441, RI Chemical Corporation), epoxy (Epolam 2017) and ortho-polyester (A-7110, Pacific Resin Inc.). Geosil fillers were precipitated by adding 10 ml sodium hydroxide (7.5%w/w NaOH) and 10 ml calcium chloride (3000 ppm $CaCl_2$) to 1 liter geothermal brine (Energy Development Corporation, Philippines) at 28°C for 1 h [13]. Specified loading of dried geosil fillers (0, 1, 3, and 5%w/w) was mixed with thermoset resin using an overhead stirrer (IKA RW20 digital) at 252 rpm for 5 min. Specific hardeners (1.2 g of methyl ethyl ketone peroxide and 0.1g of 6% cobalt octoate per 60 g of polyester; 15 g of polyoxyalkyleneamine, 3-aminomethyl-3,5,5-trimethylcyclohexylamine, 4-tert-butylphenol, and 2,2'-iminodiethylamine mixture per 50 g of epoxy) were added and mixed for 3 min. The mixture was then degassed in an ultrasonic bath (Wisd WiseBath WB-11) for 3 min and in a vacuum desiccator for 2 min to eliminate trapped bubbles. The degassed mixture was poured into metal moulds (177 × 135 × 2 mm) pre-coated with mold release agent (Durawax) and cured in a thermal press (Chemplast, S.V. Industries). Epoxy composites were cured at 80°C for 90 min and 130°C for 30 min [9], while polyester composites were cured at 50°C, 80°C, and 130°C for 60 min at each temperature [7].

Five replicates of thermoset specimens were cut into 60 × 22.5 × 2 mm dimensions and conditioned at 50°C for at least 100 h in a laboratory oven (F53 Binder GmbH). The hardness and flexural properties of specimen were measured using a Shore D durometer (Kori Seiki) and a universal testing machine (UTM Shimadzu AGS-5kNX), respectively. The cross section of specimens was characterized by scanning electron microscopy (SEM) and energy dispersive x-ray spectroscopy (EDS JEOL JSM-6510LA).

Water sorption test was done by immersing the composites in deionized water at 80°C and were monitored at 24 h interval until the weight of composites became constant. The water uptake was calculated using equation (1) [3, 7, 14]. The wet composites were dried in an oven (F53 Binder GmbH) until weight became constant. The weight loss was calculated using equation (2) [15]. Equations (1) and (2) are:

$$\text{water uptake}\,(\%) = 100 \times \frac{(W_t - W_i)}{W_i} \tag{1}$$

$$\text{weight loss}\,(\%) = 100 \times \frac{(W_i - W_{td})}{W_i} \tag{2}$$

where W_t is the weight of immersed composite at time t, W_i is the initial weight of the composite, and W_{td} is the weight of composite after drying. An optical microscope (Euromex Novex B-Plus) was used to monitor the physical changes on the surface of immersed composites.

The main factor effects were determined by computing the mean responses based on specified factor levels [16]. Statistical significance of computed main factor effects at 95% confidence level ($p < 0.05$) was determined by analysis of variance (ANOVA) using Minitab 17.

3. Results and Discussion

Figure 1 shows the SEM image of geothermal silica/ortho-polyester composite at 5%w/w geosil. Similar to the other thermoset composites, the geosil fillers agglomerated and settled at the bottom of the composite as shown. Figure 2 shows the result of EDS mapping of silicon on the cross-section of geothermal silica/ ortho-polyester composite with 5 % w/w geosil. Same with the other thermoset composites, higher concentration of silicon was observed at the bottom of the composite which confirms agglomeration and settling of the agglomerated geosil.

Figure 3a shows that decrease in flexural modulus of geosil/iso-polyester composites is observed when geosil loading is increased. In the case of geosil/epoxy and geosil/ortho-polyester composites, increase in flexural modulus is observed at low geosil loading but started to decrease at higher loading. The decrease in flexural modulus is attributed to the formation of interphase layer with higher mobility and lower modulus than the thermoset matrix, while the increase in flexural modulus could be due to the filler agglomeration that increases the constraint between thermoset chains [10, 17]. Figure 3b shows that decrease in flexural strength is observed for all geosil/thermoset composites as geosil loading is increased. The decrease in flexural strength is caused by agglomeration of geosil fillers at the surface which is an easy point of cracking during mechanical testing [5, 10].

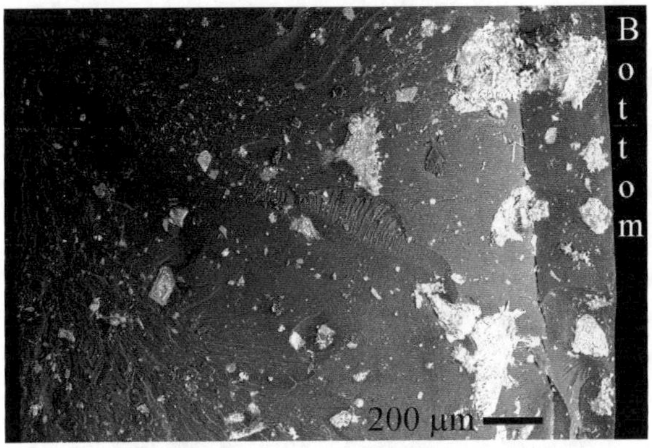

Figure 1. SEM image of geothermal silica/ortho-polyester composite 5%w/w geosil.

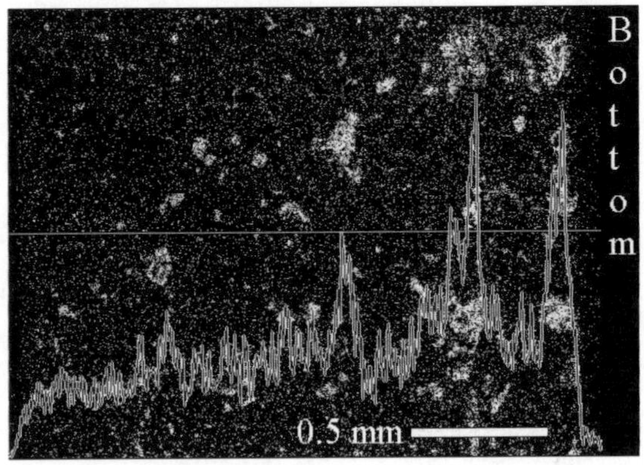

Figure 2. EDS mapping of silicon on the cross-section of geothermal silica/ ortho-polyester composite 5%w/w geosil.

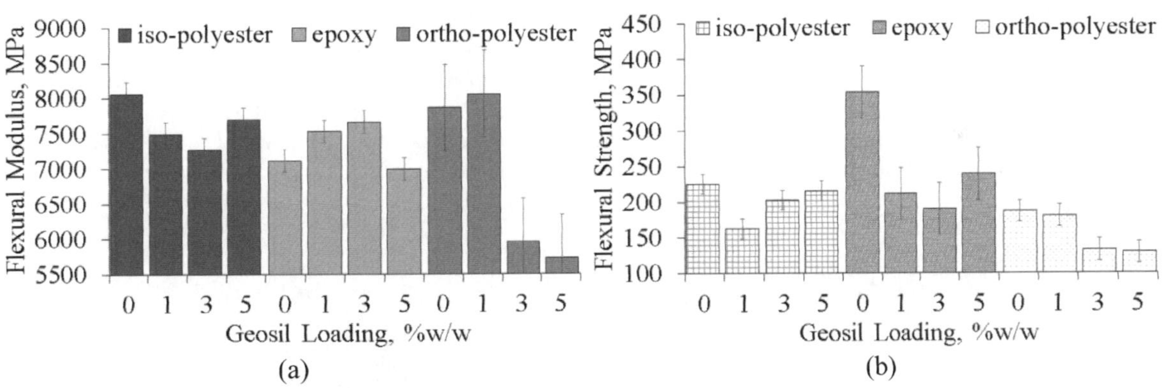

Figure 3. (a) Flexural modulus and (b) flexural strength of geosil/thermoset composites.

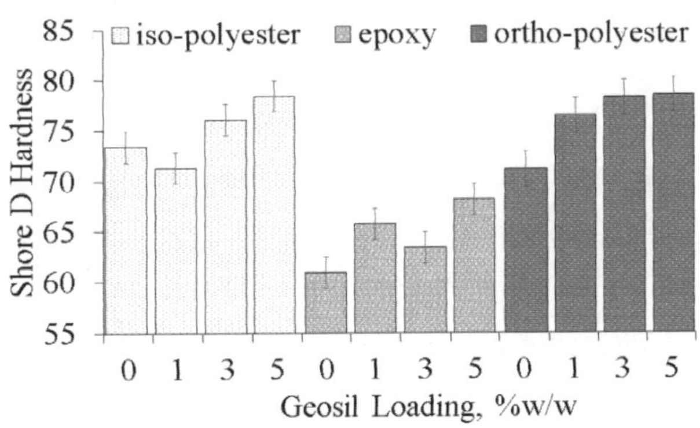

Figure 4. Shore D hardnesss of geosil/thermoset composites.

Figure 4 shows the hardness of composites to increase with geosil loading. The geosil/iso-polyester, geosil/epoxy, and geosil/ortho-polyester composites with 5%w/w geosil loading show the highest Shore D hardness of 78.4, 68.2, and 78.5, respectively. The increase in hardness of the geosil/thermoset composite is expected since silica is a material of high hardness [18]. The maximum hardness was observed at the highest geosil loading because of the higher probability that the indentation pressure during hardness test will press the thermoset matrix and geosil filler together [11].

Figure 5 shows that increase in water uptake is observed as geosil loading in geosil/thermoset composites is increased due to the hydrophilic nature of silica [8]. Also, free spaces between agglomerated geosil particles due to poor adhesion with polymer matrix offer sites for water absorption [19]. The geosil/ortho-polyester composites with 5%w/w geosil loading has the highest water uptake of 5.14% which could be due to surface cracking attributed to formation of phthalic acid by hydrolysis [20] when ortho-polyester composite was immersed in water as shown in figure 6.

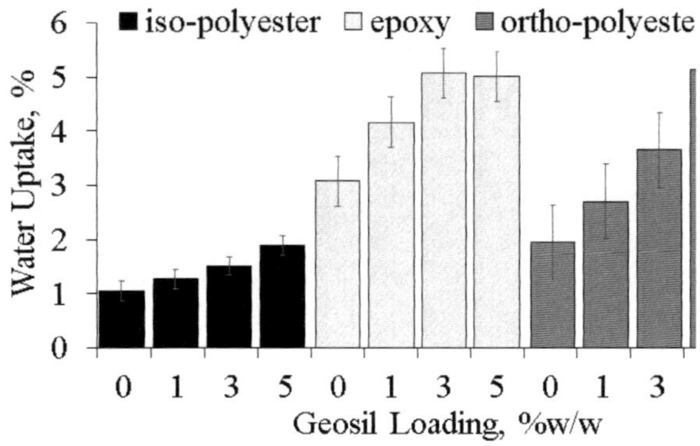

Figure 5. Water uptake of geosil composites after 264 h at 80°C.

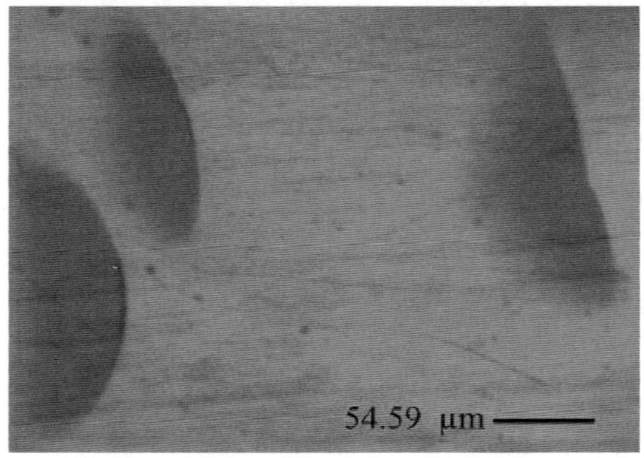

Figure 6. Surface of water-immersed ortho-polyester sample after 254 h at 80°C.

Figure 7 shows that increase in weight loss after drying is observed as geosil loading in geosil/thermoset composites increases which could be due to the aggregation of silica that resulted to filler-filler wicking, filler-matrix debonding and polymeric chain leaching [3, 13, 21]. The geosil/orthopolyester composites with 5%w/w geosil loading has the highest weight loss of 10.50%.

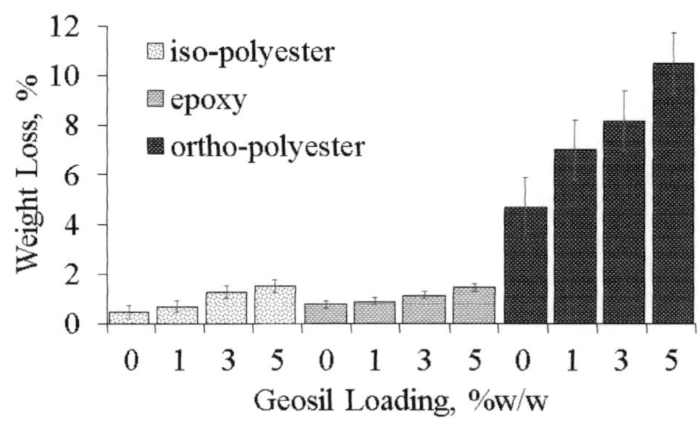

Figure 7. Weight loss of wet geosil composites after drying.

ANOVA at 95% confidence level shows that the type of thermoset significantly affects the flexural strength (p = 0.046), hardness (p = 0.000), water uptake (p = 0.001) and weight loss (p = 0.000) of geosil/thermoset composites. From the results of Tukey's test, the flexural strength of epoxy (248.8 MPa) is significantly higher than ortho-polyester (157.6 MPa), while polyesters (ortho- at 76.1; iso- at 74.8) are harder than epoxy (64.6). Epoxy composites (4.3%) have significantly higher water uptake than iso-polyesters (1.4%), while ortho-polyester composites (3.4%) are higher than its iso- counterpart. Ortho-polyester composites (7.6%) also have significantly higher weight loss than epoxy (1.1%) and iso-polyester (1.0%). Meanwhile, geosil loading significantly affects the hardness (p = 0.041) and water uptake (p = 0.024) of geosil/thermoset composites. From the results of Tukey's test, composites at

5%w/w loading (75.0) are significantly harder than at 0%w/w (68.5). Composites at 5%w/w (4.0%) also absorb more water than composites at 0%w/w (2.0%).

4. Conclusion

Epoxy has higher flexural strength than ortho- and iso-polyester, while ortho-polyester has higher hardness than iso-polyester and epoxy. Iso-polyester has lower water uptake and weight loss than epoxy and ortho-polyester. The hardness and water uptake of geosil/thermoset composites significantly increase as geosil loading increases from 0 to 5%w/w. Among the thermoset resins, iso-polyester can result to high flexural properties and hardness, low water uptake and weight loss. The 5%w/w geosil loading can result to high hardness and water uptake of geosil/thermoset composites.

Acknowledgement
The authors would like to thank the ASEAN University Network/ Southeast Asia Engineering Education Development Network (AUN/SEED-Net) under Japan International Cooperation Agency (JICA) and the Science Education Institute of the Department of Science and Technology (DOST-SEI) under the Engineering Research and Development for Technology (ERDT) scholarship for the funding of this study. Also, the authors would like to acknowledge Dr. Masatoshi Kubouchi of Tokyo Institute of Technology for SEM-EDS testing of samples.

References

[1] Sabu T, Kuruvilla J, Malhotra S K, Goda K, and Sreekala M S 2012 *Polymer Composites* vol 1, (Wiley-VCH Verlag GmbH &Co. KGaA)
[2] Sequeira, D, Mascarenhas J, Picardo D, Dias R, and Sutari O 2015 *Materials* **5** 92-95
[3] Pajarito B B, Kubouchi M, Tomita H, and Sakai T (2012) *Mater. Sci. Technol. Jpn* **49** 32-38
[4] Wang H, Bai Y, Liu S, Wu J, and Wong C P 2002 *Acta Mater.* **50** 4369-77
[5] Ahmad F N, Jaafar M, Palaniandy S, and Azizli K A M 2008 *Compos. Sci. Technol.* **68** 346-53
[6] Khankrua R, Pivsa-Art S, Hiroyuki H, and Suttiruengwong S 2013 *Energy Procedia* **34** 705-713
[7] Ollier R, Rodriguez E, and Alvarez V 2013 *Compos. Part A-Appl. S.* **48** 137-43
[8] Sun Y, Zhang Z, and Wong C P 2005 *Polymer* **46** 2297-2305
[9] Antonelli C, Serrano B, Baselga J, Ozisik R, and Cabanelas J C 2015 *Eur. Polym. J.* **62** 31-42
[10] Wichmann M H G, Cascione M, Fiedler B, Quaresimin M, and Schulte K 2006 *Compos. Interface.* **13** 699-715
[11] Sudheer M, Prabhu R, Raju K, and Bhat T 2014 *Adv. Mater. Sci. Eng.* **2014**
[12] Burton E A, Bourcier W L, Wallace A, Bruton C J,and Leif R 2003 *TrGRC* **27** 519-22
[13] Santiago D E O and Pajarito B B 2016 *Key Eng. Mater.* **705** 121-125
[14] Pajarito B B and Kubouchi M 2015 *Mech. Compos. Mater.* **50** 789-800
[15] Correlo V M, Pinho E D, Pashkuleva I, Bhattacharya M, Neves N M, and Reis R L 2007 *Macromol. Biosci.* **7** 354-63
[16] Pajarito B B, De Torres C A, and Maningding M 2014 *Sci. Diliman* **26** 21-39
[17] Al-Ajaj I A, Abd M M, and Jaffer H I 2013 *Int. J. Mining, Metall. Mech. Eng.* **1** 93-97
[18] Ibtihal A N, Ahmed A I, and Manal F H 2011 *Iraqi J. Mech. Mater. Eng.* **11** 486-506
[19] Demir H, Atikler U, Balköse D, and Tıhmınlıoğlu F 2006 *Compos. Part A-Appl. S* **37** 447-56
[20] Abeysinghe H P, Edwards W, Pritchard G, and Swampillai G J 1982 *Polymer* **23** 1785-1790
[21] Kumar D J, Singh A K, and Singh RP 2008 *Process Properties Perform Comp Mater* **4** 2652-58

IOP Conf. Series: Materials Science and Engineering **201** (2017) 012004 doi:10.1088/1757-899X/201/1/012004

Effect of degrading yellow oxo-biodegradable low-density polyethylene films to water quality

B A Requejo[1,2,a] **and B B Pajarito**[2,b]

[1] Environmental Engineering Program, University of the Philippines, Diliman, Quezon City, 1101, Philippines
[2] Polymer Research Laboratory, Department of Chemical Engineering, University of the Philippines, Diliman, Quezon City, 1101, Philippines

[a] benedict_requejo@hotmail.com, [b] bryan.pajarito@gmail.com

Abstract Polyethylene (PE) contributes largely to plastic wastes that are disposed in aquatic environment as a consequence of its widespread use. In this study, yellow oxo-biodegradable low-density PE films were immersed in deionized water at 50°C for 49 days. Indicators of water quality: pH, oxidation-reduction potential, turbidity, and total dissolved solids (TDS), were monitored at regular intervals. It was observed that pH initially rises and then slowly decreases with time, oxidation-reduction potential decreases then slowly increases with time, turbidity rises above the control at varied rates, and TDS increases abruptly and rises at a hindered rate. Moreover, the films potentially leach out lead chromate. The results imply that degrading oxo-biodegradable LDPE films results to significant reduction of water quality.

1. Introduction

The desirability of polyethylene (PE) as a packaging material stems from its good mechanical properties, low-cost production, chemical resistance, and non-toxicity [1]. To improve aesthetic appeal of PE packaging material, pigments are incorporated in the PE matrix during fabrication [2]. As a consequence of its widespread use, however, PE is also found at large amounts in bodies of water, which poses environmental, waste management and disposal issues [3,4]. PE, like other plastics, does not easily degrade, which causes detriment to the environment and harm to the relying wildlife for survival.

A prevalent way to accelerate the degradation of PE in the environment is to incorporate pro-oxidants in the PE matrix during fabrication. The presence of oxygen and either UV radiation or heat are necessary to allow the degradation of PE, with the presence of pro-oxidants catalyzing this process [1,4-7]. Pro-oxidants catalyze the oxidation of the hydroperoxides present in PE, which then form products of low molecular weights that are susceptible to microbial attack [1,4-7]. However, these plastics are not exempted from waste management issues, and many still are improperly handled and disposed.

A previous study was conducted by the authors in order to investigate the effect of transparent oxo-biodegradable PE plastic bags to water quality [8]. Similarly, this study aims to assess the impact of yellow oxo-biodegradable low density PE (LDPE) films to water quality by monitoring the following parameters with time: pH, oxidation-reduction potential (ORP), turbidity, and total dissolved solids (TDS). Results are also discussed in comparison to the transparent oxo-biodegradable PE bags previously investigated [8].

IOP Conf. Series: Materials Science and Engineering **201** (2017) 012004 doi:10.1088/1757-899X/201/1/012004

2. Experimental

Yellow oxo-biodegradable LDPE films were previously prepared using film grade LDPE pellets Lotrene® 374 by Qatar Petrochemical Corporation (Doha, Qatar), yellow colorant Yellow® 7060U by Kolora Ink and Chemical Corporation (Quezon City, Philippines), and pro-oxidant Biomate® BM-205 by First in Colours, Incorporated (Quezon City, Philippines) via melt mixing and film blowing technique [7]. Yellow colorant was analyzed by energy dispersive x-ray (EDX) fluorescence spectroscopy (Shimadzu EDX-8000) to determine pigment used. Thickness and pro-oxidant loading of yellow oxo-biodegradable LDPE films are shown in table 1. Deionized (DI) water was used as immersion medium.

Table 1. Thickness and pro-oxidant loading of yellow oxo-biodegradable LDPE films [7]

Sample	Thickness (mils)	Pro-oxidant Loading (%)
Y13	1	3
Y24	2	4
Y32	3	2

Strips of 1 × 1 cm were cut from the yellow oxo-biodegradable LDPE films using stainless steel scissors. Three grams of cut LDPE strips (see figure 1) were then placed inside 250-ml glass storage bottles. About 225 ml of DI water were added to each bottle, giving a liquid-to-solid ratio of 75 ml g^{-1}. Five replicates of each sample were prepared and tested. A control sample, i.e., bottle containing only DI water, was also prepared. Each bottle was tightly covered and sealed with the provided caps and were manually shaken to ensure that the strips were evenly dispersed in the media.

Figure 1. Y32 strips being weighed

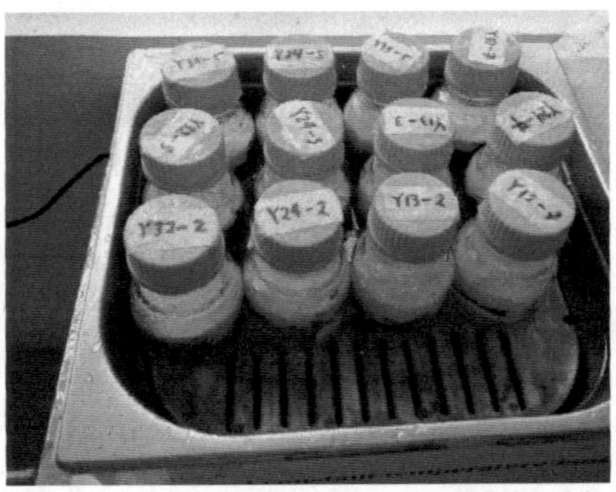

Figure 2. Samples as placed inside the water bath

Figure 3. Y13, Y24, and Y32 LDPE strips after 49 days of immersion

The glass storage bottles were submerged inside a water bath (Premiere®, HH-4) maintained at 50°C for a total of 49 days (see figure 2). The water bath was covered with a 1 in.-thick box made of polystyrene foam, whose inside walls were lined with thick cardboard and plastic to prevent water loss from the bath and minimize the exposure of LDPE samples to ambient light. Figure 3 shows the bottles containing the LDPE samples after 49 days of immersion.

Water quality parameters were measured on the 7th, 14th, 21st, 35th, and 49th days. One replicate sample was taken out per measurement, and three 30-ml aliquots were drawn for testing. Respective meters and probes for pH (Lutron Electronic Enterprise Co., LTD, PH-221 and PE-21), ORP (Lutron Electronic Enterprise Co., LTD, YK-23RP), turbidity (Lutron Electronic Enterprise Co., LTD, TU-2016,), and TDS (Lutron Electronic Enterprise Co., LTD, CD-4317SD) were used.

3. Results and Discussion

To determine the content of the yellow colorant, samples were subjected to EDX testing. Results of EDX testing indicate that the top 2 metals comprising the colorant are lead (2.744% w/w) and chromium (0.764% w/w), showing the presence of lead chromate. Lead chromate, a toxic material, is used in the industry as the main ingredient for inorganic yellow pigments [2].

The top 3 countries that contribute to waste plastics disposed into the oceans [9] are surrounded by warm tropical waters, whose temperatures can reach 32.8°C [10]. In this study, the immersion temperature was elevated to 50°C in order to increase the rate at which the LDPE degradation products are formed [1,6] but still environmentally relevant according to ASTM D6954-04. The selected immersion temperature was within 20°C from the disposal temperature of plastics and did not exceed the 70°C limit specified by the standard [11].

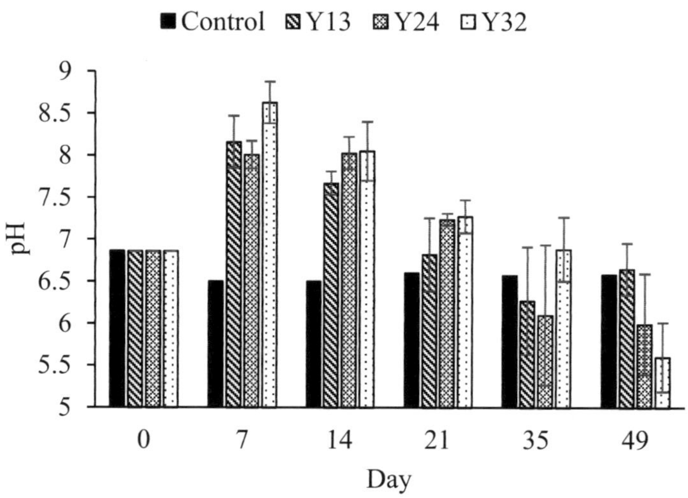

Figure 4. pH of control, Y13, Y24, and Y32 samples over time

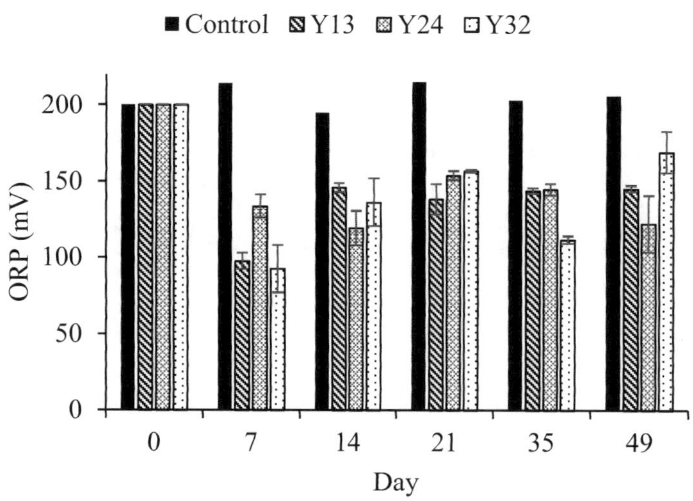

Figure 5. ORP of control, Y13, Y24, and Y32 samples over time

The pH is an indicator of the acidity or alkalinity of a solution [12]. Figure 4 shows that for all yellow oxo-biodegradable LDPE samples, pH abruptly rises after 7 days and then slowly decreases accordingly. The initial rise in pH could indicate the presence of inhibitors that are alkaline in nature, such as hydroxyl- or hindered amines, in the composition of oxo-biodegradable plastics [13] and yellow colorant [7]. As degradation progresses, acidic hydroperoxides are formed from LDPE [14], consuming the inhibitors and decreasing the pH of the media. Minor changes in pH are observed in the control.

The ORP of a solution shows its oxidizing or reducing ability, i.e., the more positive the ORP of a solution is, the more it exhibits an oxidizing property [12]. As shown in figure 5, the ORP of the water media containing the yellow oxo-biodegradable LDPE samples were observed to decrease after 7 days and then slowly increase on the succeeding days. The initial decrease in ORP of the water media is attributed to the consumption of oxygen, an oxidizing agent, dissolved in the water media during the degradation of PE [8]. Consequently, the increasing presence of radicals causes ORP to rise gradually thereafter. Meanwhile, the ORP of the control exhibits minimal changes.

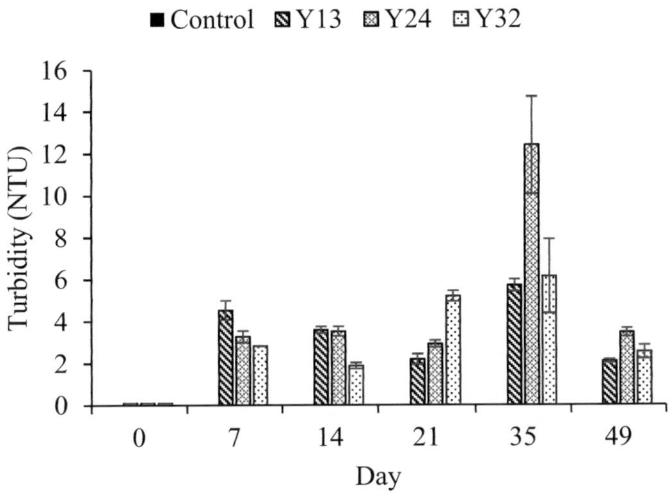

Figure 6. Turbidity of control, Y13, Y24, and Y32 samples over time

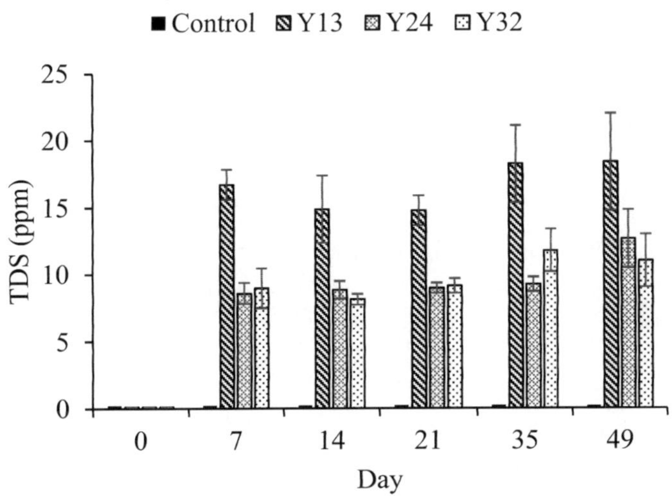

Figure 7. TDS of control, Y13, Y24, and Y32 samples over time

Turbidity indicates the presence of suspended particles or colloidal matter that obstruct the transmission of light through water [15]. It is observed that the media containing yellow oxo-biodegradable LDPE films increase in turbidity for 7 days but then the trend varies (see figure 6). Media containing Y13 and Y24 films decrease in turbidity from 7 to 21 days, maximizes at 35 days, and then decreases at 49 days. Meanwhile, media containing Y32 decreases in turbidity from 7 to 14 days, increases from 21 to 35 days, then decreases at 49 days. The increase in turbidity is due to the increasing concentration of low-molecular weight LDPE fragments due to the physical degradation of plastic. However, because LDPE degradation is a random process, the rate may differ for all samples [4].

TDS is the measurement of the sum of inorganic salts and small amounts of organic matter in a solution [15]. TDS of the water media containing yellow oxo-biodegradable LDPE films increases abruptly after 7 days and then slowly during the succeeding days (see figure 7). Media containing Y13 samples attained the highest TDS, while Y24 and Y32 results are comparable. Pro-oxidants and small LDPE fragments leaching onto the water cause the initial increase in TDS, but this increase is hindered possibly due to the protective role of aromatic amine stabilizer, which interferes with auto-oxidation of LDPE, present in the yellow colorant [7].

A similar study was conducted previously by the authors, in which transparent PE-blend plastic bags composed by weight of 88% HDPE, 11% LDPE, and 1% pro-oxidant, with thicknesses of 0.39 mils and 0.51 mils, were employed [8]. In comparison, pH, ORP, and turbidity changes were more profound for the LDPE films in this study, which can be due to greater extent of branching than HDPE, causing the former to degrade at a greater extent [16]. Meanwhile TDS values were much higher for the PE-blend films than the LDPE films in this study due to lower diffusion resistance exhibited by the thinner PE-blend films [8].

This study therefore proposes the extent of LDPE degradation, a surface phenomenon [4], can be presumed directly from pH, ORP, and turbidity readings. Similarly, the extent of leaching, a bulk phenomenon, can be deduced from TDS results. Ideally, pro-oxidant loading enhances the rate of degradation and smaller film thickness encourages leaching. A synergistic effect may exist between these 2 factors, which will be the subject of a future study.

4. Conclusion

The immersion of yellow oxo-biodegradable LDPE films has an effect on the quality of water where they are immersed. The yellow films show a potential for the leaching of lead chromate, a toxic compound. The pH of the water media containing the films increases abruptly then slowly decreases with time. The ORP of the water media decreases initially then slowly increases. Furthermore, both their turbidity and TDS increase but at a varied and hindered rate. The results suggest that the prolonged presence of yellow oxo-biodegradable LDPE films in aquatic environment results to a decreasing water quality and even a potential for toxicity.

Acknowledgements

The authors would like to thank Dr. Himawan Tri Bayu Murti Petrus from the Chemical Engineering department of the Universitas Gadjah Mada in Yogyakarta, Indonesia for EDX testing and the Engineering Research and Development for Technology (ERDT) Program of the Department of Science and Technology (DOST) of the Philippines for their financial support in this research.

References

[1] Benitez A 2013 *Polym. Degrad. Stab.* **98** 490
[2] Tolinski M 2015 *Additives for Polyolefins* (Elsevier) pp 105-106
[3] Rios L M, Moore C, Jones P R 2007 *Mar. Pollut. Bull.* **54** 1230
[4] Gewert B, Plassmann M, and MacLeod M 2015 *Environ. Sci.: Processes Impacts* **17** 1513
[5] Roy P K, Hakkarainen M, and Albertsson A C 2014 *J. Polym. Environ.* **22** 69
[6] Jakubowicz I 2003 *Polym. Degrad. Stab.* **80** 39
[7] Maalihan R and Pajarito B 2015 *J. Plast. Film Sheeting* **32** 124
[8] Requejo B and Pajarito B 2016 *Mater. Sci. Forum* submitted for publication
[9] Jambeck J R et al 2015 *Science* **347** 768
[10] NOAA OSPO, Sea Surface Temperature Contour [Online image]. Retrieved November 29, 2016 from http://www.ospo.noaa.gov/Products/ocean/sst/contour/

[11] ASTM International 2013 D6954-04
[12] Banhidi M 2007 *Met. Finish.* **105** 518
[13] Downie R H 2002 US Patent 6482872 B2
[14] Henry J, Ruaya N, and Carton A 1992 *J. Polym. Sci. A Polym. Chem.* **30** 1693
[15] World Health Organization 2011 *Guidelines for drinking-water quality* 4th ed.
[16] Park, J W et al 2000 *Korean J. Chem. Eng.* **17** 489

IOP Conf. Series: Materials Science and Engineering **201** (2017) 012005 doi:10.1088/1757-899X/201/1/012005

Effect of type and loading of surface-modifying agent on mechanical properties of modified geothermal scale powder/stereolithography polymer composite

A. C. Tilendo and B. B. Pajarito

Polymer Research Laboratory, Department of Chemical Engineering, University of the Philippines, Diliman, Quezon City 1101 Philippines

amierson.tilendo@gmail.com, bryan.pajarito@gmail.com

Abstract. This study investigated the effect of stearic acid (SA), glycerol monostearate (GMS) and 3-trimethoxysilylpropyl methacrylate (TSPM) at varied loadings on the hardness and flexural properties of modified geothermal scale powder (GSP)/stereolithography (SLA) polymer composite. TSPM-modified GSP/SLA composite has the highest value of hardness due to increased filler dispersion and crystallinity. Hardness of GSP/SLA composite increases with loading of surface-modifying agent due to increase filler dispersion. Pronounced effect of surface modification to flexural modulus is observed. While low loading of SA and GMS leads to reduction of flexural modulus, increasing loading enhances the said property. Further increase of SA deteriorates the property. TSPM-modified GSP enhances the modulus due to increased crystalline phase of the system owing to TSPM copolymerization. Likewise, addition of SA and GMS increases flexural strength due to efficient reduction of filler agglomerates. However, unreacted TSPM produces weak interfaces and poor adhesion between GSP and SLA matrix.

1. Introduction

In the advent of solid freeform fabrication techniques, stereolithograpy (SLA) is continuously evolving, fabrication costs are decreasing and properties of the manufactured parts are becoming better [1]. However, photosensitive monomer resin alone produces products of limited mechanical properties [2] that becomes critical in producing functional parts for demanding end-use applications [1]. This provides routes to further advancement on material properties of SLA products such as reinforcement of various fillers [1-4]. Fillers are inorganic particles incorporated into polymer; however, it is very difficult to well disperse the fillers because particles of high surface energy tend to agglomerate [5]. Silica is one of the most important reinforcing fillers used mainly in composite SLA owing for its optical compatibility [3]. However, silica provides lower degree of reinforcement because it contains a large number of silanol (-SiOH) groups on the surface making it highly polar and less compatible with many non-polar photopolymers. In addition, the presence of -SiOH groups leads to strong filler–filler interaction giving rise to a high tendency for filler agglomeration in the matrix [6] resulting to poor interfacial interactions and weak SLA composites [5].

Effective reinforcement of fillers to polymer can be achieved by surface modification of fillers to minimize filler/filler interaction and to enhance filler/matrix interaction. Also, surface modification increases the hydrophobicity of the filler that enhances the interfacial adhesion either by entanglement

IOP Conf. Series: Materials Science and Engineering **201** (2017) 012005 doi:10.1088/1757-899X/201/1/012005

or chemical bonding. It has also been proved that increasing the number of filler/matrix interfaces leads to unusual performance [5,7]. Surface modification of fillers can be physical or chemical. Physical modification uses surfactants which are chemicals containing one or more polar groups and a long aliphatic chain. The preferential adsorption of polar group onto the high energy surface of the fillers by electrostatic interaction produces encapsulation of inorganic fillers [5]. Two surfactants were used in this study as surface-modifying agents of fillers. Stearic acid (SA) is an amphiphilic non-ionic surfactant that has been proved to enhance compatibility and dispersion of filler in polymer matrix [5,8]. Glycerol monostearate (GMS) is another non-ionic surfactant that is an ester of stearic acid with hydrophilic group of a non-dissociable type [9, 10] previously used to improve organo-layer silicate intercalation [11]. In comparison, chemical treatment is based on covalent attachment of modifying agent to avoid desorption from filler surface. Better filler distribution and compatibility between filler and matrix have been achieved by the use of different coupling agents [5,6, 12-14]. The reactive group of coupling agent helps construct chemical bonding between fillers and polymer. Among the different coupling agents, organo-silane ($R\text{-}Si\text{-}(OR')_3$) is the most effective, cheap, readily available and is widely used for modifying surface properties to improve the reinforcing ability of silica-based fillers. Also, silanes have alkoxysilane groups, which after hydrolysis are capable of reacting with hydroxyl-rich surfaces [14].

It is expected that surface modification of geothermal scale powder (GSP), a silica-based filler derived from waste of geothermal power plants, will improve its chemical compatibility and distribution throughout the SLA resin, thus providing better mechanical properties than the untreated (uT) filler. This study investigates the effect of type and loading of surface-modifying agent on hardness and flexural properties of modified GSP/SLA polymer composite.

2. Experimental

GSP was obtained from geothermal wastes generated by Energy Development Corporation (EDC). Geothermal silica scale was grinded in a planetary ball mill (Pulverisette 6, Fritsch) using 10-mm tempered steel balls for 5 min. Ground silica scale was treated with 6 M sulfuric acid (Univar, Ajax Finechem) solution, washed with distilled water, filtered and dried in a vacuum oven (LVO-2030, Labtech) at 80 °C for 48 h [15]. Surface modifying agents utilized are SA (RTC Laboratory Services & Supply House), GMS (Cerin GMS, Chemrez Technologies, Inc.) and a silane coupling agent 3-trimethoxysilylpropyl methacrylate (TSPM, 440159, Aldrich Chemistry) with loading varied at 1, 2 and 3 wt% of GSP. Modification of GSP using surfactants were carried out by dry-blending at 100 °C for 15 min [16]. For silane treatment, GSP was modified with a mixture of TSPM and acetone (AR1003-G4L, RCI Labscan) using a magnetic stirrer (WiseStir, Wisd) for 30 min, followed by acetone evaporation in the vacuum oven for 24 h. GSP was sieved using 270 mesh (No. 270, WS Tyler) before and after surface modification. Fillers were characterized by attenuated total reflectance – Fourier transform infrared spectroscopy (ATR-FTIR, Nicolet 6700) to determine the presence of surface-modifying agents. Commercial SLA resin (Clear Resin, MoonRay) was blended with GSP at 10 vol% using an overhead stirrer (RW20, IKA) for 5 min, sonicated for 5 min in an ultrasonic bath (WUC-D30H, Wisd) and degassed for 30 min using a vacuum desiccator. The degassed GSP/SLA mixture was poured into a 1-mm thick glass mould pre-coated with silicon oil, covered with glass plate, and was placed on top of a 356 × 60 mm UV exposure unit (300-02, Mega Electronics) for 125 min. The unit has 4 × 15 W UVA bulbs with peak output wavelength of 350 nm and average intensity of 316 $\mu W\ cm^{-1}$ as measured by a UV light meter (UV-340A, Lutron). Filler dispersion in cured GSP/SLA composites was assessed using an optical microscope (Novex B Series, Euromex), while water contact angles were measured using a digital microscope (AM7915MZT, Dino-Lite). Composites were cut into dimensions specified by ASTM D790-03 for determination of flexural properties using universal testing machine (AGS-5kNX, Shimadzu). Shore D hardness of composites was measured using a durometer (GS-702N, Teclock).

3. Results and Discussion

IOP Conf. Series: Materials Science and Engineering **201** (2017) 012005 doi:10.1088/1757-899X/201/1/012005

GSP at 50 μm magnification has particle size range of 1-15 μm and irregular shape due to high impact of milling [15]. Presence of surface-modifying agents in GSP is confirmed in the ATR-FTIR spectra as shown in figure 1. In figure 1a, the small peaks at 2915 and 2855 cm^{-1} are due to symmetric and asymmetric stretching vibration of aliphatic $–CH_2–$ groups; the small peak at 1697 cm^{-1} is due to C=O band stretching, confirming the presence of SA in the modified filler [10]. The ATR-FTIR spectrum of GMS modified filler in figure 1b exhibits peaks at 3 positions which are 2915, 2930 and 2848 cm^{-1}. These peaks are due to $-CH_2-$ stretching vibrations confirming the presence of GMS in the modified filler [10]. In figure 1c, small peaks at 2937 and 2838 cm^{-1} are assigned to the symmetric of C–H stretch. Peaks at 1716, 1635 and 1456 cm^{-1} correspond to the stretching vibration of C=O, methylene C-H and vinyl C-H bending vibration of TSPM; peak at 1394 cm^{-1} is attributed by O-H bending derived from hydrolysis [17]. Presence of these peaks could be associated to inefficient bonding between the O-H group and C=O group from TSPM with O-H group from GSP. Thus, neighboring silanols are condensed making TSPM unreacted to GSP and consequently, the vinyl group of TSPM is polymerized by free radicals. The cured TSPM are not affected by SLA monomer and thus copolymerization occurs at the coupling agent-resin interface [18]. In addition, increase of absorption band at 930 to 1028 cm^{-1} corresponding to Si-O stretching is observed after the silane modification of GSP [17].

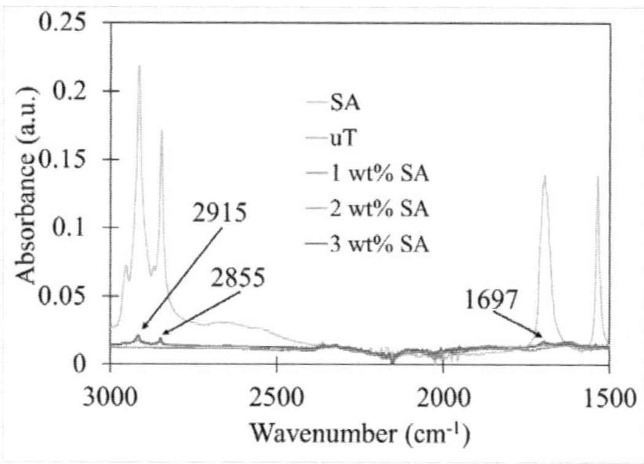

(a)

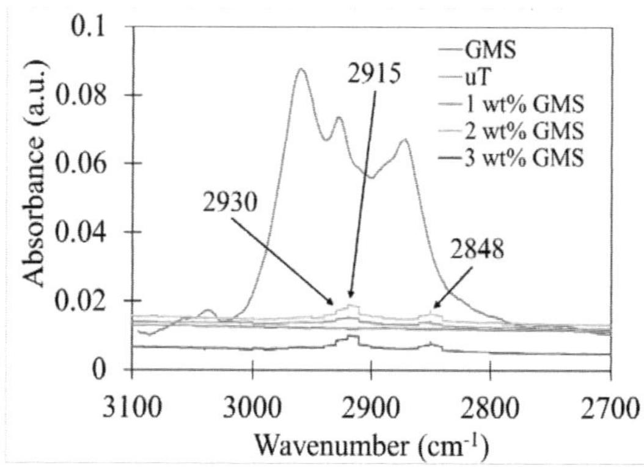

(b)

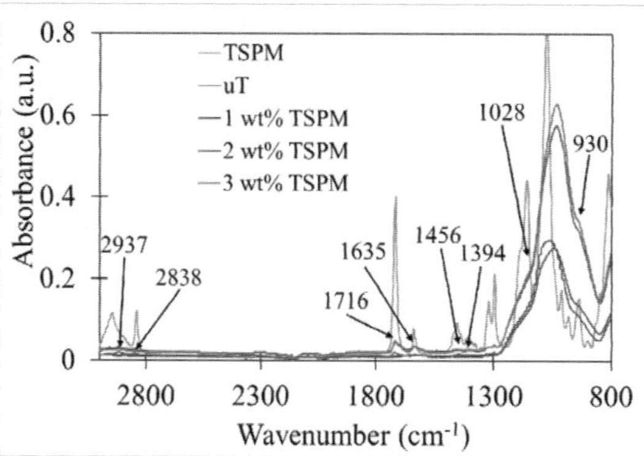

(c)

Figure 1. ATR-FTIR spectra of surface modified GSP: (a) SA, (b) GMS, (c) TSPM.

Dispersion of modified GSP in the SLA matrix was evaluated using optical microscopy. The unmodified GSP forms agglomerates in the SLA matrix. However, with SA-coated GSP, the distribution of modified fillers is not homogeneous owing to limitation in critical micelle concentration [19]. As more SA is added, equilibrium shifts with surfactants leaving the filler surface to start forming micelles. In consequence, filler is no longer repulsed and favors reagglomeration.

GMS effectively encapsulates the filler and reveals homogeneous distribution throughout the matrix. Increasing the amount effectively coats the fillers for homogenized dispersion. Similarly, TSPM-modified fillers are homogeneously distributed owing to the significant reduction of filler–filler interaction and better filler dispersion [6]. Moreover, the crosslink density is found to increase with increasing TSPM loading [6] that results to increase crystalline phase of the system [20] as exhibited by hazy white blend shown in figure 2.

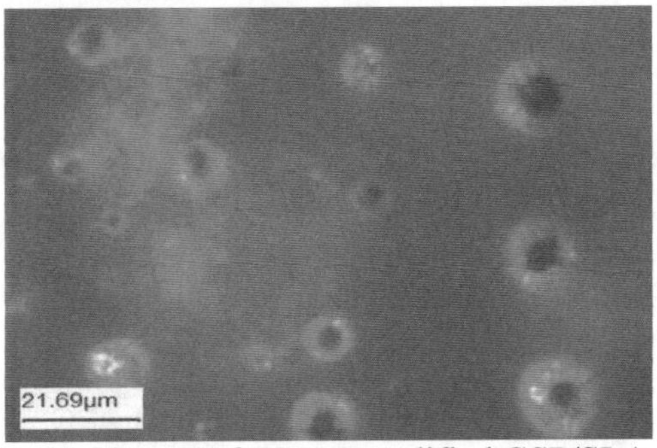

Figure 2. Optical image of TSPM-modified GSP/SLA composite.

Wetting and hydrophobicity of composite are characterized by measuring the contact angle formed by liquid water droplet over the surface of composite using ImageJ software. Figure 3 reveals that addition of modifying agents to the GSP increases the contact angle. Furthermore, increase in loading of modifying agents also increases the contact angle indicating increase in hydrophobicity of the composite. GSP modified with 3 wt% SA reveals the highest hydrophobicity due to localized micelles formed in the matrix; hydrophobic tails are directed outward the micelle for compatibility with hydrophobic polymer. Also, TSPM is likely to contribute to the improvement of filler–matrix compatibility by better wetting of SLA polymer matrix to GSP surface [14].

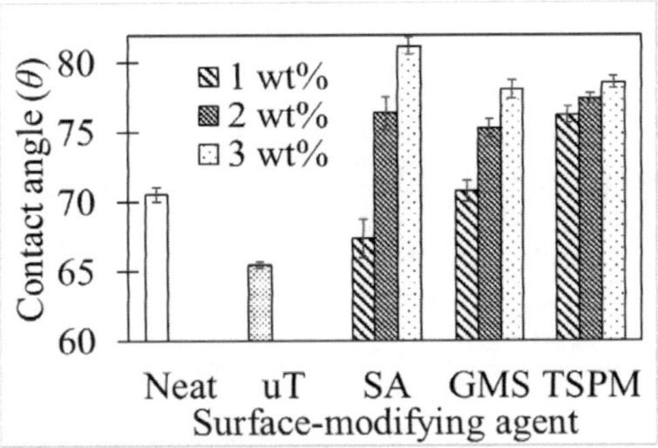

Figure 3. Contact angle of GSP/SLA composite.

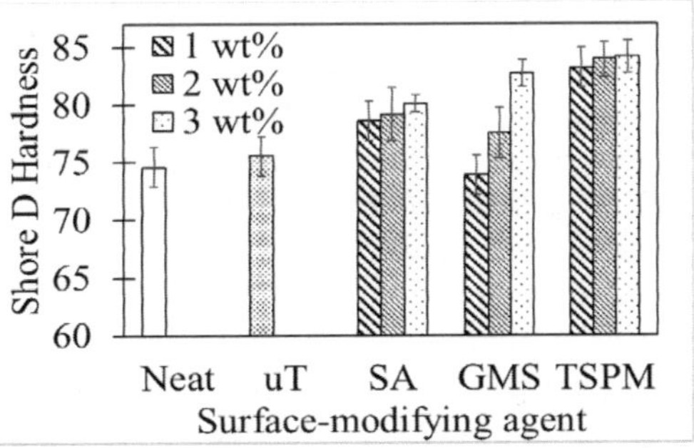

Figure 4. Hardness of GSP/SLA composites.

Figure 4 shows the improvement in hardness of GSP/SLA composites after use of surface-modifying agents. Furthermore, increasing the loading of modifying agents also enhances the hardness. Due to increased hydrophobicity, the modified fillers are more dispersed in SLA matrix than the untreated GSP. This increases the chance for the durometer indenter to hit the hard filler during measurement. For TSPM-modified fillers, C=C functional groups could covalently bond via copolymerization with the polymer matrix [21] resulting to increased crosslink density and crystalline phase of the system [6, 19]. Thus, modified GSP acts as chemical junction in the network of harder polymer.

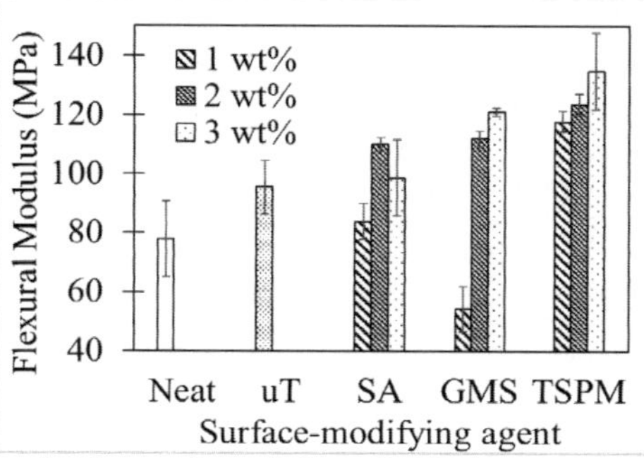

Figure 5. Flexural modulus of GSP/SLA composites.

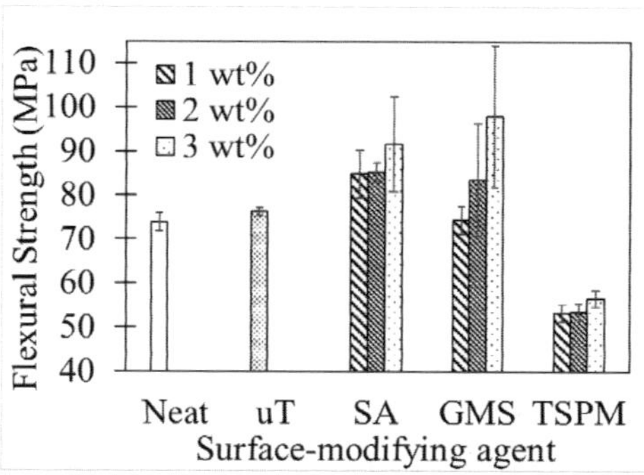

Figure 6. Flexural strength of GSP/SLA composites.

Figure 5 shows that at low loading of SA, incomplete surfactant coating leads to inadequate bonding with the organic matrix and consequently reduces the flexural modulus. Increasing up to 2 wt% SA augments the modulus due to improvement in adhesion between the filler and polymer matrix and reduces the interaction between fillers. However, modulus deteriorates with high amount of SA due to decrease in interactions with excess surface coverage [22]. The decrease in the moduli of composites can also be explained by the plasticizing effect of the surface-modifying agent, which is adsorbed on the surface of the GSP as a multilayer [8]. Behavior of SLA composite containing GMS-modified GSP is similar to SA at low loading; however, 3 wt.% GMS effectively coats the fillers for homogenized dispersion. TSPM, on the other hand, enhances filler dispersion and crystalline phase of the system due to polymerization of silane at the surface of GSP as shown in figure 2. Polymerized TSPM reinforces the SLA matrix and improves the stiffness of the composite [17]. Figure 6 shows the improvement in flexural strength upon addition of SA and GMS. This is attributed to the efficient reduction of filler agglomerates and increase in effective stress transfer [7] between the filler and polymer. Increasing surfactant loading increases the number of fillers effectively modified resulting to improved number of strong interfaces. Increased number of strong interfaces increases the effective stress transfer. Although TSPM improves filler dispersion and hydrophobicity of GSP/SLA composite, significant reduction of flexural strength is observed due to copolymerization of TSPM. TSPM, instead of reacting with the GSP, condenses with neighboring TSPM and polymerizes on its own which creates weak interfaces and poor adhesion between GSP and SLA matrix [18]. This polymerization of methylene and vinyl groups of TSPM produces weak polymeric interfaces which promote crack growth when stress is applied.

4. Conclusion

Various surface modifying agents compared in this study influence the hardness and flexural properties of GSP/SLA composites differently. TSPM-modified GSP/SLA composite has the highest value of hardness due to increased filler dispersion and crystallinity. Hardness of GSP/SLA composite increases with loading of surface-modifying agent due to increase filler dispersion. Flexural modulus is greatly affected by the type and amount of surface modifying agent. While low loading of SA and GMS leads to reduction of modulus, high loading enhances the said property. However, further increase of SA deteriorates flexural modulus. TSPM-modified GSP significantly improves the flexural modulus due to better filler dispersion and increased crystalline phase of the system. Improvement of flexural strength is observed at high loading of SA and GMS due to increased filler dispersion and hydrophobicity. However, all loadings of TSPM significantly reduce the flexural strength due to possible polymerization of silane at the GSP surface, generating weak polymeric interfaces and poor adhesion between GSP and SLA matrix. Results shows that 3 wt% TSPM is appropriate treatment to enhance hardness and flexural modulus of GSP/SLA composite while 3 wt% GMS is recommended to improve flexural strength.

Acknowledgement

The authors gratefully acknowledge the Department of Science and Technology (DOST) – Engineering Research and Development for Technology (ERDT) for the financial support and 3D2Go Philippines for the supply of commercial SLA resin.

References

[1] Melchels F, Feijen J, Grijpma D 2010 *Biomaterials* **31** (2010) 6121-30
[2] Bartolo P J, Gaspar J 2008 *CIRP Ann-Manuf. Techn.* **57** 235–38
[3] Wozniak M, Graule T, de Hazan Y, Kata D and Lis J 2009 *J. Eur. Ceram. Soc.* **29** 2259–65
[4] Zak G, Sela M N, Park C B and Benhabib B 1997 J. *Manuf. Sci. Eng* **3** 17-24
[5] Rong M Z, Zhang M Q and Ruan W H 2006 *Materials Science and Technology* 22 **7** 787-96
[6] Sae-oui P, Sirisinha C, Thepsuwan U and Hatthapanit K 2006 *Eur. Polym. J* **42** 479–86
[7] Fu S, Feng X, Lauke B and Mai Y (2008) *Composites: Part B* **39** 933–61
[8] Ahn S, Kim S and Lee S 2004 *J. Appl. Polym. Sci.* **94** 812–18
[9] Salager J L *Surfactants Types and Uses* (Venezuela: University de Los Andes) Booklet # E300-A
[10] Hussain T, Saeed T, Mumtaz A, Javaid Z, Abbas K, Awais A and Idrees H a. 2013, *Acta. Pol. Pharm.- Drug Research* **70** 749-57
[11] Liza,C , Soegijono B, Budianto E, Alinasiri J and Jayatin 2012 *Proc. Chem.* **4** 47–52
[12] Sabzi M, Mirabedini S M, Zohuriaan-Mehr J and Atai M 2009 *Prog. Org. Coat.* **65** 222–28
[13] Yoshida K and Greener E H 1994 *J. Dent.* **22** 57-62
[14] Girones J, Mendez J A, Boufi S, Vilaseca F and Mutje P 2007 *J. Appl. Polym. Sci* **103** 3706-17
[15] Tilendo A, Pajarito B 2016 *Mater. Sci. Forum* submitted for publication
[16] Santiago D, Pajarito B, Mangaccat W F, Tigue M R and Tipton M 2016 *AIP Conf. Proc.* **1736** 020047
[17] Tham W, Chow W and Mohd Ishak Z 2010 *J. Appl. Polym. Sci* **118** 218-28
[18] Vallittu P K 1997 *J. Oral Rehabil.* **24** 124-30
[19] Wan Y, Tang L, Yan D, Zhao L, Wu L, Jiang J and Lai G. 2013 *Compos. Sci. Technol.* **82** 60-68
[20] Khalil R, Chryss A G, Jollands M and Bhattacharya S 2007 *J. Mater. Sci.* **42** 10219–27
[21] Hong R, Fu H, Zhang Y, Liu L, Wang J, Li H, Zheng Y 2007 *J. Appl. Polym. Sci* **105**, 2176–84
[22] Dányádi L, Móczó J and Pukánszky B 2010 *Composites: Part A* **41** 199–06

Development of porous structured polyvinyl alcohol/zeolite/carbon composites as adsorbent

J A Laksmono[1], M Sudibandriyo[1*], A H Saputra[1], A Haryono[2]

[1] Department of Chemical Engineering, Faculty of Engineering, Universitas Indonesia, Depok – Java, Indonesia, 16424.

[2] Research Center for Chemistry, Indonesian Institute of Sciences, Kawasan PUSPIPTEK Serpong, Tangerang Selatan – Banten, Indonesia, 15314.

*Email: msudib@che.ui.ac.id

Abstract. Adsorption is a separation process that has higher energy efficiency than others. Analyzing the nature of the adsorbate and the selection of suitable adsorbent are key success in adsorption. The performance of the adsorbent can be modified either physically or chemically to obtain the efficiency and effectiveness of the adsorption, this can be facilitated by using a composite adsorbent. In this study, we have conducted the preparation process of a polyvinyl alcohol (PVA)/zeolite/carbon composites. The resulting adsorbent composites are dedicated for ethanol – water dehydration proposes. The composites were prepared using cross-linked polymerization method followed by supercritical fluid extraction (SFE) to obtain the porous structured upon drying process. The characterization of the functional groups and morphology were performed by using Fourier Transform Infra-Red (FTIR) and Scanning Electron Microscopy (SEM), respectively. The FTIR analysis showed that composite prepared by SFE method formed hydrogen bonding confirmed by the appearance of peaks at 2950 – 3000 cm^{-1} compared to composite without SFE method, whereas, the results of SEM study showed the formation of three layered structures. On basis of the obtained results, it can be shown that PVA/zeolite/carbon has high potential to be develop further as an adsorbent composite.

***Keywords*:** adsorption, adsorbent, polymerization, composite, polyvinyl alcohol, zeolite, carbon.

1. Introduction

Energy use in the processes, particularly in separation processes *i.e* distillation, azeotropic distillation, membrane pervaporation, adsorption, extraction, and so on (1, 2) is becoming a major concern of researchers in the past few decades (3-5). However, adsorption process provide higher energy efficiency compared to other techniques (6). Adsorption is a process that occurs when a fluid, liquid or gas as adsorbate, is bound to a solid or a liquid as the adsorbent to finally form a molecular or atomic layer (7). Adsorption term used to describe the presence of a particular substance either liquid or solid with a higher concentration at the surface than in its bulk phase medium. The adsorption process occurs as a result of molecular physical bonds of the adsorbate on the surface of the adsorbent (8). The key factor to the success of the adsorption process is related to the adsorbent performance (9). The bond among the adsorbate and adsorbent will be enhanced whether designing higher surface area of the adsorbent or expanding the contact area among them (10).

In addition, improving interaction among adsorbent and adsorbate can also be done chemically on the adsorbent surfaces (11). Surface modification can be performed by adding the chemical compound to the adsorbent (12-14). Polymeric substances are potential as an adsorbent composite in terms of their good mechanical rigidity, adjustable surface chemistry (15), and good selectivity toward organic adsorbate (16). However, their performance is dependent on the type of functional groups used *i.e.* extent of chemical activation and modification, nature and degree of functionality grafted on the polymeric surfaces.

In this study, PVA is added to the zeolite and activated carbon to form a composite with rich hydroxyl groups. Hydroxyl groups is expected to form hydrogen bonds with water so that the potential adsorbent composite increase the water adsorption capacity of the adsorbent. In this case, certain condition is required in which the PVA polymer can bind to zeolite and activated carbon. The best condition can be achieved by using CO_2 at supercritical condition. Under these conditions, the viscosity is lowered so that the polymer can be more reactive and enhance the diffusivity of the polymer to other materials (17). The aim of this study is to develop a PVA/Zeolite/Carbon adsorbent composite with porous structured and possessing good chemical bonding among the three substances forming the composite.

2. Materials and Method

2.1. Materials

In this study, the main materials used are polyvinyl alcohol (MW 140,000 purchased from Emerck), natural zeolite (obtained from Gunung Kidul, Yogyakarta), wood activated carbon (obtained from West Java), glutaraldehyde (purchased from Emerck as crosslinker), CO_2 UHP (purchased from Air Liquide), furfural alcohol (purchased from Sigma Aldrich), Toluene (purchased from Emerck) and distilled water.

2.2. Methods

2.2.1. Composite preparation. We have conducted the pretreatment and activation process of natural zeolites by washing using distilled water, acidified and then dried in a furnace at 300°C for 3 hours. Zeolite and activated carbon were crushed using a laboratory ball mill for 72 hours, separated, and then screened to obtain consistent particle size. The activated carbon, furfural alcohol, and solvent toluene were dispersed in a 500 mL glass beaker and stirred using a magnetic stirrer. For the impregnation method, zeolite was inserted slowly into the dispersion mixture and stirred continuously. After 1 hour, the mixture was transferred into an incubator shaker and agitated at 25°C for 24 hours Afterwards, the solids were filtered and then transferred to a crucible. The next step is carbonization at 150°C to remove moisture, then increased to 800°C with rate 10°C min^{-1} then detained for 8 hours. This treatment produced a zeolite/carbon composite. The amount of PVA was added into a 500 mL glass beaker and dissolved using distilled water at a temperature of 80°C for 3-4 hours and stirred using a magnetic stirrer continuously until the PVA clearly dissolved. The amount of composite zeolite/carbon was added to the PVA mixture slowly and stirred continuously to reach a homogeneous mixture. In order to reach the crosslinking, 4 wt% glutaraldehyde was added to the mixture and stirred overnight. The sample was dried using CO_2 Supercritical Fluid Extraction (SFE) units in the pressure 2,000 psi and temperature 50°C for 4 hr.

2.2.2. Material characterization. In accordance with the aim to find the composite functional groups and formation of hydrogen bonding, we have carried out the FTIR (Shimadzu IR Prestige – 21) analyses with ATR as a method of measurement and KBr was used as a carrier. The next step was SEM to analysis morphological and pore structured of the composite. We have conducted SEM (JEOL JSM-6510LA) analysis in 20 kV with a gold coated sample previously in the current 30 mA for 55 sec. Hydroxyl number was used to obtain the amount of hydroxyl groups in the composite. In accordance with ASTM D1957 and E222-10, 30 mL of acetic acid anhydride is dissolved in 350 mL of pyridine. Acetylation of the hydroxyl group involves reacting 2 mL sample into 4 mL of acetic acid anhydride. The reaction was performed at a temperature of 90°C for 2 hours. After 2 hours of reaction, 4 mL of distilled water was added to the solution at room temperature. Determination of hydroxyl value can be conducted by titrating reaction resulting solution diluted with 1N KOH solution. The hydroxyl value was calculated following the equation:

$$HV = \left\{ \frac{(56{,}1)(N)(V_B - V_{acet})}{W_{acet}} \right\} + AV$$

Where HV is hydroxyl value (mg KOH/g sample); MW of KOH is 56.1 g/mole; N is normality of titer; V_B is amount of potassium hydroxide solution (mL) required to titrate the blank; V_{acet} is amount of potassium hydroxide solution (mL) required to titrate the acetylated sample; W_{acet} is amount of sample (g) required for acetylation; AV is acid value of the sample (mg KOH/g sample).

3. Results and Discussion

In this study, the PVA used is in higher hydrolyzed grade. This type of PVA has a high degree of crystallinity (40% to 50%), and can increase the adhesion on a hydrophilic surface. In addition, PVA has higher hydroxyl groups in the molecule structure. Zeolites are aluminosilicate tetrahedral framework structure that provides cavities to be occupied by the cations and water molecules. While the other substance of the composite is activated carbon. It is composed of a random arrangement of micro-crystallites and with a strong crosslink among them, then formed a good pore structure. The main characteristics of these three substances plays an important role to undertake the preparation of composite materials. Based on these characteristics, there are several active groups of each substance that can be engineered through chemical bonds, where chemical bond can mix the three substances into a composite.

Analysis of composite formation can be showed from the FTIR study as seen in Fig 1. Several important chemical bonds were formed during the composite preparation process. The IR spectra showed that the wavelength range 400-450 cm^{-1} appear to form the group C - O - C composite. While in the wavelength range 750-850 cm^{-1} appear to form the stretching Si - C aliphatic which is the bond within zeolite and carbon. In the wavelength range 1050 - 1100 cm^{-1} corresponds to the Si - O - C group which is the result of the bond between PVA and zeolite. At a wavelength range of 2900 - 3000 cm^{-1} corresponds to a hydrogen bond and the OH group stretching. This indicates that the hydrogen from polyvinyl alcohol bounds to zeolite, while the hydroxyl groups on the composite PVA/zeolite/carbon has only in compound polyvinyl alcohol, that is making possibility the occurrence of hydrogen bonds OH groups on poly vinyl alcohol.

According to the hydroxyl value analysis, the amount hydroxyl group of PVA can be determined 1,220.85 mg KOH/g sample. After the composite was formed, the hydroxyl value decreased to 620.94 mg KOH/g sample. From the FTIR analysis, the hydrogen bonds formation on the composite was confirmed. The hydrogen bonding formation in this composite have an important role in changing the properties of the material, particularly on the improvement of the material mechanical strength. Improvement of mechanical properties is very influential to the resistance ability of the adsorbent during the adsorption process takes place. As previously explained, that each initial substance has specific properties. In this case, the PVA is essentially insoluble in water, however it is needed as a OH group donor in the adsorbent. With the change in mechanical properties, it is expected that the glass temperature, Tg, to be increased, thus PVA will be insoluble in water and remain as OH groups donor although the hydroxyl value was decline.

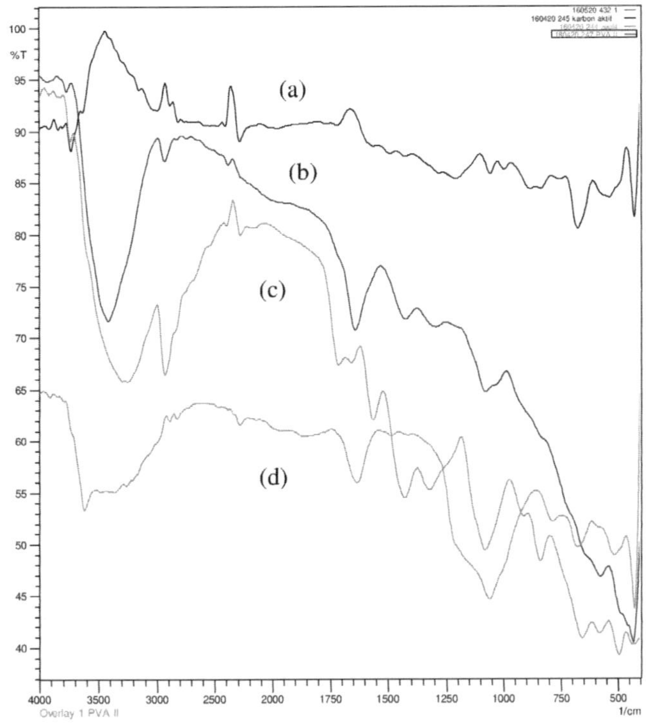

Figure 1. The IR spectra that show the formation of PVA/zeolite/carbon composite and compared with the initial constituent substances. The spectra are (a) activated carbon; (b) PVA/zeolite/carbon; (c) PVA; (d) zeolite.

The morphology of the composite can be analyzed using SEM. In addition, the affinity of the initial substances can be seen as well in order to understand the formation of pore structure and occurring the composite layer specifically as seen in Fig. 2. SEM analysis was performed on two section, first in the interface section to see the morphology and the formation of pores in the top layer, second in the cross section to analyzethe composite layer formation and pore structure. SEM analysis indicate that in the interface was formed uniquely pores of the composite material. These pores become a very important characteristic in the adsorption process, while it would improve a higher surface area. In other words, it would be possible that contact among the adsorbate and adsorbent will increase which will impact to the effectiveness and efficiency in the adsorption process.

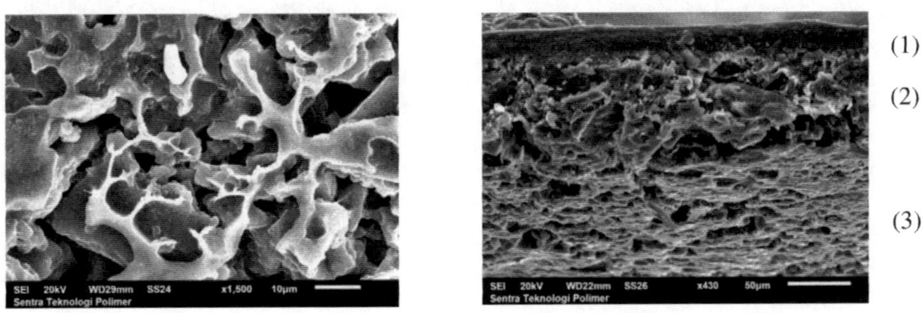

(a) (b)

Figure 2. SEM micrograph analysis of the PVA/zeolite/carbon composite (a) interface section with magnificent 1,500x; (b) cross section with magnificent 430x while in this part consist of three layers (1) PVA; (2) zeolite; (3) carbon.

Furthermore, it appears that the upper layer was smooth that shows the polymer layer. The cross section shows more clearly that the polymer is in the top layer, then zeolites and carbon, respectively. The fact that the first layer is a polymer becomes important considering that PVA is a OH groups donor in the adsorbent, therefore chemisorption process will occur in the first layer adsorption thereafter it will be followed by the process of physisorption on zeolites and carbon layers sequentially. SEM analysis also showed that their regularity and pore structure caused by the formation of hydrogen bonding between PVA with zeolites and zeolite with carbon.

4. Conclusion

In this study, we examined the characteristics of composite materials primarily through supercritical fluid extraction method. We confirmed that the method was instrumental to the unique pore formation and chemical bonding of the three starting substance, in this case hydrogen bonds to form a composite. Composite was expected consisting of three layers namely PVA, zeolites and carbon respectively therefore it has fulfilled characteristics and potential to be used as adsorbent.

References

[1] A K Frolkova V M Raeva. *Theoretical Foundations of Chemical Engineering*. 2010;**44**(4):545 - 56.
[2]. H J Huang, S Ramaswamy, U W Tschirner, B V Ramarao. *Separation and Purification Technology*. 2008;**62**:21.
[3] M Balat, H Balat, C Oz. *Progress in Energy and Combustion Science*. 2008;**34**:23.
[4] S I Mussatto, G Dragone, P M R Guimarães, J P A Silva, L M Carneiro, I C Roberto, A Vicente, L Domingues, J A Teixeira. *Biotechnology Advances*. 2010;**28**:14.
[5] S K Maity. *Renewable and Sustainable Energy Reviews*. 2015;**43**:19.
[6] T Yamamoto, Y H Kim, B C Kim, A Endo, N Thongprachan, T Ohmori. *Chemical Engineering Journal*. 2012;**181 - 181**:6.
[7] J Huang, Y Zhou, K Huang, S Liu, Q Luo, M Xu *Journal of Colloid and Interface Science*. 2007;**316**:9.
[8] C Long, A Li, H Wu, Q Zhang. *Colloids and Surfaces A: Physicochemical and Engineering Aspects*. 2009;**333**:6.
[9] S Kumar, N Singh, R Prasad. *Renewable and Sustainable Energy Reviews*. 2010;**14**:15.
[10] J Kuhn, J M C-Sanchez, J Gascon, S Calero ,D Dubbeldam, T J H Vlugt, F Kapteijn, J Gross. *J Phys Chem C*. 2009;**113**:12.
[11] K Zhang, R P Lively, J D Noel, M E Dose, B A McCool, R R Chance, W J Koros. *Langmuir* 2012;**28**:10.
[12] İ Y İpek, N Kabay, M Yüksel, D Yapıcı, Ü Yüksel. *Desalination*. 2012;**306**:5.
[13] M S Islam, M S Rahaman, J H Yeum. *Colloids and Surfaces A: Physicochemical and Engineering Aspects*. 2015;**484**:10.
[14] M Kozlov, M Quarmyne, W Chen, T J McCarthy. *Macromolecules*. 2003;**36**:6.
[15] N M Mahmoodi, F Najafi, S Khorramfar, F Amini, M Arami. *Journal of Hazardous Materials*. 2011;**198**:8.
[16] E I Unuabonah, A Taubert. *Applied Clay Science*. 2014;**99**:10.
[17] S G Kazarian. *Polymer Science, Ser C*. 2000;**42**(1):24.

IOP Conf. Series: Materials Science and Engineering **201** (2017) 012007 doi:10.1088/1757-899X/201/1/012007

Antimicrobial and water-triggered release characteristics of a copper sulfate-polyvinyl acetate adhesive composite

A. P. O. De Jesus, R. M. L Roxas-Villanueva, M. U. Herrera

Institute of Mathematical Sciences and Physics, University of the Philippines, Los Baños, Laguna, Philippines

aodejesus@up.edu.ph, rrvillanueva3@up.edu.ph, muherrera@up.edu.ph

Abstract. Water-triggered release of antimicrobial solutions is advantageous in inhibiting the growth of bacteria and fungi in moist and wet environments. In this study, we fabricated a composite, by mixing polyvinyl acetate adhesive with copper sulfate solution, which exhibits antimicrobial activities against bacteria. Polyvinyl acetate adhesive serves as the binder and water soluble substance while copper sulfate serves as the antimicrobial agent. The composite was coated in an acetate film and air-dried. To monitor the rate of release of copper ions, the composite was submerged in water and the conductivity was measured. The conductivity saturation time was determined. The composite showed antimicrobial activity against *Escherichia coli* (Gram-negative) and *Staphylococcus aureus* (Gram-positive).

1. Introduction

Wet environments are conducive for the growth of microorganisms. The dependence of microorganisms in water makes it easier for them to thrive in these types of environment. For instance, their number rapidly increases in water-exposed surfaces (e.g., walls, woods, trees) after rainfall. Inhibiting the growth of microorganisms is essential in order to prevent possible causes of diseases that may harm humans, animals, and even plants.

Antimicrobial agents have been incorporated in different materials such as papers, textiles, fibers, films, and others [1-8]. However, if the intention is to release these agents on events wherein the environment becomes wet (a condition where microorganisms greatly multiply), the fabrication of materials that release antimicrobial agents in the presence of water is advantageous. This type of materials could have two integral components: (1) the water-sensitive component and (2) the antimicrobial component. The intention is that when the water-sensitive component interacts with water, the component releases the antimicrobial agents. Polymers have been made to interact with water for studies concerning ion release [9] and shape memory alloys [10]. Polyvinyl acetate (PVA) is a candidate material for the water-sensitive component. It is water-soluble; thus, it can release the encapsulated antimicrobial agents when dissolved in water. Furthermore, it is low-cost, widely available and its viscosity makes it easily be drawn into coatings and films.

Copper-based materials [11-14] have antimicrobial properties. Compared to silver metal which is a well-known antimicrobial agent, copper-based materials are cheaper. Copper sulfate [15-18] is a copper-containing salt which easily dissolves into ions in aqueous media. It can be used as molluscicide for the destruction of snails and slugs. It has antifungal applications and it can be used to inhibit algal growth in plant pots and swimming pools.

In this study, we incorporate copper sulfate (CuSO4) in water-soluble matrix (polyvinyl acetate). We characterize its color and antimicrobial activity and determine its rate of release in aqueous solution.

2. Methodology

2.1 Preparation of Samples

Copper sulfate pentahydrate ($CuSO_4{\cdot}5H_2O$) salts were dissolved in water to prepare a 0.5 M copper sulfate solution. Varying ratios of polyvinyl acetate adhesive and copper sulfate solution were fabricated – 100:0, 95:5, 90:10, 80:20, 70:30, and 60:40. Each mixture was coated in an acetate film and was air dried. The copper sulfate pentahydrate serves as the antimicrobial agent while the polyvinyl acetate serves as a binder and water soluble substance in the sample.

2.2 Characterization

Transmission spectroscopy was done using a USB spectrometer to describe the color of each sample. The antimicrobial activity was determined through Disc Diffusion Method using *Staphylococcus aureus* (Gram-positive) and *Escherichia coli* (Gram-negative) as test microorganisms. The samples were prepared in discs of 10.00mm in diameter. The test reference for this method was "US Pharmacopeia 30-NF 25, 2007".

For the conductivity measurement, acetate film was submerged in a beaker filled with water. Conductivity of water through time was monitored using OAKTON PC 510 pH/conductivity bench meter to determine the rate of release of copper ions.

3. Results and Discussion

We were able to fabricate a blue, translucent, paste-like composite containing polyvinyl acetate adhesive and copper sulfate. Transmission spectroscopy within the blue wavelength was done to characterize the color of these composites deposited on acetate films.

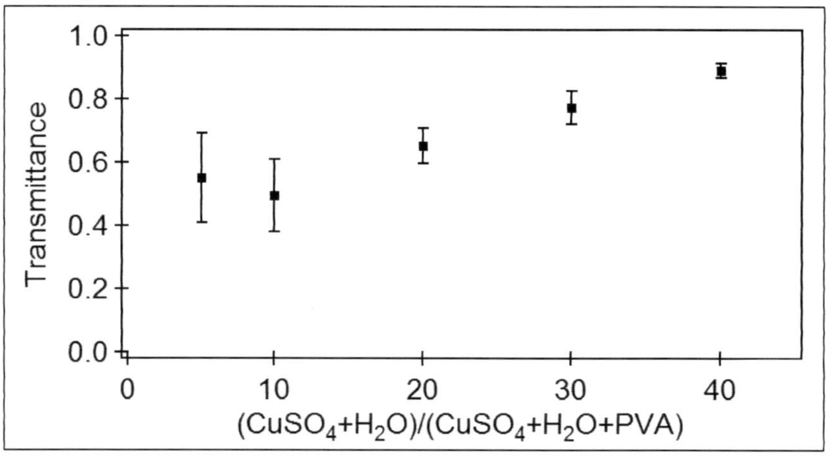

Figure 1. Transmittance of $CuSO_4$-PVA composite on film containing 5%, 10%, 20%, 30%, and 40% $CuSO_4$+H_2O.

Figure 1 shows the plot of the transmittance per amount of copper sulfate. Wavelengths 450-495 nm were selected as the range of the blue color that will be accounted for the embedded copper sulfate in each sample. Based from the graph, the intensity of the blue color generally increases as the amount of copper sulfate increases. However, there is a slight difference between the sample containing 5% and 10% copper sulfate which can be attributed to the topography of the sample. The distribution of mixture in the surface of the air-dried sample was not uniform i.e., a portion of the sample appears to be bluer than the other portion.

The antimicrobial activity of the samples was also determined as shown in Table 1. Cefuroxime was used as positive control for *E. coli* while that of *S. aureus* is Clindamycin.

Table 1. Antimicrobial activity of samples against *Escherichia coli* and *Staphylococcus aureus.*

Sample/Control	*Escherichia coli*		*Staphylococcus aureus*	
	Zone of inhibition (mm)	Inhibitory activity	Zone of inhibition (mm)	Inhibitory activity
$CuSO_4$-PVA composite on film	19.41	complete	14.76	complete
Positive control: Cefuroxime 30 μg	17.28	complete	N/A	N/A
Positive control: Clindamycin 2 μg	N/A	N/A	19.68	complete

The composite containing polyvinyl acetate adhesive containing copper sulfate exhibited antimicrobial activity for both test organisms. The sample tested against *E. coli* (Gram-negative) produced a complete inhibitory activity within the zone of inhibition (ZOI) of 19.41mm which is higher than the positive control. The sample tested against *S. aureus* (Gram-positive) also produced a complete inhibitory activity within the zone of inhibition of 14.76 mm which is lower than the positive control. Gram-negative bacteria possess a thin peptidoglycan, a part cell wall which is responsible for bacterial survival. On the other hand, Gram-positive bacteria possess a thick peptidoglycan. Copper ions may adhere on the cell wall causing the rupture and killing of bacteria which possibly explains why the ZOI of *E. coli* is larger than that of *S. aureus.*

Conductivity of water containing the sample through time was also measured to determine the rate of release of copper ions. Figure 2 shows the plot of conductivity over saturation conductivity (σ/σ_{sat}) versus time over saturation time (t/t_{sat}) of samples containing 5% and 40% copper sulfate solution.

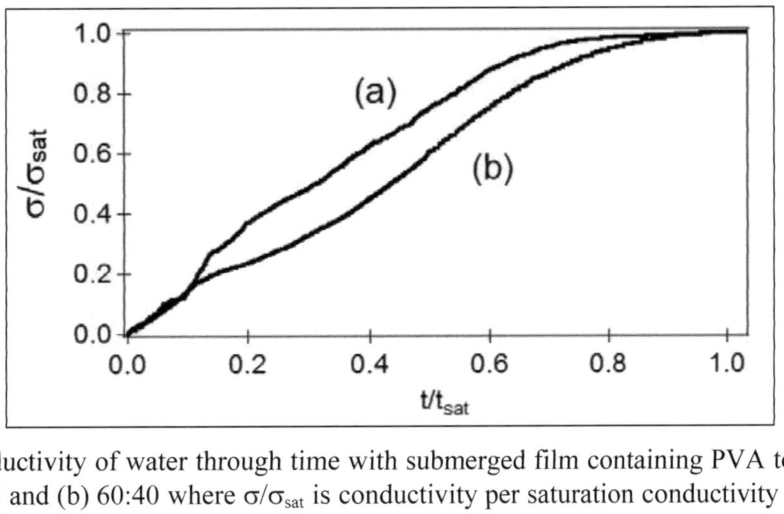

Figure 2. Conductivity of water through time with submerged film containing PVA to $CuSO_4 + H_2O$ ratio of (a) 95:5 and (b) 60:40 where σ/σ_{sat} is conductivity per saturation conductivity and t/t_{sat} is time per saturation time.

PVA molecules were dissolved which led to the dissociation of copper sulfate salts in water. As soon as the sample comes in contact with water, the PVA adhesive which acts as a binder starts to dissolve. Since the dissolution of PVA molecules is not abrupt, the copper ions are released in water gradually as seen in Figure 2. The conductivity of water continues to increase as the copper ions are released. After reaching a certain time, saturation occurs when all the copper ions are released in water resulting to an almost constant conductivity reading.

Table 2 shows the summary of the saturation time and saturation conductivity of the two samples. It can be seen that the sample with 40% copper sulfate solution has a significantly higher saturation conductivity and took longer time to saturate than the sample with 5% copper sulfate solution. This is due to larger amounts of copper ions released in water. The amount of copper ions is essential for its role to adhere and destroy organisms of interest.

Table 2. Comparison of saturation time and saturation conductivity of $CuSO_4$-PVA composite containing5% and 40% $CuSO_4 + H_2O$.

$CuSO_4 + H_2O$ amount	Saturation time (mins.)	Saturation conductivity (μS/cm)
5%	869	9.84
40%	2117	33.04

4. Summary and Conclusion

Samples with different ratios of PVA to copper sulfate were prepared and deposited in film. The copper sulfate served as the antimicrobial agent while the PVA served as a binder for the fabrication of the sample. Between the two samples, the sample with 40% copper sulfate exhibited higher transmittance in the range of the blue wavelength. Antimicrobial activity against *Staphylococcus aureus* (gram-positive) and *Escherichia coli* (gram-negative) were also evaluated resulting to more antimicrobial resistance by the latter organism. The conductivity of water increased through time as the $CuSO_4$-PVA composite on film is submerged in water. Also, the sample with 40% copper sulfate solution showed a higher saturation conductivity than the sample with only 5% copper sulfate solution. The fabricated mixture of PVA and copper sulfate can be smeared on wet surfaces for possible prevention of bacterial growth.

Acknowledgments

The authors would like to acknowledge the financial support of the **STAMP: Standards and Testing Automated Modular Platform** Project funded by DOST-PCIEERD.

References

[1] Dickerson M, Sierra A, Bedford N, Lyon W, Gruner W, Mirau P and Naik R 2013 Keratin-based antimicrobial textiles, films, and nanofibers *J. Mater. Chem.* B **1** 5505-14
[2] Shalini G and Anitha D 2016 A Review: Antimicrobial property of textiles *IJSR* **5** 766-8
[3] Teli M D, Sheikh J 2013 Modified bamboo rayon-copper nanoparticle composites as antibacterial textiles *Int. J. Biol. Macromolec.* **61** 302-7
[4] Booshehri A Y, Wang R and Xu R 2015 Simple method of deposition of CuO nanoparticles on a cellulose paper and its antibacterial activity *Chem. Eng. J.* **262** 999-1008
[5] Tankhiwale R and Bajpai S K 2009 Graft copolymerization onto cellulose-based filter paper and its further development as silver nanoparticles loaded antibacterial food-packaging material *Colloids Surf.* B **69** 164-8
[6] Wang J, Liu W, Li H, Wang H, Wang Z, Zhou W and Liu H 2013 Preparation of cellulose fiber- TiO_2 nanobelt-silver nanoparticle hierarchically structured hybrid paper and its photocatalytic and antibacterial properties *Chem. Eng. J.* **228** 272-80
[7] Ling Y, Luo Y, Luo J, Wang X and Sun R 2013 Novel antibacterial paper based on quaternized

carboxymethylchitosan/organic montmorillonite/Ag NP nanocomposites *Ind. Crops Prod.* **51** 470-9

[8] Nassar M and Youssef A 2012 Mechanical and antibacterial properties of recycled carton paper coatedby PS/Ag nanocomposites for packaging *Carbohydr. Polym.* **89** 269-74

[9] Radheshkumar C and Mnstedt H 2006 Antimicrobial polymers from polypropylene/silver composites – Ag^+ release measured by anode stripping voltammetry *React. Func. Polym.* **66** 780-8

[10] Gu X and Mather P 2013 Water-triggered shape memory of multiblock thermoplastic polyurethanes (TPUs) *RSC Adv.* **3** 15783-91

[11] Grass G, Rensing C and Solioz M 2011 Metallic Copper as an Antimicrobial Surface *Appl. Environ. Microbiol.* **77** 1541-7

[12] Subramanian P, Anu Priya K, Thanka Rajan S, Dhandapani P and Jayachandran M 2014 Antimicrobial activity of sputtered nanocrystalline CuO impregnated fabrcs *Mater. Lett.* **128** 1-4

[13] Das D, Nath B C, Phukon P, Dolui S K 2013 Synthesis and evaluation of antioxidant and antibacterial behavior of CuO nanoparticle *Colloids Surf.* B **101** 430-3

[14] Sonia S, Jayram N D, Kumar P S, Mangaralaj D, Ponpandian N, Viswanathan C 2014 Effect of NaOH concentration on structural, surface and antibacterial activity of CuO nanorods synthesized by direct sonochemical method *Superlattice Microst.***66** 1-9

[15] Sierra M, Sanhueza A, Alcantara R and Sanchez G 2013 Antimicrobial evaluation of copper sulfate (II) on strains of *Enterococcus faecalis*. In vitro study *J. Oral Res.* **3** 114-8

[16] Sun Q, Hu K and Yang X 2014 The efficacy of copper sulfate in controlling infection of *Saprolegnia parasitica J. World Aquacult. Soc.* **45** 220-5

[17] Straus D, Farmer B, Ledbetter C, Beck B, Williams R, Clark M and Freeze M 2016 Use of copper sulfate to control egg saprolegniasis at a commercial sunshine bass hatchery *N. Am. J. Aquacult.* **78** 243-50

[18] Halpern M, Gasith A, Teltsch B, Porat R and Broza M 1999 Chloramine and copper sulfate as control agents of planktonic larvae of *chironomus luridus* in water supply systems *J. Am. Mosquito Contr.* **15** 453-7

IOP Conf. Series: Materials Science and Engineering **201** (2017) 012008 doi:10.1088/1757-899X/201/1/012008

Electrically conductive carbon fibre-reinforced composite for aircraft lightning strike protection

Andrzej Katunin[1], Katarzyna Krukiewicz[2], Roman Turczyn[2], Przemysław Sul[3], Marcin Bilewicz[4]

[1]Institute of Fundamentals of Machinery Design, Silesian University of Technology, Konarskiego 18A, 44-100 Gliwice, Poland
[2]Department of Physical Chemistry and Technology of Polymers, Silesian University of Technology, M Strzody 9, 44-100 Gliwice, Poland
[3]Institute of Theory of Electrical Engineering, Measurement and Information Systems, Warsaw University of Technology, Koszykowa 75, 00-662 Warsaw, Poland
[4]Institute of Engineering Materials and Biomaterials, Silesian University of Technology, Konarskiego 18A, 44-100 Gliwice, Poland

E-mail: andrzej.katunin@polsl.pl

Abstract. Aircraft elements, especially elements of exterior fuselage, are subjected to damage caused by lightning strikes. Due to the fact that these elements are manufactured from polymeric composites in modern aircraft, and thus, they cannot conduct electrical charges, the lightning strikes cause burnouts in composite structures. Therefore, the effective lightning strike protection for such structures is highly desired. The solution presented in this paper is based on application of organic conductive fillers in the form of intrinsically conducting polymers and carbon fabric in order to ensure electrical conductivity of whole composite and simultaneously retain superior mechanical properties. The presented studies cover synthesis and manufacturing of the electrically conductive composite as well as its characterization with respect to mechanical and electrical properties. The performed studies indicate that the proposed material can be potentially considered as a constructional material for aircraft industry, which characterizes by good operational properties and low cost of manufacturing with respect to current lightning strike protection materials solutions.

1. Introduction

Increasing demands to aircraft structures cause that during the last decades more and more aircraft elements started to manufacture using polymeric composites, and elements of modern airplanes, especially the exterior fuselage elements, are manufactured from such materials, primarily from carbon fibre-reinforced plastic (CFRP). The application of CFRP allows for significant increasing of strength-to-mass ratio with respect to its metallic analogues, which entails significant economic benefits related to lower fuel consumption, corrosion resistance, etc. However, manufacturing of exterior elements from polymeric composites implies a problem when a lightning strikes such a structure since practically it is insulating to electrical current. This fact causes that during lightning strikes aircraft exterior composite structures are subjected to serious damage caused by heating and very high temperature in the surrounding of the strike location. Considering a fact that appearance of such damage implies a necessity of element repair and further testing of repaired structure or even replacement of a damaged element, which significantly increase operational costs of aircraft, the effective lightning strike protection (LSP) is highly desired.

LSP solutions which already found an application in aircraft structures of the newest and the biggest passenger aircraft are based on integration of a typical CFRP structure with highly conductive metallic foils or meshes. Recent studies [1-4] confirmed the high effectiveness of such integrated structures during lightning strikes, i.e. the resulting damage characterizes by significantly lower extent in comparison to structures not integrated with metallic layers. Besides the great advantages of the described solutions, the integration process considerably hampers the manufacturing process, and additionally, due to the weak adhesion between polymer and metal initial manufacturing damage may occur. Finally, the complication of the manufacturing process adversely affects the manufacturing costs.

In order to overcome the reflected problems on the metal-polymer interface and simplify the manufacturing process it is suitable to develop a structure which has a capability of conducting electrical current and simultaneously retain mechanical properties of traditional aircraft structures. The alternative LSP solutions developed to date that meet these requirements are based on dispersion of electrically conductive fillers inside an insulating matrix. This solves the problem of manufacturing complexity. In the case of dispersion of metallic or metal-coated particles [5,6] the adhesion problems remains. Moreover, in order to ensure appropriate electrical conductivity properties of a structure the content of metallic particles should be significant, which additionally increases mass of a structure. Other LSP solutions are based on using of carbon nanostructures (CNS) as a conductive filler [7,8]. Such material solutions constitute a valuable alternative to the foil- or mesh-protected composites, however, considering considerably high content of dispersed particles in a resulting structure, the manufacturing costs are too high to consider this approach for industrial manufacturing conditions of aircraft structures.

The development of an electrically conductive composite, which characterized by considerably high electrical conductivity and simultaneously retain mechanical properties of traditional CFRP structures, which is free of manufacturing flaws resulting from internal structural architecture (interface cracking) and considerably inexpensive in manufacturing was the goal of the authors. The detailed concept of such a material was presented in [9]. The developed material is fully organic with good electrical and mechanical properties. The paper describes synthesis and characterization of a conductive matrix as well as the manufactured carbon-reinforced conductive composite, including the high-voltage tests which allow for evaluation of resistance to the simulated lightning strike of the developed material.

2. Synthesis and characterization of a conductive matrix

Following the idea of development the fully organic conductive composite with LSP properties, as a conductive filler the polyaniline (PANI) was considered. As previous studies show [10], this polymer is characterized by electrical conductivity of the semiconductors level, which allow for transport electrical charge through the structure and thus can be potential candidate as a conductive filler. The preliminary numerical simulations [11] as well as analysis of chemical stability and availability of various intrinsically conducting polymers (ICPs) [12] show that PANI has the best properties for LSP applications. However, due to the poor mechanical properties of PANI, an appropriate amount of this filler in insulating polymeric matrix with good mechanical properties (in manufacturing of aircraft composite structures it is usually epoxy resin) should be determined in order to obtain electrically conductive composite with good mechanical properties. For determination of optimal content of PANI numerical simulations were performed. The simulation results show [13] that the desirable properties can be obtained by addition of 30%vol. of PANI into the insulating epoxy.

In order to increase electrical conductivity of PANI it was doped with camphorsulfonic acid (CSA). The morphology of the obtained mixture was homogeneous and the resulting conductivity at the level of unities of S/cm was observed during the bipotentiostatic measurements [12]. The CSA-doped PANI was then used to prepare PANI/epoxy composites. Further studies on characterization of their physical properties show that considering the processibility properties the amount of 30%vol. of PANI in the resulting composite is an optimal value. The performed measurements of electrical conductivity of the resulting PANI/epoxy composite show a decrease of its values with respect to CSA-doped PANI to tenth of S/cm due to the observed agglomeration of PANI particles [12]. Mechanical tests of the obtained PANI/epoxy composite show ca. three times lower ultimate tensile strength with respect to pure epoxy, however it was still acceptable for structural application since the loading will be carried out by the reinforcement added at the next stages.

The performed high-voltage and high-current tests which allow simulating conditions close to those appeared during lightning strike reveal satisfactory results. During high-voltage testing the specimens of 2 mm thickness were subjected to the voltage impulse of 1.2/50 μs with a maximal voltage value of ca. 32 kV using the generator of lightning voltage impulses. The average value of the avalanche breakdown current equaled 1062 A at the constant charge voltage of the generator of 15.5 kV. The concentration of electric current during a strike, which is accompanied by local rapid increase of temperature causes degradation of the tested specimens in the form of burnouts and crumbling of a composite in the area of a strike location. The observed damage with extent of ca. 2-3 mm was observed (see [12] for details). The high-current tests were performed using the generator of lightning current impulses at the constant charge voltage of 11.1 kV for generation of an impulse of 1.2/50 μs. The registered current conducted through the specimen has the maximum value of 8.04 kA. During the test the disintegration of the specimen was observed due to the occurrence of the acoustic wave which acted mechanically on the tested specimen. However, the obtained results confirmed the ability of conducting high-current changes by the developed material.

3. Carbon fibre-reinforced conductive composite

In order to enhance mechanical characteristics of the developed conductive matrix its reinforcing was necessary. Considering the strength requirements to aircraft structures the most exterior fuselage elements of aircraft are manufactured from CFRP. Considering also that carbon fibre (CF) is electrically conductive, which can be considered not only as a reinforcement, but also as a linking medium between conducting particles, it was decided to reinforce the developed conductive matrix by this fibre.

The CF-reinforced composite was manufactured by the plain weave carbon fabric. The composite sheets of 2 mm in thickness were manufactured by manual lamination method and contained 4 layers of carbon fabric. The low molecular epoxy resin (Epidian 6, 80g) was mixed with CSA-doped polyaniline (10%.vol, 16 g) and cured with amine hardener (IDA, 40 g). The cross-linking process was carried out in room temperature for 24 hours.

The prepared composite was then subjected to mechanical tests in order to evaluate the improvement of mechanical properties with respect to the conductive matrix as well as to compare the properties of the obtained structure to those of typical aircraft CFRP structure. The mechanical testing was performed on the specimens prepared for tensile tests with fabric orientation of 0° and 45°, which allows to evaluate both Young and Kirchhoff moduli. The obtained results show that the averaged Young modulus is of 7.185 GPa and the averaged Kirchhoff modulus is of 4.444 GPa.

In order to evaluate the resistance of the obtained composite to high-current impulse discharges, which simulate lightning strike conditions a series of laboratory tests were performed using the generator of lightning current impulses. The testing setup is presented in Figure 1, where the specimen is connected to the flat electrode (bottom side) and a wire electrode (top side). During these tests, the constant charge voltage was increased to 25 kV and the impulse characteristic of 1.2/50 μs remains the same as during the tests of the PANI/epoxy composite. Obtained results indicated that current conducting through the specimen has aperiodic character which is characteristic to the lightning-type discharges with a maximum value of 8.5-9 kA.

(a) (b)

Figure 1. Experimental setup for high-current testing.

Due to the reinforcement presence the tested specimens were not subjected to disintegration after the high-current impulse. However, the energy accumulation accompanied with high temperature around the wire electrode and the resulting acoustic wave, which influences mechanically on the structure the structural damage may be observed. The exemplary damaged structures made of the developed composite are presented in Figure 2.

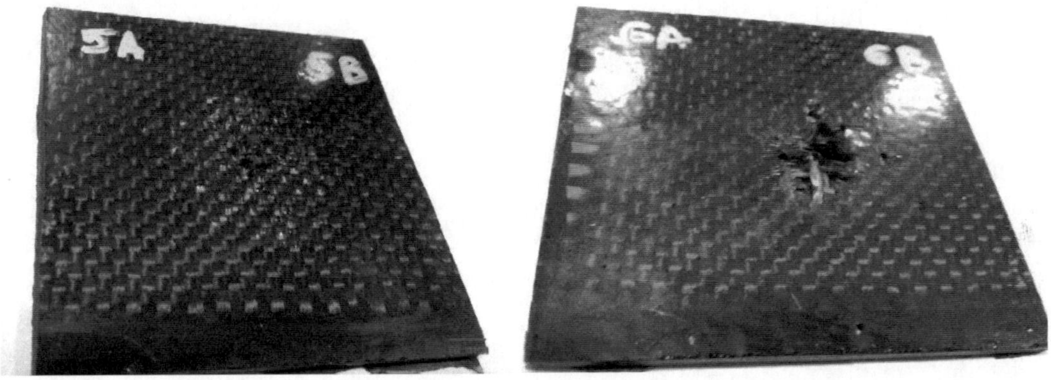

(a) (b)

Figure 2. Exemplary specimens damaged by high-current impulse simulating lightning strike conditions.

One can observe that besides the local lacks of a matrix, which is decomposed and evaporated due to the high temperature, the damage of the fabric can be noticed (see especially Figure 2b). One should also notice that the resulting damage has a surface character and the depth of such a damage is low. The through the-thickness damage was not observed during testing the specimens of the developed composite which confirm that the mechanisms of dissipation of electrical charge and thermal energy work properly.

4. Conclusions

In this study, the authors presented the novel fully organic electrically conductive composite material which can be considered as a inexpensive and effective alternative to currently applied LSP solutions of aircraft structures. The obtained results show that the proposed LSP solution has all the desirable properties including good electrical conductivity on the level of semiconductors as well as excellent mechanical properties comparable with classical CFRP aircraft structures. High-current tests show that the developed composite can successfully carry high-current charges with comparatively low damage. Due to the application of ICP as a conductive filler in the developed composite the technological processes of manufacturing of aircraft structures can be significantly simplified with respect to their analogues with integrated metallic meshes and foils, which additionally influences on decreasing of manufacturing costs of such structures. The presence of carbon fabric, among improving mechanical properties improve carrying electrical charge during lightning strike as well as thermal energy by increasing thermal conductivity of the composite. These factors additionally influences on minimization of damage occurring during lightning strikes.

References

[1] Feraboli P and Miller M 2009 *Compos. A* **40** 954–967
[2] Hirano Y, Katsumata S, Iwahori Y and Todoroki A 2010 *Compos. A* **41** 1461–1470
[3] Kawakami H and Feraboli P 2011 *Compos. A* **42** 1247–1262
[4] Gagné M and Therriault D 2014 *Prog. Aerosp. Sci.* **64** 1–16
[5] Rupke E J 2002 *Lightning Direct Effects Handbook* (Pittsfield: Lightning Technologies Inc.)
[6] Chakravarthi D K, Khabashesku V N, Vaidyanathan R, Blaine J, Yarlagadda S, Roseman D, Zeng Q and Barrera E V 2011 *Adv. Func. Mater.* **21** 2527–2533
[7] Gou J, Tang Y, Liang F, Zhao Z, Firsich D and Fielding J 2010 *Compos. B* **41** 192–198
[8] Han J, Zhang H, Chen M, Wang D, Liu Q, Wu Q and Zhang Z 2015 *Carbon* **94** 101–113
[9] Katunin A, Krukiewicz K, Herega A and Catalanotti G 2016 *Adv. Mater. Sci.* **16** 32–46
[10] Krukiewicz K and Katunin A 2016 *Synth. Met.* **214** 45–49
[11] Katunin A and Krukiewicz K 2015 *J. Polym. Eng.* **35** 731–741
[12] Katunin A, Krukiewicz K, Turczyn R, Sul P, Łasica A and Bilewicz M 2017 *Compos. Struct.* **159** 773–783
[13] Katunin A 2016 *J. Appl. Math. Comput. Mech.* **15** 59–69

Acknowledgements

The publication is supported within the framework of the Rector's grant in the area of scientific and development research. Silesian University of Technology, 10/060/RGJ17/0052.

IOP Conf. Series: Materials Science and Engineering **201** (2017) 012009 doi:10.1088/1757-899X/201/1/012009

Study of the mechanical properties of hybrid composite basalt / alumina / shells for brake lining pads

I K. Adi Atmika[1,2,4], IDG.Ary Subagia[2], I W. Surata[2,4] and I N. Sutantra[3]
[1] Doctoral Student of Engineering Science of Udayana University, Denpasar, Bali, Indonesia
[2] Mechanical Engineering Department of Udayana University, Street of Bukit Jimbaran, Badung, Bali, Indonesia
[3] Mechanical Engineering Department of Institut Teknologi Sepuluh Nopember, Surabaya, Indonesia
[4] tutadi2001@yahoo.com, iwasura@gmail.com

Abstract. Brake lining pad as one of the active safety components in motor vehicles has been studied thoroughly. Asbestos is the main material forming the brake in addition to other alloy materials that have a negative impact on health and the environment. This paper explain the behavior of hybrid composites phenolic resin with basalt/alumina/clamshell powder reinforced on brake lining pad. This materials has been manufactured use compaction and sintering process through any steps, that an emphasis of 2,000 kg for 30 minutes at a constant temperature of 150° C. The research aims to investigate hardness characteristic of hybrid composite that test using the vickers according to standard ASTM E-384. The reinforced materials and phenolic resin composition is 60%: 40%. The results show for the average hardness VHN to 24.18, 25.11, 26.34, 27.21 and 28.83. The average hardness hybrid composite shows the hardness harder than asbestos materials.

1. Introduction

The braking system is one part of a vehicle that has a very important role for the comfort and safety of the rider. Brake serves to decelerate or stop the vehicle by means of friction between the brake drum system or disc system. Brake lining is an important part of the braking system in a motor vehicle and so deserve more attention by government in order to protect consumers and reduce the percentage of the causes of accidents on the highway.The brake lining pad is the brake components that directly rub against the rotating parts in the brake system corresponds to the type and brakes used. In general, the brake lining material made of asbestos because of its superior in temperature resistance up to 800C [1] and is also good at sound absorption and low in water absorption [2]. However, asbestos material has been shut down because it has carcinogenic properties that have a negative impact on the environment and human health [3]. Then, asbestos fiber material is replaced with aramid fiber, Kevlar, Twaron, rockwool, fiberglass, potasiumtitanate, carbon fiber, graphite, cellulose, vermiculate, steel fiber, BaSO4, resin, Nitrile butadiene rubber.

Utilization of hybrid composite materials on motor vehicle components has become popular because it effectively increases the stress and strain due to the imposition of impact as well as reduce the cost of reinforcing material [4]. Hibridisasi is the process of forming a composite material with new traits which made merging two or more materials that have characteristics which differ in a

IOP Conf. Series: Materials Science and Engineering **201** (2017) 012009 doi:10.1088/1757-899X/201/1/012009

matrix binder (thermoplastic or thermosetting) or otherwise of the reinforcing fibers bonded with two or more types / phase matrix of different [5]. Nowadays a lot of research on hybrid composite material with epoxy basalt reinforced that has been done. Czigany [6], has conducted research on the characteristics of epoxy composite with basalt fiber reinforcement, and the following year a hybrid composites reinforced basalt fiber, polypropylene matrix has been produced, in which the strength properties of the hybrid composites due to increased surface treatment after mechanical tests and microscopic analysis. Besides it also studied the characteristics of basalt fiber hybrid composite laminates due to the imposition of tensile and flexural [7]. Liu, et al [8] conduct research of mechanical properties of basalt fiber composite with the transport system where the basalt is very possible material applied to the transportation environment.
In the contect in order to develope newly material for lining brake pads, basalt as new member of organic material was introduced. Basalt is a natural fiber results volcanic eruption that has a heat resistance of up to above 1500C [9], resistant to corrosion, low water absorption and is resistant to chemical treatments and non-toxic [10]. In the last five years of research related to the characteristics of basalt has been done with the conclusion that the material of basalt have the physical and mechanical properties very well, high ductility, as well as the resilience of the high wear [11], and may replace fiber glass [12]. Then the properties of basalt most important is to have a low thermal conductivity. These properties are used to address potential brake problems today.

In the context of the development of the brake lining material was developed using hybridization method powder basalt) with other natural materials,powder shells and alumina are tied with epoxy as a composite material. The hybrid composite have been manufactured using the five kind of weight fraction ratio. The research aims to investigate hardness characteristic of hybrid composite that test using the Vickers according to standard ASTM E-384.

2. Materials and Method

Basalt powder has been implemented as a reinforcement of phenolic resin composite material as a motor vehicle brake linings. Basalt as described above, are derived from the hardening of lava mountain (figure 1). The grain size basalt, alumina and clamshell powder used in this study was 0.0074 millimeter. Besides, the mechanical properties of basalt is shown as table 1. Clamshell powder material produced by burning the shells are pulverized, the powder can be used as a mixture or in addition to the manufacture of brake linings. The clamshell powder material consists of 66.70% CaO, 22.28% MgO, 7.88% SiO_2, 1.25% Al_2O_3 and 0.03% Fe_2O_3. Phenolic resin material (Sumiliteresin) which is one type of resin in the form of powder. Where the phenolic resin used in this study coded PR -51 510 I usually white or yellow.

The brake lining pads in this study was manufactured by mixing, compacting and sintering as well as cutting the specimen (figure 2). The composition of the composite is shown in Table 2 is based on the standard ASTM D 3171-09. In the process of making the emphasis is done with the loading of 2000 N for approximately 30 minutes for each veriasi composite brake lining. Furthermore, the composite sintering at a temperature of 150°C for each variation.

Figure 1. Material basalt

Table 1. Mechanical properties of basalt material

Properties of Basalt	Value (unit)
Density	2600-2630 (kg/m^3)
Tensile strength	500k-550k (psi)
Sintering Temperatur	1050 (^oC)
Operation Temperatur	-265-+700 (^oC)
Modulus of elastisitas	9100-1100 (kg/mm^3)
Mohs Hardness @20^oC	5-9
Melting point	1170(^oC)
Heat resistance	700-1000 (deg.C)
Elongation at break	3.15(%)

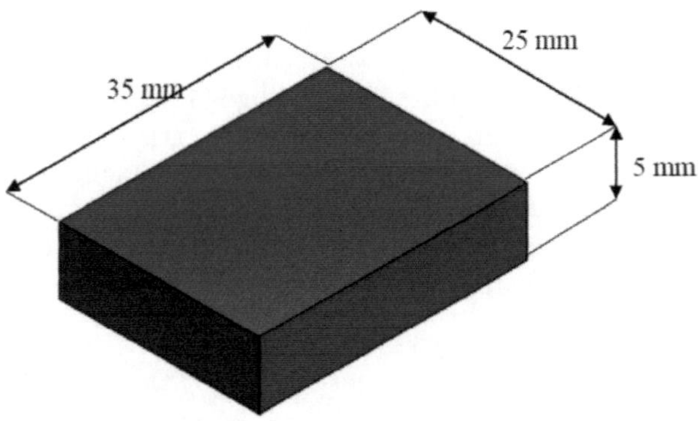

Figure 2. Specimen geometry

Table 2. The weight fraction ratio of hybrid composite

Composition	Basalt powder	Clamshell powder	Alumina powder	Phenolic Resin
I	45	5	10	40
II	40	10	10	40
III	35	15	10	40
IV	30	20	10	40
V	25	25	10	40

Hardness test use vickers method according ASTM E384-99 (Standard test method for micro indentation hardness of materials).

3. Result and Discussion

The average result of VHN (Vickers Hardness Number) of each composition and comparison with the hardness of the brake lining pads that asbestos material (X) are shown in figure 3.

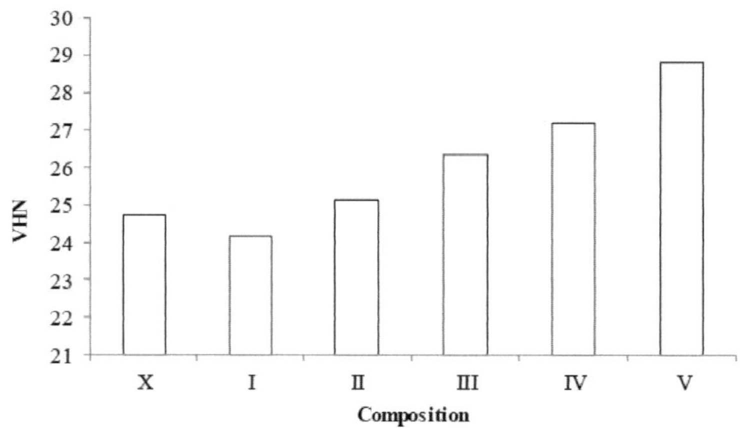

Figure 3. Vicker Hardness Number oh the hybrid composite and asbestos

Figure 3 shows the characteristic VHN diagram of brake lining pads using hybrid composite. It was seen comparison between the asbestos material. It is a connection diagram of the composite mixture composition variation VHN where the decreasing amount of content on the specimen basalt greater hardness value. From these test results obtained hardness value composition I is 24.72 with the content of basalt 45%, the hardness value composition II is 25.40 with the content of basalt 40%, the hardness composition III is 25.67 with the content of basalt 35%, the hardness composition IV is 25.74 with the content of basalt 30%, and HK5 hardness value is 26.55 with basalt content of 25%. From the results of these tests show that the brake lining asbestos has a hardness value is at the middle of the range only between compositio I and composition II. According to the result the conclusion is the composition the number of basalt powder has a significant influence on the rate of hardness. Beside, basalt/clamshells are a natural alternative material that can be used as an alternative material for asbestos brake pad replacement.

4. Conclusion

This study concludes that the weight fraction at each variation of composite brake lining against violence is a reduction in the percentage of basalt and increase the percentage of powder shells, as well as a constant percentage of aluminum oxide in any composition produces a hardness value higher. The comparison between the brake lining asbestos with composite brake then obtained hardness value is not much different, therefore it can be concluded that the brake lining Hybrid Composite potential as a replacement brake lining asbestos material with excellence environmentally friendly.

5. References

[1] Sivarao, M. Amarnath, M.S.Rizal, and Kamely, A. 2009. An investigation toward development of economical brake lining wear alert system. International Journal of Engineering & Technology IJET-IJENS vol. 09, No. 09, pp. 63-68

[2] Aigbodion, V. S. , Akadike, Hassan, U.S.B., Asuke, F., and Agunsoye, J.O.. 2010. *Development of asbestos - free brake pad using bagasse*, Tribology in industry, vol. 32, No.1, pp. 12-18.

[3] Louis St, 2004. U.S. Survey shows imports of asbestos brake materials increasing. St. Louis.
[4] Kedar S Pandya, Ch. Veerraju, and N. K. Naik. 2011. Hybrid composites made of carbon and glass woven fabrics under quasi-static loading. Materials and Design, vol. 32 pp. 4094–4099.
[5] Ashby, M. F. and Y. J. M. Brechet,. 2003. Designing hybrid materials. Acta Materialia, vol. 51, pp. 5801-5821.
[6] Czigany, T. 2005. Basalt fiber reinforced hybrid polymer composites. Materials Science, Testing and Informatics Ii, vol. 473-474, pp. 59-66.
[7] Ary Subagia, I.D.G., and Y. Kim. 2013. A study on flexural properties of carbon-basalt/epoxy hybrid composites. Journal of Mechanical Science and Technology, vol. 27 (4), p. 987~992.
[8] Qiang Liu, Montgomery T. Shaw, and Parnas, R. S. 2006. Investigation of basalt fiber composite mechanical properties for applications in transportation. POLYMER COMPOSITES, vol. 27, pp. 41-48.
[9] Singha, K. 2012. A short review on basalt fiber. International Journal of Textile Science, vol. 1 (4), pp. 19-28.
[10] Fiore, V. G. Di Bella, and Valenza, A. 2011. Glass–basalt/epoxy hybrid composites for marine applications, Materials and Design, vol. 32, pp. 2091–2099.
[11] Dalinkevich A.A., Gumargalieva K.Z, Marakhovsky S.S, and Soukhanov A.V. 2009. Modern basalt fibrous materials and basalt fiber-based polymeric composites. Journal of Natural Fibers, vol. 6, pp. 248-271.
[12] Lopresto, V.C. Leone, and I. De Iorio,. 2011. Mechanical characterisation of basalt fibre reinforced plastic. Composites Part B: Engineering, vol. 42, pp. 717–723.

IOP Conf. Series: Materials Science and Engineering **201** (2017) 012010 doi:10.1088/1757-899X/201/1/012010

Synthesization and characterization of poly(lactic-*co*-glycolic acid) / calcium phosphate bone cement from crab shells

M R Abdul Hanan[1], N M Daud[1], L H Ismail[1] and S Saidin[1,*]
[1]Faculty of Biosciences & Medical Engineering, Universiti Teknologi Malaysia, 81310 UTM Johor Bahru, Johor, Malaysia

redzuanhanan@gmail.com

Abstract. An injectable calcium phosphate (CaP) bone cement has been widely used for musculoskeletal and bone disorder due to its biocompatible and osteoconductive properties. In this study, CaP was successfully synthesized from crab shells by a wet chemical route. Poly(lactic-*co*-glycolic acid) (PLGA) microspheres which have been produced through a double emulsion technique were incorporated into the CaP mixture for the purpose of bone cement solidification. The ratio of both compounds, CaP and PLGA, were set at 8:2. The CaP and PLGA/CaP bone cement were analyzed by ATR-FTIR, FESEM-EDX and contact angle analyses. The bone cement was composed of CaP and PLGA where the micro-powders of CaP were agglomerated on the PLGA microspheres. Addition of the PLGA has increased the hydrophilicity of the bone cement which will be beneficial for materials degradation and bone integration.

1. Introduction

Natural products such as chitin, chitosan and calcium phosphate (CaP) derived from waste marine products have received much attention by researchers [1]. Waste marine products including crab shell and shrimp shell which composed of calcium carbonate are converted into CaP derivatives such as hydroxyapatite (HA) or biphasic calcium phosphate (BCP) through a wet chemical route [2]. These CaPs are biocompatible, bioactive and osteoconductive which have great potential to be substituted as a bone cement in orthopedic, maxillofacial and reconstructive applications [3]. It is also has been used as a scaffold for tissue material engineering due to its main composition of calcium and phosphorus that similar to bone minerals [4]. However, CaP powders are difficult to be casted into block or solid structure before or upon implantation as it is brittle [5]. Therefore, CaP should be fixed inside a paste or injected materials for the purpose of filling bone defects [3].

Despite of its excellent biocompatibility, CaP (mostly HA) has low degradation rate which can limits bone growth onto biomaterials [6]. Thus, incorporation of natural or synthetic polymers such as poly(lactic-*co*-glycolic acid) (PLGA) into a CaP bone cement may increase its degradability [6]. PLGA has been approved by Food and Drug Administration (FDA) and widely used due to its biodegradable and biocompatible properties. Degradation of PLGA microspheres, hydrolytically, will produces acidic nature which composed of monomers lactic and glycolic acid that may combine with poorly degrade CaP bone cement [6].

Therefore, the purpose of this study was to synthesis CaP powders from crab shells and incorporate the powders into PLGA microspheres to form a bone cement composite. The characterization analyses are concerning the composition, morphology and hydrophilicity of the bone cement using attenuated total reflectance-Fourier transform infrared spectroscopy (ATR-FTIR), field emission scanning electron microscopy-energy dispersive X-ray spectroscopy (SEM-EDX) and contact angle instruments, respectively.

2. Methodology

2.1. Sample preparation

2.1.1 Synthesization of calcium phosphate from crab shells. Waste crab shells were collected from local seafood stores and restaurants. They were crushed into small pieces, washed and pre-treated through an ultrasonic cleaning for 15 minutes to remove contaminants and residues. The crab shells were then rinsed twice with deionized water and dried overnight. The crab shells were treated with 1.0 M NaOH at 80°C for 1 hour under stirring condition. Then, the crab shells were rinsed twice with deionized water. Calcium carbonate that was contained in the crab shells were heat-treated in a muffle furnace (BSK-1700X-S, Berkeley Scientific, USA) at 1000°C for 5 hours. The calcium oxide (CaO) obtained from the sintering process was diluted with deionized water to form 0.6 M calcium hydroxide (CaOH).

The CaP was synthesized by titrating 0.6 M of 100 mL CaOH into 0.2 M of 100 mL diammonium phosphate, ($(NH_4)_2HPO_4$) at 40°C under stirring condition. The pH was adjusted and controlled at 9, by adding either ammonium (NH_4) or nitric acid (HNO_3). The CaP mixture was further stirred at 40°C for 4 hours. The homogenized CaP mixture was leaved at room temperature for 24 hours to allow complete reaction and precipitation. Afterward, the CaP precipitates were filtered using a filter paper and dried in an oven at 110°C for 5 hours. Further heat treatment at 1000°C for 4 hours was performed for the calcination purpose. Finally, the CaP was crushed using a mortar and pestle to form fine CaP powders.

2.1.2 Preparation of PLGA microspheres. Microspheres of PLGA were prepared by an established method of double emulsion technique [5]. 1 g of PLGA was dissolved in 4 mL of dichloromethane (DCM) inside a 50 mL conical tube, followed by addition of 500 μL of deionized water. The mixture was emulsified at 8000 rpm for 90 seconds. A total of 6 mL of 0.3% polyvinyl alcohol (PVA) solution was added into the mixture and it was emulsified again at similar speed and duration. Then, the emulsified mixture was transferred into a beaker that contained 394 mL of 0.3% PVA solution under stirring condition. 400 mL of 2% isopropyl nitrate (IPN) solution was immediately added into the beaker and the mixture was stirred for 1 hour. Then, the mixture was allowed to rest for 15 minutes. The clear top supernatant was decanted and the left suspension was centrifuged at 1500 rpm for 15 minutes. The microspheres were collected, freeze-dried for 24 hours and stored at -20°C.

2.1.3 Preparation of PLGA/CaP bone cement. The composition of bone cement was set at 80% CaP and 20% PLGA. 0.8 g of CaP and 0.2 g of PLGA microspheres were added into a plastic syringe. Then, 350 μL of 2% sodium phosphate dibasic (Na_2HPO_4) solution was added into the composite and it was mixed vigorously. The mixture was then injected into a plastic mold with a diameter of 1.1 cm and a thickness of 0.4 cm. Finally, the casted bone cement was left at room temperature for 24 hours.

2.2. Surface characterization

The chemical functionalities of the samples were characterized by ATR-FTIR (Nicolet iS5-IR Spectrometer, Thermo Scientific, USA). The obtained spectra were analyzed at a frequency interval of 500 cm^{-1} to 1000 cm^{-1}. The number of scanning was set at 36 with a resolution of 4. Meanwhile, the chemical composition and morphology of the samples were examined by FESEM (SU8020, HITACHI, Japan) equipped with EDX (SU8020, HITACHI, Japan). Each sample was sputter coated with an ultrathin platinum film to enhance image resolution and visualization.

For the contact angle measurement, 1.0 g of CaP powders were compressed for 60 seconds at 1000 Psi by using a pellet compressor (E-Z Press, International Crystal Laboratories, USA) to form a pellet (diameter = 1.4 cm; thickness = 0.3 cm). The hydrophilicity of the samples was then investigated by using a video contact angle instrument (VCA, Optima, AST products, Inc., USA) at 6 μL/min. The angle of dropped 0.5 mL deionized water was captured within 10 seconds. The average data of both samples were determined by measuring the angle of water drop at three difference spots.

3. Results and discussion

3.1 ATR-FTIR analyses

Figure 1 shows ATR-FTIR spectra of PLGA/CaP bone cement and CaP powders derived from crab shells. The phosphate peaks of $V_1 PO_4^{3-}$ (961 cm^{-1}), $V_3 PO_4^{3-}$ (1020 cm^{-1} and 1085 cm^{-1}) and $V_4 PO_4^{3-}$ (603 cm^{-1} and 625 cm^{-1}) were presented in both samples, clarifying the composition of phosphate group in the CaP powders and the PLGA/CaP bone cement. These three vibration peaks show a tendency of HA formation as stated in other study [7]. However, further analyses such as X-ray diffractometer (XRD) should be performed to confirm the formation of HA compound since CaP has several derivatives such as HA, TCP, TTCP, OCP, etc [8]. After the incorporation of PLGA microspheres into the bone cement, the carbonyl group, C=O was noticed at 1655 cm^{-1}. The peaks at 560 and 1120 cm^{-1} were assigned to C–O bond stretching vibration of PLGA. While the peak in between of 3200 and 3540 cm^{-1} was corresponded to the hydroxyl group, OH^- in the PLGA/CaP bone cements [9]. No traces of acid phosphate (HPO_4^{2-}) peak at 530 - 540 cm^{-1}, demonstrating a complete phase of apatite formation [10].

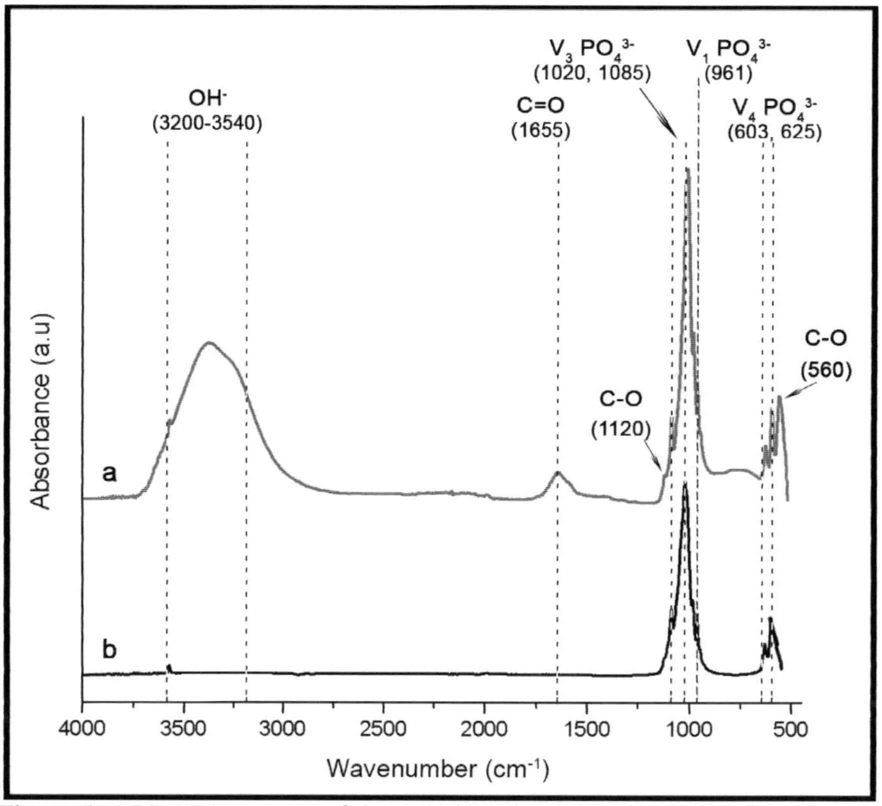

Figure 1. ATR-FTIR spectra of a) PLGA/CaP bone cement and b) CaP powders

3.2 FESEM-EDX analyses

The elemental composition of the CaP powders and the PLGA/CaP bone cement was analyzed and expressed as atomic percentage (Table 1). The Ca/P ratio was calculated to be 1.59 and 0.61 for the CaP powders and the PLGA/CaP bone cement, respectively. The Ca/P ratio of the bone cement was extremely reduced due to the utilization of ($(NH_4)_2HPO_4$) as a hardener in the bone cement which increased the composition of phosphorous element. Figure 2 shows morphology of CaP powders and PLGA/CaP bone cement. The diameter of the Ca/P powders derived from the crab shells was noticed to be less than 1 μm. Nano size of CaP is important to improve the mechanical properties of the materials and may aid in protein adsorption and cell adhesion as compared to micron size of CaP [11, 12]. The incorporation of PLGA into the CaP mixture has caused the agglomeration of CaP powders on the surface of PLGA microspheres indicating good physical contact between these two compounds.

Table 1. Elemental composition of CaP powders and PLGA/CaP bone cement

Sample	Atomic percentage (wt%)			
	C	O	Ca	P
CaP powders	17.89	55.09	15.33	9.64
PLGA/CaP bone cement	22.96	52.19	8.43	13.86

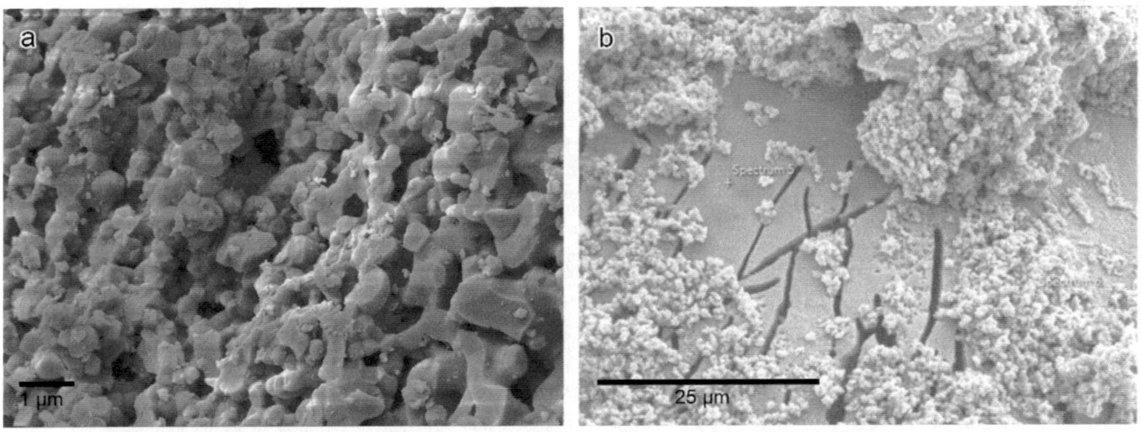

Figure 2. Surface morphology of a) CaP powders and b) PLGA/CaP bone cement

3.3 Contact angle analyses

A sessile drop method was used to determine the hydrophilicity of the samples (Figure 3). The static contact angle of the CaP pellet was found to be 16.63±3.00° revealing the nature of CaP which is hydrophilic [10]. Upon the PLGA incorporation, the hydrophilicity of the bone cement was increased with the contact angle measurement of 14.13±3.94°. These results are in parallel with the FTIR analyses which demonstrated the presence of OH^- group in the PLGA/CaP bone cement. Improvement in hydrophilicity property of the bone cement will aid in tolerate degradation of the bone cement while allowing osteoblast cells to attach and integrate within the bone defect area [13]. Furthermore, this property is crucial for bone growth and bone development upon restoration [13]. According to Habraken *et al.* [14], the hydrolyzed PLGA will increase the porosity of bone cement which is useful for water intake and cell attachment. Moreover, it will also contribute to extra initial strength of the composite as it results in equal distribution of relatively small and polydisperse microspheres [14].

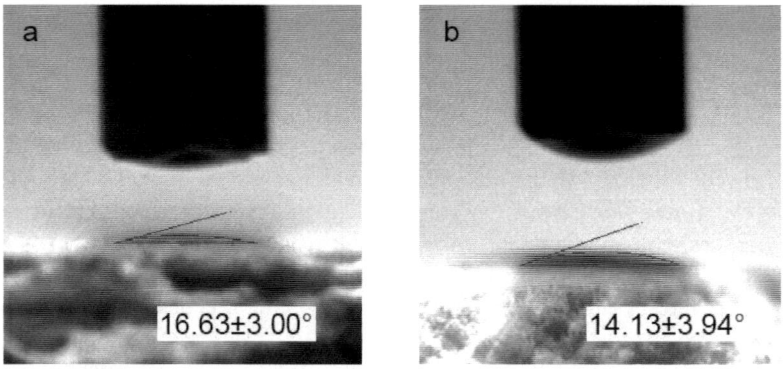

Figure 3. Contact angle measurement on a) CaP pellet and b) PLGA/CaP bone cement

4. Conclusion

The CaP powders were synthesized from waste product of crab shells. The incorporation of CaP powders and PLGA microspheres are necessary to fabricate an injectable bone cement which is preferable for the filling of bone defect area. Identical composition of CaP and PLGA were noticed in the PLGA/CaP bone cement with the agglomeration of CaP powders on the surface of PLGA microspheres. Increased in the hydrophilicity of the bone cement upon the addition of PLGA may improve the degradation rate of the bone cement for bone cells integration.

References

[1] Shavandi A, Bekhit A El-D A, Ali M A, Sun Z 2015 *Int. J. Biol. Macromolec.* **80** 445
[2] Qian J, Xu W, Yong X, Jin X, Zhang W 2014 *Mater. Sci. Eng., C* **36** 95
[3] Khairoun I, Magne D, Gauthier O, Bouler J M, Aguado E, Daculsi G, Weiss P 2002 *J. Biomed. Mater. Res.* **60** (4) 633.
[4] Ge H, Zhao B, Lai Y, Hu X, Zhang D, Hu K 2010 *J. Mater. Sci.-Mater. Med.* **21** (6) 1781.
[5] Habraken W J E M, Wolke J G C, Mikos A G, Jansen J A 2008 *J. Biomater. Sci., Polym. Ed.* **19** (9) 1171.
[6] Lanao R P F, Leeuwenburgh S C G, Wolke J G C, Jansen J A 2011 *Acta Biomater.* **7** (9) 3459.
[7] Nordin J A, Prajitno D H, Saidin S, Nur H, Hermawan H 2015 *Mater. Sci. Eng., C* **51** 294.
[8] Xie Y, Liu X, Chu P K, Ding C 2006 *Surf. Sci.* **600 (3)** 651.
[9] Nenad I, Victoria W, Zorica A, Tatjana M K, Vuk U, Dragan U 2016 *Mater. Sci. Eng., C* **60** 357.
[10] Gergely G, Wéber F, Lukács I, Tóth A L, Horváth Z E, Mihály J, Balázsi C 2010 Ceram. Int. **36** (2) 803.
[11] Wei G and Ma P X 2004 *Biomaterials* **25** (19) 4749.
[12] Wang H, Li Y, Zuo Y, Li J, Ma S, Cheng L 2007 *Biomaterials* **28** (22) 3338
[13] Chang H-I. and Wang Y 2011 Cell responses to surface and architecture of tissue engineering scaffolds *Regenerative Medicine and Tissue Engineering - Cells and Biomaterials (*InTech) chapter 27 pp 569-588.
[14] Habraken W J E M, Wolke J G C, Mikos A G, Jansen J A 2006 *J. Biomater. Sci., Polym. Ed.* **17** (9) 1057.

Acknowledgement

The professional service charge, chemical reagents and materials used in this study were supported by Ministry of Higher Education (MOHE), Malaysia through Tier 2, Research University Grant (Q.J130000.2645.11J68).

Effect of Mineral Filler type on Strength of Roller Compacted Rubbercrete for Pavement Applications

Musa Adamu[1]*, Bashar S Mohammed[1], Nasir Shafiq[1]

Department of Civil & Environmental Engineering, Universiti Teknologi Petronas, 32610 Bandar Seri Iskandar, Perak, Malaysia

Corresponding author: musa.adamu_g03310@utp.edu.my

Abstract. This paper investigated the possibility of using crumb rubber as partial replacement to fine aggregate in roller compacted concrete for pavement applications where fine aggregate was replaced with crumb rubber at 0%, 10%, 20% and 30% to produce roller compacted rubbercrete (RCR). In order to achieve a combined aggregate gradation similar to the one recommended by ACI 211.3R and US Army Corps of Engineers method, fly ash was used as mineral filler. Several trials were done to achieve the combined grading, and finally a combination of 55% fine aggregate, 40% coarse aggregate and 5% fly ash as mineral filler was used. Silica fume was then used as mineral filler instead of fly ash and the fresh density and compressive strength were compared. A target flexural strength of 4.5 MPa and the mix design was developed using the geotechnical concept according to ACI 211.3R/CRD-C161-92. The Results showed that fresh density and compressive strength decreases with increase in partial replacement of fine aggregate with crumb rubber. RCR containing silica fume showed lower fresh density and higher compressive strengths than that with fly ash. However, all the mixes achieved a strength higher than the design target strength at 28 days except for 30% crumb rubber containing fly ash mineral filler, while 10% crumb rubber achieved target strength even at 3 and 7 days for silica fume and fly ash.

1. Introduction

As the world population keeps increasing, likewise the number of vehicle usage due to the rapid demand in transportation which is a basic necessity of life. This leads to increase in waste tire generation [1]. Disposal of waste tires is becoming a serious issue to most countries as the world is now focusing towards sustainability of environment. In most highly technological advanced countries it was estimated that every one person generate one waste tire annually and the total number generated annually stands to about 1 billion, with future estimation to reach 1.2 billion by 2030. Globally approximately 4 billion numbers of waste tires are discarded as landfills yearly where more than half of it is dumped without any pre-treating [2-5]. In the US, the waste tire generation increased rapidly from 3.8 million tons in 2013 to more than 4.1 million tons in 2015 where only about 7% are used for civil engineering applications [6, 7]. These waste tires constitute parts of solid waste where its management is becoming a major concern globally as it is non-biodegradable, it will only be stockpiled when disposed as landfills causing aesthetic and health hazards to the environment by providing shelter and breeding grounds for rodents, mosquitoes, snake etc. [8]. Several attempts of waste tire management are been practiced worldwide, some methods of recycling or reusing waste tires include; production of fuel for cement kiln, in asphaltic pavement, production of carbon black, in marine as artificial reef, in pyrolysis i.e. production of tire derived fuel [4, 9, 10]. The construction industry keeps expanding due to rapid development in technology with concrete been the most used construction material and the most used substance in the world after air and water, one of the major challenges facing the world of construction is the sustainability of the natural materials mostly aggregate as their production involves pollution to the environment [2, 4, 11]. One of the possible ways to reduce the effect of shortage in construction materials, sustainability of environment and waste tire disposal issues is by utilizing the waste tire as partial replacement to natural aggregate in concrete to produce rubberized concrete [12], and can be used in roller compacted concrete to produce roller compacted rubbercrete (RCR) [13]. This will also reduce the cost of concrete by reducing the amount of costly aggregate used [4, 9, 14, 15]. Waste tires are grounded into smaller particles ranging from 75μm to 4.75mm with the steel, fiber, dust and any contaminant removed, and used as crumb rubber to partially replaced fine aggregate in concrete [9].

In this paper the effect of partial replacement of fine aggregate with crumb rubber on roller compacted rubbercrete (RCR) for pavement applications was studied and the effect of replacing fly ash with silica fume as mineral filler was also studied. Crumb rubber (CR) was used from 0% to 30% in multiples of 10%, where fly ash and silica fume were used as mineral fillers at 5%.

2. Materials and Methods

This section describe the materials used and the experimental procedures adopted to obtain the results of the study

2.1. Materials

In this study, Portland cement of grade 43 with specific gravity of 3.15 is used, natural river sand was used as fine aggregate with specific gravity of 2.65, fineness modulus 2.86 and water absorption 1.24%. Two sizes of coarse aggregate of NMSA 19 mm and 10 mm were used as coarse aggregate with specific gravities 2.66 and 2.55 respectively. Three sizes of crumb rubber mesh 30, 1-3 mm and 3-5 mm with specific gravities 0.95, 0.9 and 0.94 respectively were used in proportions 40%, 40% and 20% respectively. Figure 1 shows the grading of fine aggregate and crumb rubber. While the properties of fly ash and silica fume used are shown in Table 1.

Table 1: Properties of fly ash and silica fume

Oxides (%)	**SiO_2**	**Fe_2O_3**	**CaO**	**Al_2O_3**	**MgO**	**K_2O**	**SO_3**	**TiO_2**	**P_2O_5**	**LOI**	**S.G**
Fly ash	34.5	23.6	19.0	12.8	2.27	2.08	1.49	1.46	1.27	-	2.3
Silica Fume	91.6	1.68	0.91	1.69	-	1.19	0.33	-	-	2.75	2.26

2.2 Methods

In order to evaluate the performance of roller compacted rubbercrete (RCR) pavement with partial replacement of fine aggregate with crumb rubber, 8 mixes were prepared by varying the percentage of CR content at 0%, 10%, 20% and 30% by volume of fine aggregate for 5% fly ash and 5% silica fume as mineral fillers. The mixes with 0% CR for 5% fly ash were taken as the control. The combined aggregate grading curve as shown in Fig 2 was obtained by combing 55% fine aggregate, 20% 19 mm NMSA, 20% 10 mm NMSA, and 5% mineral filler

A target flexural strength of 4.5 MPa which is equivalent to strength C30/37, 37MPa cube strength (BS EN 1992-2-1) was selected, and the mix design was developed using the geotechnical concept according to ACI 211.3R/CRD-C161-92, the water and cement contents were obtained using the moisture content-dry density relationship according to ASTM D1557. The cement content used was 13% by weight of dry aggregate. The fresh density was determined according to ASTM C1170; while compressive strengths were obtained for 3, 7 and 28 days of curing according to ASTM C 1435, and the compaction was done using a vibrating hammer.

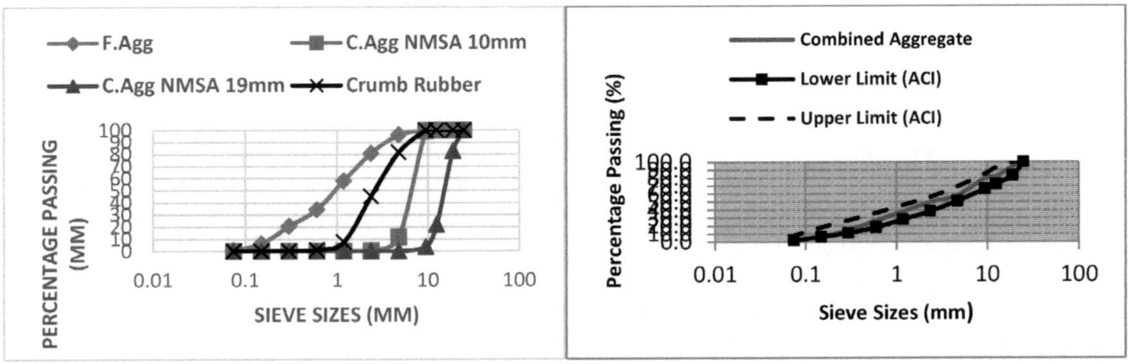

Figure 1 Sieve Analysis (a) Aggregates (b) Combined aggregate gradation

3. Results and Discussions

3.1 Effect of Crumb Rubber on Fresh Density of Roller Compacted Rubbercrete

The fresh density of RCR mix decreases with increase in partial replacement of fine aggregate with CR as shown in Fig 3. However, as shown in Fig 4 the reduction in was more pronounced when silica fume was used as mineral filler compared to fly ash. This is due to lower specific gravity of silica fume compared to fine aggregate with silica fume having more water absorption capacity. The reduction in fresh density due to crumb rubber addition can mainly be attributed to lower specific gravity and density of the crumb rubber compared to fine aggregate it replaced, and due to hydrophobic and non-polar nature of crumb rubber making it to entrap air on its surface thus reducing its fresh density

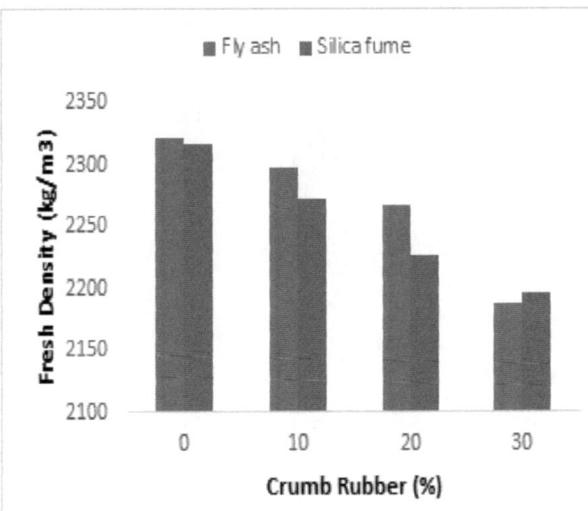

Figure 3: Fresh density of RCR

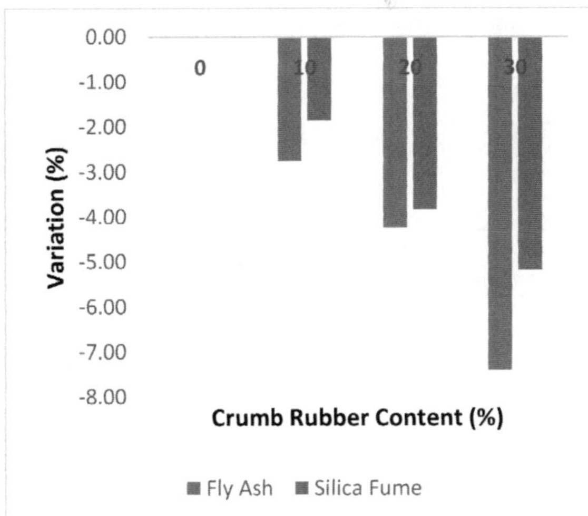

Figure 4: Variation of fresh density with reference to the control mix.

3.2 Compressive Strength

The compressive strength of RCR with either silica fume or fly ash as mineral filler decreases with increase in partial replacement of fine aggregate with crumb rubber as shown in Figs 5 and 6. This reduction in strength can be due to the following reasons; poor bonding between crumb rubber particles and cement paste, lower specific gravity, density and stiffness of crumb rubber in relation to fine aggregate, high porosity in the hardened RCR mix due to entrapped air on the fresh mix. However, all the mixes achieved a strength higher than the target strength at 28 days expect 30% crumb rubber containing 5% fly ash mineral filler, and 10% crumb rubber achieved a strength more than the target strength even at 3 and 7 days. This is due to the ability of the mineral fillers to contribute to strength development through their pozzolanic reactions. Though, the reduction in strength was lower when silica fume was used as mineral filler compared to fly ash as seen in Fig 6. This is due to the higher pozzolanic reactivity of silica fume compared to fly ash making it to consume more $Ca(OH)_2$ and producing more C-S-H gel which is responsible for strength development, with fly ash having a very low pozzolanic reactivity at early ages of curing.

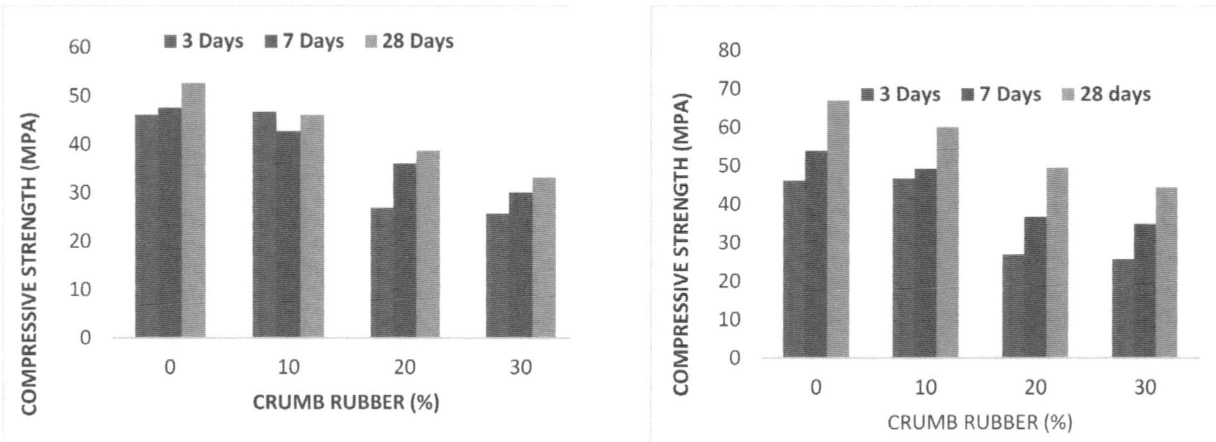

Figure 5: Compressive strength of (a) Fly ash mineral filler (b) Silica fume mineral filler

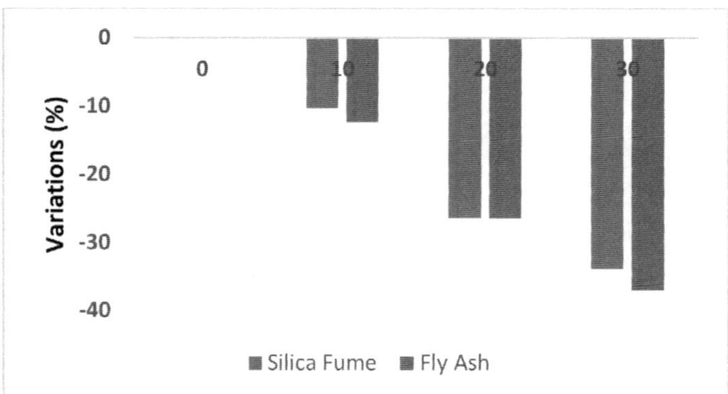

Figure 6: Variation of compressive strength with reference to the control mix

4. Conclusions

This paper investigated the possibility of using crumb rubber as partial replacement to fine aggregate in roller compacted concrete for pavement applications, and compares the use of fly ash and silica fume as mineral filler. The results obtained showed that crumb rubber addition decreases the fresh density and compressive strength. However, all the mixes achieved strength higher than the design target strength at 28 days except for 30% crumb rubber containing fly ash mineral filler. Conversely, when silica fume was used as mineral filler the reduction in fresh density was higher compared to that of fly ash. However the reduction in compressive strength was higher for fly ash mineral filler than silica fume which was mainly due to the slower pozzolanic reaction of fly ash. Replacing fly ash with silica as a mineral filler was successful in mitigating loss in compressive strength for up to 20% CR.

Acknowledgement

The authors wish to acknowledge the research grant provided by Fundamental Research Grant Scheme (FRGS 0153AB-K04) that leads to the realization of this work and also Universiti Teknologi PETRONAS (UTP) for supporting this work.

References

[1] L. He, Y. Ma, Q. Liu, and Y. Mu, "Surface modification of crumb rubber and its influence on the mechanical properties of rubber-cement concrete," *Construction and Building Materials,* vol. 120, pp. 403-407, 2016.

[2] B. S. Thomas, R. C. Gupta, P. Mehra, and S. Kumar, "Performance of high strength rubberized concrete in aggressive environment," *Construction and Building Materials,* vol. 83, pp. 320-326, 2015.

[3] WBCSD, "Managing End-of-Life Tires," World Business Council for Sustainable Development, Washington DC; USA2008.

[4] B. S. Thomas, R. C. Gupta, and V. J. Panicker, "Recycling of waste tire rubber as aggregate in concrete: durability-related performance," *Journal of Cleaner Production,* vol. 112, pp. 504-513, 2016.

[5] F. Azevedo, F. Pacheco-Torgal, C. Jesus, J. B. de Aguiar, and A. Camões, "Properties and durability of HPC with tyre rubber wastes," *Construction and building materials,* vol. 34, pp. 186-191, 2012.

[6] RMA, "Scrap tire markets in the United States," Rubber Manufacturers Association, Washington DC, USA2009.

[7] RMA, "Scrap tire markets in the United States," Rubber Manufacturers Association, Washington DC, USA2016.

[8] A. A. Aliabdo, A. E. M. A. Elmoaty, and M. M. AbdElbaset, "Utilization of waste rubber in non-structural applications," *Construction and Building Materials,* vol. 91, pp. 195-207, 2015.

[9] R. Siddique, *Waste materials and by-products in concrete*: Springer Science & Business Media, 2007.

[10] A. Ayanoğlu and R. Yumrutaş, "Production of gasoline and diesel like fuels from waste tire oil by using catalytic pyrolysis," *Energy,* vol. 103, pp. 456-468, 2016.

[11] A. Meddah, M. Beddar, and A. Bali, "Use of shredded rubber tire aggregates for roller compacted concrete pavement," *Journal of Cleaner Production,* vol. 72, pp. 187-192, 2014.

[12] B. S. Mohammed, A. B. Awang, S. San Wong, and C. P. Nhavene, "Properties of nano silica modified rubbercrete," *Journal of Cleaner Production,* vol. 119, pp. 66-75, 2016.

[13] M. Adamu, B. S. Mohammed, and N. Shafiq, "Nano silica modified roller compacted rubbercrete – an overview," in *Engineering Challenges for Sustainable Future - Proceedings of the 3rd International Conference on Civil, offshore and Environmental Engineering, ICCOEE 2016*, 2016, pp. 483-488.

[14] B. S. Thomas and R. C. Gupta, "Properties of high strength concrete containing scrap tire rubber," *Journal of Cleaner Production,* 2015.

[15] T. Gupta, S. Chaudhary, and R. K. Sharma, "Mechanical and durability properties of waste rubber fiber concrete with and without silica fume," *Journal of Cleaner Production,* vol. 112, pp. 702-711, 2016.

IOP Conf. Series: Materials Science and Engineering **201** (2017) 012012 doi:10.1088/1757-899X/201/1/012012

Rheological and rutting evaluation of composite nanosilica/polyethylene modified bitumen

Nura Bala[1], Ibrahim Kamaruddin[1], Madzlan Napiah[1], Nasiru Danlami[1]

[1] Department of Civil & Environmental Engineering, Universiti Teknologi PETRONAS, 32610 Bandar Seri Iskandar Perak Malaysia

nura.bala_g03311@utp.edu.my

Abstract. In this research, composite nanosilica/polyethylene modified binder samples were prepared at varying concentration of nanosilica. The modified binder samples are prepared by adding 1%, 2% and 3% nanosilica by weight of bitumen in to 6% linear low density polyethylene (LLDPE) polymer modified binder. Effects of nanosilica on physical and rheological properties of composites were investigated through penetration, softening point, temperature susceptibility, dynamic mechanical analysis and rutting resistance evaluations. Results shows that, nanosilica affects the rheological properties of LLDPE modified binder due to decrease in penetration and increase in softening point observed with increase in nanosilica content, this enhances composite binders hardness and resistance to temperature. Temperature susceptibility shows that, composite modified binders are less susceptible to temperature compared to LLDPE modified binder. DSR rheological analysis shows that, nanosilica enhances the composite modified binders properties at intermediate to high temperatures. Also rutting parameter evaluation indicates that composite modified binders have high resistance to rutting deformations due to increase in elastic behavior and stiffness of the composites.

1. Introduction

Bitumen consists of complex chemical composition, it often exhibits both elastic and viscous behaviors which largely depends on loading time and temperature[1, 2]. Bitumen is generally known as aggregates binder applied for flexible pavements constructions due to its unique viscoelastic properties [3, 4]. However, current increase in axle loads globally together with varying climatic conditions leads to rapid deterioration of pavements, to overcome such effects, pavement demands binder with higher performance than normal grade bitumens applied [5]. Consequently, now a day more and more modified bitumens are applied for pavement constructions. Recently, over the past ten years, composite polymer/inorganic nanomaterials modification becomes more interested for construction of pavements as it improves both physical and rheological properties of binders [6, 7].

Nanomaterial are material having one of its external dimensions or internal structure in order of 100 nm or less, and this material can exhibit novel properties that the same material without nanoscale structure cannot [8]. However, there is limited research conducted on composite polymer/nanomaterial modified binder performance properties characterization. Therefore, in this study the performance characteristics of composite LLDPE/Nanosilica modified binder were evaluated.

2. Materials and modification process

2.1 Materials

The binder modified is penetration grade 80/100 bitumen supplied by PETRONAS refinery Melacca, Malaysia. Polymer used in pellet form is linear low density polyethylene (LLDPE) produced by Etilinas polyethylene Sdn Bhd in Kerteh, Malaysia. Nanosilica supplied by Benua Sains chemical Sdn Bhd Malaysia is used in this research.

2.2 Modification process

The composite LLDPE/nanosilica modified binder samples were preferred by first melting the binder at temperature of 150 °C. Upon reaching temperature of 150 °C it becomes sufficiently fluid , a weighted 1%, 2% and 3% amounts of nanosilica by weight of bitumen binder were gradually added to 6% LLDPE polymer modified binder. The composites were mixed using a desktop propeller blade multimix high shear mixer at high blending rate of 4000 rpm for 2 hours to obtain a homogenous composite mixture.

3. Experimentation methods

3.1 Penetration Test

An empirical needle Penetration test (ASTM D5-13) is used to establish binders consistency at a temperature of 25°C under a load of 100g and time period of 5 seconds which produces penetration depth in units of 1/10 mm on the tested sample. The depth of penetration in units of 0.1 mm is recorded as the penetration value of the binders.

3.2 Softening Point Temperature Test

Ring and ball test (ASTM D36-12) is used to characterize and establishes the temperature at which bitumen binder softens. Softening point is taken as the temperature under which bitumen binder sample cannot support steel ball of 3.5g weight, when heated at uniform rate of temperature of 5°C/ min.

3.3 Temperature susceptibility

Temperature susceptibility for composite nanosilica modified binders was evaluated based on penetration index (PI) parameter. Estimation of PI is described by Shell Bitumen Hand book [9] as shown by Eq. 1.

$$PI = \frac{1952 - 500\log P_{25} - 20T_{R\&B}}{50\log P_{25} - T_{R\&B} - 120} \quad (1)$$

where PI is the penetration index, P_{25} is the binder penetration at temperature of 25 ^{0}C, and $T_{R\&B}$ is the softening point temperature.

3.4 Dynamic shear rheometer test

A Malvern Kinexus rheometer was used to determine the rheological characteristics of composite LLDPE/nanosilica modified binders. Temperature sweep test was conducted under controlled strain loading mode over a temperature range of 20-60 °C and constant frequency of 10 rad/s. A heated sample of modified binder about 1.0 g weight was placed on top of the DSR lower plate. The upper parallel plate was then lowered to a height of 50 μm plus the required testing gap geometry of 1mm or 2mm and excess binder around the plates were trimmed out and the final gap was adjusted to a required 1mm or 2mm testing gap geometry.

4. Results and discussions

4.1 Penetration Test

Fig. 1 shows the effect of nanosilica on penetration of LLDPE and composite LLDPE/nanosilica modified binders. It can be seen that, nanosilica causes reduction in penetration values of the LLDPE modified binders, and the reduction is observed to increase with increase in nanosilica content. This may result in an enhancement of the modified binders resistance against temperature defects there by increasing its durability and performance during service life stages. Also, large decrease in penetration observed within nanocomposite modified binders indicates that nanosilica addition makes the binder to be more consistent and harder, this might increase the binders resistance against rutting

4.2 Softening point test

Fig. 2 shows the softening point temperatures for LLDPE and composite LLDPE/nanosilica modified binders. It can be seen that nanosilica addition causes an increase in softening point temperatures of the modified binders. The increment in softening point temperature is observed to be increase with increase in nanosilica, binder containing 3% nanosilica shows the highest softening point increment of 64 °C. The increase in softening point temperature for the composites indicates a reduction in their ability to soften easily under influence of temperature, and this will results in binders higher resistance against rutting.

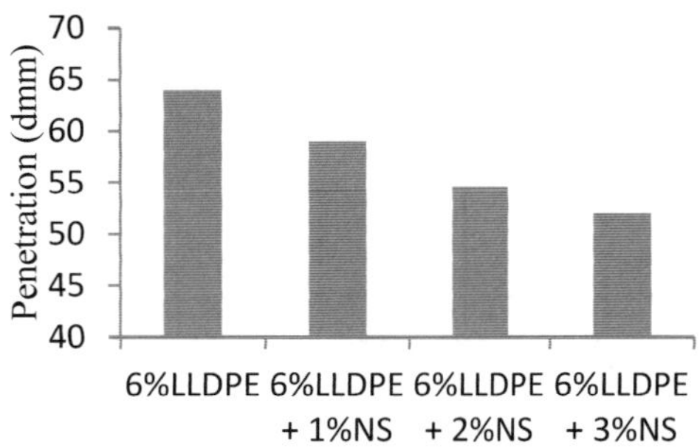

Figure. 1 Effects of penetration on modified binders

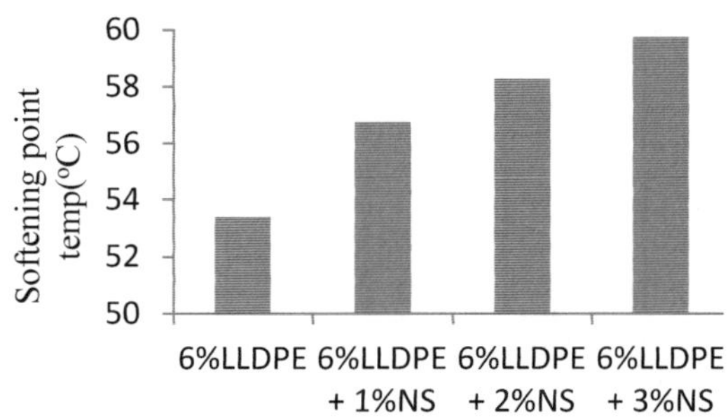

Figure. 2 Softening point temperature of modified binders

4.3 Temperature susceptibility

Fig. 3 shows penetration index values as an indication of temperature susceptibility for LLDPE and composite modified binders. It can be observed that, the initial PI value for LLDPE modified binder is 1.2, but with addition of nanosilica up to 3%, the PI value increases to 1.8, this shows that, composite modified binders exhibit high temperature resistance compared LLDPE modified binder, and the resistance increase with increase in nanosilica content. Increase in PI values observed in composites modified binders with increase in nanosilica can be attributed to exfoliated structure formed due to nanosilica addition which increases service temperature range and resistance to low temperature cracking and permanent deformation [10].

4.4 Rheological evaluation

Fig. 4, 5 and 6 are isochronal plots of complex modulus, complex viscosity and phase angle of LLDPE and composite modified binders against temperature. It can be seen that, composite modified binders complex modulus (Fig. 4) and complex viscosity (Fig. 5) increases with increase in nanosilica, however, phase angle (Fig. 6) a different phenomenon were observed where by phase angle decreases with an increase in nanosilica content. From isochronal plot of complex modulus in Fig. 4, it can be seen that, for composite modified binders, there is slight increase in G* at lower concentrations of nanosilica, but at high temperatures, higher increment was observed, this indicates that, addition of nanosilica enhances complex modulus which increases binders resistance to permanent deformation.
From isochronal plot of phase angle in Fig. 5 it can be observed that, nanosilica addition to form composites reduces phase angle values of LLDPE modified binders at intermediate to high temperatures. However, the decrease in phase angle values as nanosilica was added especially at intermediate temperatures confirms the formation of exfoliated nanostructure which enhances the binders performance [10].

From isochronal plot of complex viscosity in Fig. 5, it is observed that, there was increase in viscosity of the composites; this implies that, composite modified binders will results in formation of strong bond between aggregates and binder due to thicker film of binder that will surround the aggregates.

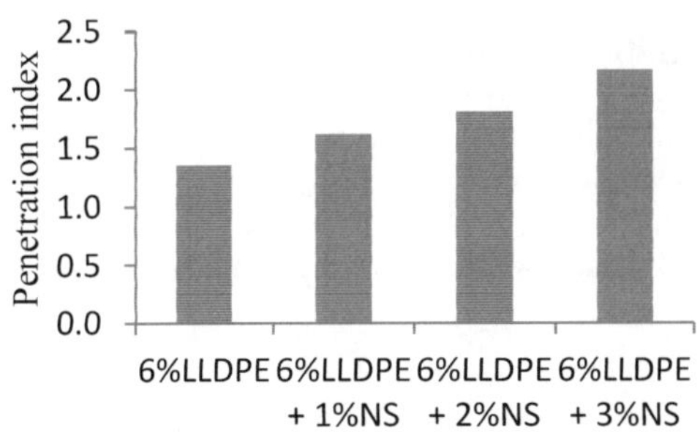

Figure 3 Temperature susceptibility of modified binders

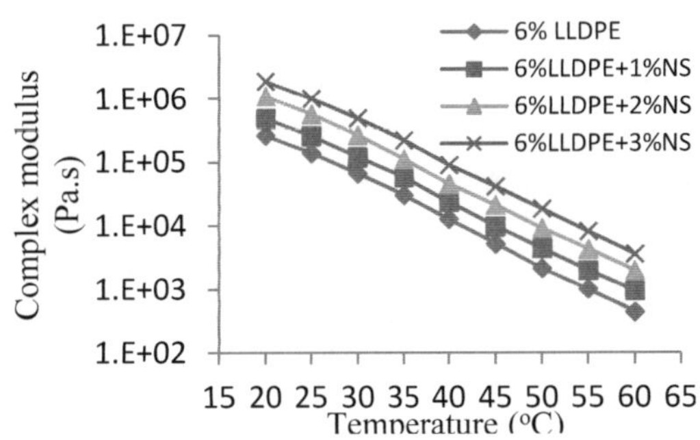

Figure. 4 Isochronal plot of complex modulus versus temperature

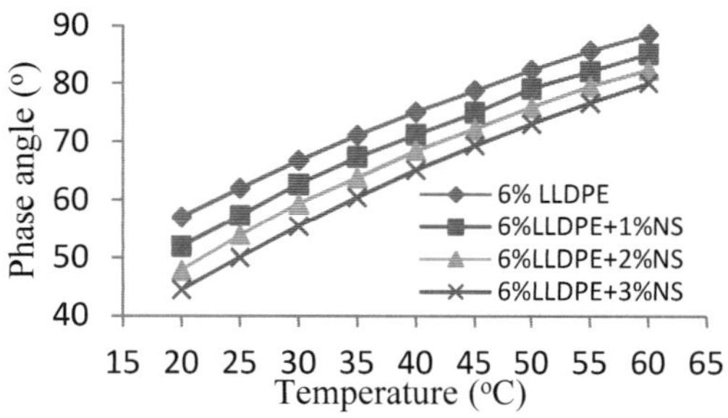

Figure. 5 Isochronal plot of phase angle versus temperature

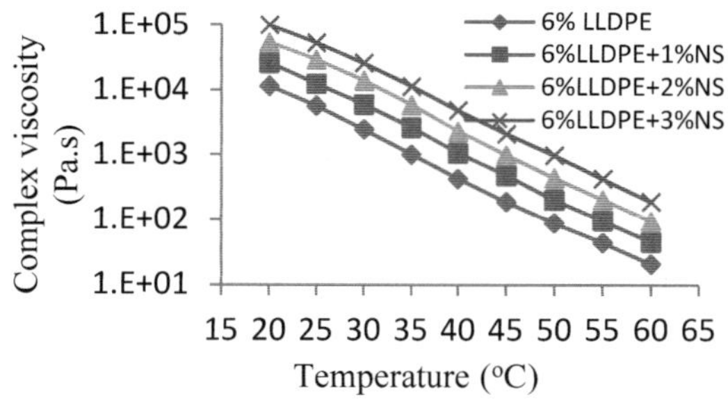

Figure 6 Isochronal plot of complex viscosity versus temperature

4.5 Rutting resistance

For evaluation of bituminous binder resistance to rutting, a G/ sin δ ratio is used as specified by Superpave specifications. Superpave specification sets a limit of G*/sin δ ≥ 1 kPa [11] as the minimum requirement for rutting resistance of unaged binders. The G*/sin δ ratios for modified binders at varying temperature ranges are presented in Fig. 7. It can be observed that, LLDPE polymer modified binder have the minimum G*/sin δ ratio, and composite modified binders have the highest G*/sin δ ratio.

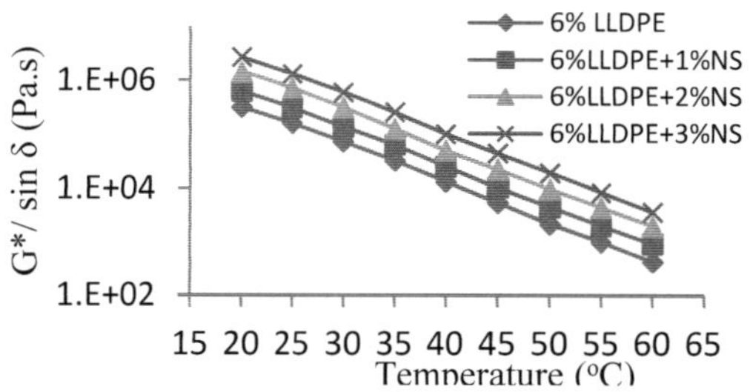

Figure. 7 Effect of temperature on rutting resistance

However, it can also observed that, at intermediate temperatures of 20 °C to 40 °C all modified samples have the minimum required value of 1.0 kPa for rutting resistance, while for high temperatures above 50 °C only 2% and 3% NS composite modified binders achieved the required minimum value of 1.0 kPa. This shows that, composite modified binders can be used to achieve higher enhancement in resistance to permanent deformation at high temperatures.

5. Conclusions

Based on observations in this research for composite modification, the results shows that, the composite penetration decreases and softening point temperature increases which enhances the binders hardness and better resistance to temperature effects. Temperature susceptibility evaluation indicates that, composite nanosilica modified binders have less susceptibility to temperature, due to higher PI values observed in composites compared to LLDPE modified binder. A rheological evaluation shows that, nanosilica enhances the rheological performance of the composite modified binders at intermediate to high service temperatures. Rutting estimations, shows that, composite modified binders are more resistant to deformations, this makes them more suitable for paving applications compared to LLDPE modified binder. From analysis of the outcome, 3% NS can be recommended as the optimum content for composite modification.

References

[1] S. Peters, T. Rushing, E. Landis, and T. Cummins, "Nanocellulose and microcellulose fibers for concrete," *Transportation Research Record: Journal of the Transportation Research Board,* pp. 25-28, 2010.

[2] Y. Yildirim, "Polymer modified asphalt binders," *Construction and Building Materials,* vol. 21, pp. 66-72, 2007.

[3] L. Loeber, G. Muller, J. Morel, and O. Sutton, "Bitumen in colloid science: a chemical, structural and rheological approach," *Fuel,* vol. 77, pp. 1443-1450, 1998.

[4] S.-p. Wu, L. Pang, L.-t. Mo, Y.-c. Chen, and G.-j. Zhu, "Influence of aging on the evolution of structure, morphology and rheology of base and SBS modified bitumen," *Construction and Building Materials,* vol. 23, pp. 1005-1010, 2009.

[5] J. Lamontagne, P. Dumas, V. Mouillet, and J. Kister, "Comparison by Fourier transform infrared (FTIR) spectroscopy of different ageing techniques: application to road bitumens," *Fuel,* vol. 80, pp. 483-488, 2001.

[6] Z. You, J. Mills-Beale, J. M. Foley, S. Roy, G. M. Odegard, Q. Dai, *et al.*, "Nanoclay-modified asphalt materials: Preparation and characterization," *Construction and Building Materials,* vol. 25, pp. 1072-1078, 2011.

[7] E. H. Fini, P. Hajikarimi, M. Rahi, and F. Moghadas Nejad, "Physiochemical, Rheological, and Oxidative Aging Characteristics of Asphalt Binder in the Presence of Mesoporous Silica Nanoparticles," *Journal of Materials in Civil Engineering,* vol. 28, p. 04015133, 2015.

[8] D. Williams, M. Amman, H. Autrup, J. Bridges, F. Cassee, K. Donaldson, *et al.*, "The appropriateness of existing methodologies to assess the potential risks associated with engineered and adventitious products of nanotechnologies," *Report for the European Commission Health and Consumer Protection Directorate General by the Scientific Committee on Emerging and Newly Identified Health Risks. Brussels,* 2005.

[9] J. Read and D. Whiteoak, *The shell bitumen handbook*: Thomas Telford, 2003.

[10] S. S. Galooyak, B. Dabir, A. E. Nazarbeygi, and A. Moeini, "Rheological properties and storage stability of bitumen/SBS/montmorillonite composites," *Construction and Building Materials,* vol. 24, pp. 300-307, 2010.

[11] T. W. Kennedy, G. A. Huber, E. T. Harrigan, R. J. Cominsky, C. S. Hughes, H. Von Quintus, *et al.*, "Superior performing asphalt pavements (Superpave): The product of the SHRP asphalt research program," *Strategic Highway Research Program, National Research Council, Report No. SHRP-A-410,* 1994.

Optimisation of nano-silica modified self-compacting high-Volume fly ash mortar

Bitrus Emmanuel Achara, Bashar S. Mohammed, and Muhd Fadhil Nuruddin

Deparment of Civil and Environmental Engineering, Universiti Teknologi PETRONAS, 32610 Bandar Seri Iskandar, Perak Darul Ridzuan, Malaysia

emmachara@yahoo.com, bashar.mohammed@petronas.com.my

Abstract. Evaluation of the effects of nano-silica amount and superplasticizer (SP) dosage on the compressive strength, porosity and slump flow on high-volume fly ash self-consolidating mortar was investigated. Multiobjective optimisation technique using Design-Expert software was applied to obtain solution based on desirability function that simultaneously optimises the variables and the responses. A desirability function of 0.811 gives the optimised solution. The experimental and predicted results showed minimal errors in all the measured responses.

1. Introduction

The need to develop a durable and cost-effective high-performance concrete has led to the utilisation of industrial by-products termed cement replacement materials' or 'supplementary cementitious materials' such as ground granulated blast furnace slag (GGBS), silica fume (SF), fly ash (FA) or limestone powder. Numerous researches have been reported utilizing by-products or waste for cement replacement in high performance concretes [1], and [2].

Nano-silica has been widely accepted by many researchers due to its contribution to improving performance in a cementitious matrix. It has been reported that nano-silica densify matrix in cementitious composites; it reacts faster than other supplementary cementitious materials such as FA, GGBS, etc. it utilizes calcium hydroxide released during cement hydration to produce more calcium silicate hydrates (C-S-H) gel for further increase in strength of concrete [3] and [4]. This is due to nano size effect, which contributes to pore filling and subsequent packing density of cementitious matrix [5], and [6]; high surface area and nucleation effect, which aid faster reaction. Silica units from cement particles during cement hydration [7] and C-S-H gel produced during the first stage are attracted to nano-silica particles [5] and form nano-silica particles as a nucleus. The products of hydration on both the surface of cement grain and newly formed nucleation sites are well dispersed in the pore solution, creating a more dense matrix [6]. The resultant effect of accelerated hydration of cement by nano-silica by its nucleation effect leads to the production of more calcium hydroxide. The two processes, nucleation effect and hydration effect, through consumption of calcium hydroxide, leads to boosting a pozzolanic value of nano-silica [3]. In effect, the presences of nano-silica shorten the dormant period of cement hydration as nano-silica consumption of calcium hydroxide stimulates the further release of hydration products in order for the hydration reaction to attain a balanced reaction [8]. Nanosilica has demonstrated faster pozzolanic reactivity compared to other active powders, particularly at initial phases [9].

Despite the numerous benefits associated with the use of nano-silica in cementitious composites some drawbacks have been highlighted: The range of optimum quantity of nano-silica recommended for improvement of mechanical properties in the literature have not been consistent. Two sources have been identified in relation to this effect. Firstly, the source of the nano-silica is of different origins, which shows differences in types, particle size, specific surface or methods of production. Secondly, the difficulty in uniform dispersion of nano-silica in the fresh cement paste, grout, mortar or concrete [10], is an issue that cannot be exposed completely on a mechanistic basis. Even in their original form, slurry or hydrosol, if well dispersed, they have the potential of aggregating once in pore solution of a cement-based material due to the availability of ions such as Ca^{2+}, Na^{+} and K^{+} released during cement reaction with water [3]. These ions have a bridging effect, agglomerating silica units in the form: (SiO^{-})-(Ca^{2+})-(SiO^{-}) [8], which leads to disrupt the dispersion of nano-silica particles in the pore fluid and restraint their uniform dispersion. Similarly, the utilization of nano-silica restraint compatibility between the cement, superplasticizer and any other chemical admixture or mineral additions [10] and [11]. This results in a negative effect on the rheology/flow of cement paste and mortar. Consequently, increasing nano-silica content lead to significantly increasing yield stress values [11]. Identifying the threshold for both nano-silica and superplasticizer balances the threat pause by their incorporation in cement paste and mortar. The varied optimum content of nano-silica reported by numerous researchers ranges between 5% to 2% weight of cement [10] and [12]. The wide disparity in the optimum content reported indicates difficulty claiming maximum nano-silica content required to improve properties of cement composites.

2. Research Significance

The study evaluates the influence of nano-silica and superplasticizer quantities on compressive strength and mercury intrusion porosimetry (MIP) of self-compacting high-volume fly ash cement mortar. Response surface and multiobjective optimization are used to develop a solution of most effective nano-silica and SP amount based on the set objectives. This is essential to harmonise differences in the literature regarding the optimum nano-silica and superplasticizer amounts in nano-silica modified mortars.

3. Constituent Materials

Washed river sand of about 600 μm average diameter, ordinary Portland cement type 1, tap water, Class F fly ash with low Calcium less than 15% (ASTM C618), commercially available nano-silica particles ordered from Dongshen Petrochemical, China, with silica purity 99.8% and average diameter 20nm and polycarboxylate-based superplasticizer sika viscocrete R 5-700® were used in all the mortar mixes.

3.1. Mixture proportion

Some of the ingredients used in the mortar preparation are fixed to minimize over-fitting in the responses. The constituents that have been fixed are presented as shown in Table 1.

Table 1. Fixed constituents of mortar mixes

W/C	Water (Kg/m^3)	Cement (Kg/m^3)	Sand (Kg/M^3)	Fly ash (Kg/m^3)
0.32	187	583	467	700

3.2. Sample preparation

A mortar mixer of 4.5 L capacity was used to prepare all mortars at an average of 8 min. Water and SP were mixed together and half the quantity is poured into the mixer containing measured dry quantities of sand, cement, and FA for about 2 min, the remaining half quantity of mixed water and SP was further added to the mixture after another 2 min until the mixture attained homogeneity, which took about 4 min and the mixer stopped. The mortars are tested for workability on a mini mortar flow. The mortars are cast in 50 mm cube moulds for compressive strength testing and MIP assessment of porosity. The compressive strength of mortar cubes was tested at 28th day after curing at lab temperature of 22 ± 2°C in accordance to ASTM C39/C39M-04 A. The samples for porosity assessment was cut from the mid-center of cube samples and tested in a MIP equipment in accordance with the requirements of ASTM C1202.

4. Method and Experimental Design

Design-Expert software was used for the response surface and optimization. The user-defined option was utilised for the experimental design to accommodate the range of values of trial mixes. The values of the responses were inputted and the model for each response was established after verification with the statistical requirement. Afterwards, a multiobjective optimisation was conducted utilising the models to obtain solutions of responses and variables that give a fair treatment to all the variables based on the rating in terms of weights/ranking inputs. Usually, model overfitting occurs when it is made excessively complex, such as having too many parameters relative to the number of observations [13]. Three parameters are selected in the present investigation for optimisation of the response variables of the mortars and other parameters are fixed. Trial mixes were also conducted to obtain a realistic range of variables and ensure further avoidance of overfitting. The range of variables of nano-silica (0%, 1%, 2%, 3%, 4%), and SP (0.65%, 0.7%, 0.8%, 1.0%, 1.2%).

5. Results and Discussion

5.1. *Development of Compressive Strength.* The experimental design and results obtained by a user-defined option in RSM are shown in Table 2. The model expression for compressive strength, f_c, of mortars at the age of 28 days is shown in equation (1). The relationship between the variables A, nano-silica and B, superplasticizer has R-squared = 0.788:

$$f_c = +8.97 - 0.22A + 0.11B + 0.35AB - 1.28A^2 \quad (1)$$

Response surface plot in fig.1 shows a quadratic trend depicted by equation (1). It illustrates the effects of nano-silica and SP on the compressive strength of mortars. From the plot, nano-silica maximise the compressive strength of mortars at about 2%. Nanosilica is more significant than SP, though, the interaction of the two showed significant effect as well.

Table 2. User-defined experimental design and responses

Run	Factor1 A: NS (%)	Factor2 B: SP (%)	Response1 Comp. strength (MPa)	Response2 Porosity (%)	Response3 Slump flow (mm)
1	3	0.65	50.5	8.1	175
2	4	1.2	60.4	10.95	215
3	4	0.8	52	12.5	195
4	1	1	72.15	6.8	230
5	4	0.65	50.55	13.5	170
6	2	0.8	89.5	5.45	210
7	1	0.8	74.5	5.9	220
8	2	1.2	81.2	7.1	228
9	0	0.7	64.09	7	219
10	2	1	87.15	6.1	220
11	1	0.65	71.5	6.1	208
12	2	0.65	80.15	7.4	192
13	4	1	56	11.55	205

14	1	1.2	70.4	7.05	240
15	2	0.7	87.6	5.85	205
16	0	0.8	62.5	7.85	223
17	0	1	60.75	8.2	227
18	3	0.8	65	7.2	195
19	0	0.65	65.77	6.88	215
20	3	1.2	68.08	8.85	227
21	3	1	70.3	6.97	210
22	1	0.7	73.65	5.85	217
23	3	0.7	60	7.5	186
24	4	0.7	51.75	12.2	184
25	0	1.2	56.25	9.15	240

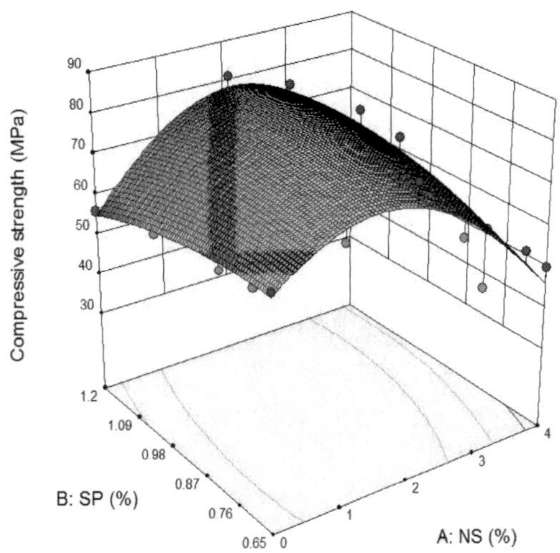

Figure 1. Response surface plot compressive strength vs nano-silica amount and superplasticizer dose.

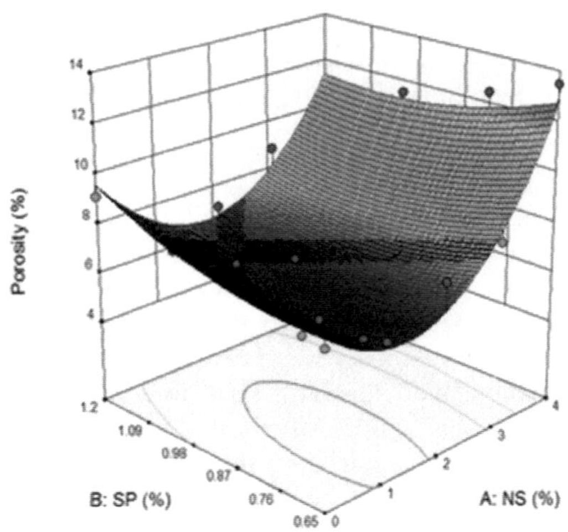

Figure 2. Response surface of porosity vs NS and SP.

5.2. *Porosity.* The experimental results of porosity based on experimental design are shown in Table 2 while the *response* surface plot based on equation (2) is shown in fig.2. The effect of nano-silica on porosity is more significant (R-squared = 0.95) while SP insignificant. However, the interaction effect of nano-silica amount and SP dose showed a significant effect. The nano-silica amount minimises porosity at about 2%, the same value as required to maximise compressive strength.

$$Porosity = +5.85 + 1.82A + 0.24B - 0.92AB + 3.74A^2 + 0.69B^2 \quad (2)$$

5.3. *Slump flow.* Equation (3) has been used to plot response surface of slump flow as shown in fig.3. The SP dosage is statistically significant with regard the slump flow (Sf) (R-squared =0.98). Contrary to this, the slump flow decreases with increasing nano-silica amount. The SP dosage is in percentage while the slump flow is in mm. looking at equation (3) and figure 3, it can be observed that the effect of SP was not truly linear, and the interaction between the two variables, the nano-silica and SP also affected the slump flow. Consequently, variation in slump flow due to SP dose was dependent on the nano-silica amount.

$$Sf = 216.83 - 23.26A + 10.24B + 5.44AB - 2.28A^2 - 4.16B^2 - 2.23A^2B - 5.12AB^2 + 11.87A^3 + 9.41B^3 \quad (3)$$

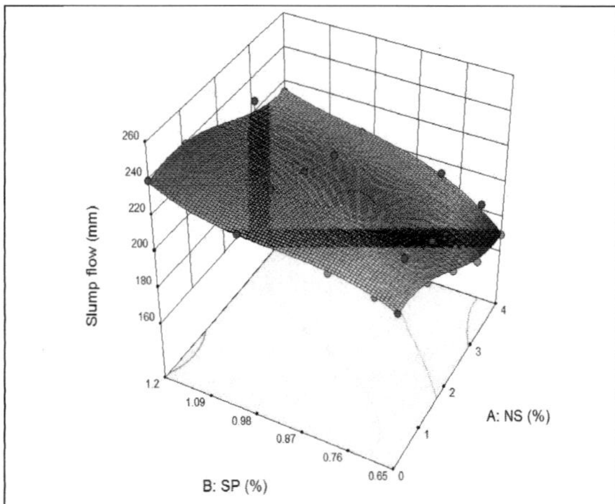

Figure 3. Response surface of slump flow vs NS and SP

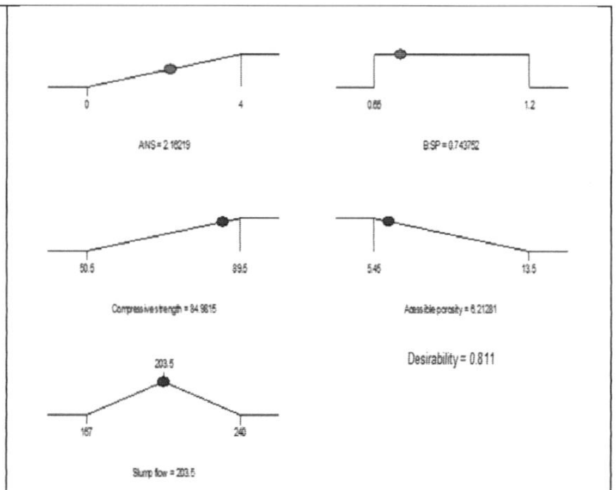

Figure 4. Rams of multiobjective optimisation of variables and responses

6. Multiobjective Optimization

The appropriate required amount of nano-silica and superplasticizer that is needed to maximise compressive strength may not necessarily be the same with the amount required to minimise porosity and maintain slump flow on target. Therefore, there is a need to determine the amount of nano-silica and dosage of superplasticizer that will satisfy all the responses. The multiobjective optimisation process is necessary to determine the most suitable solutions using desirability coefficient. Multiobjective optimisation utilises the model of all the responses and simultaneously develop solutions variables and responses values that will meet the requirements of all the responses. The rams in fig.4 illustrate the multiobjective solution based on desirability function. Using the solution variables and the fixed constituents in the mortar mix shown in Table 1, the new mixture was prepared with a slight adjustment of superplasticizer to attain the required range of flow. The experimental and predicted results are as shown in Table 3 based on desirability value of 0.811. The error calculated is minimal which indicates the reliability of the optimisation technique.

Table 3. Experimental and prediction results for optimised solution

A-NS (%)	B-SP (%)	y1 (Mpa) Exp.	y1 (Mpa) pred.	Error (%)	y2 (%) Exp.	y2 (%) pred.	Error (%)	y3 (mm) Exp.	y3 (mm) pred.	Error (%)
2.2	0.75	81.5	84.98	4.27	6.05	6.21	2.64	209	203.5	2.63

7. Conclusion

To optimise the compressive strength and MIP at a set slump flow of 203.5 mm of nano-silica modified self-compacting high-volume fly ash mortar, response surface and multiobjective optimisation were applied. The optimised mortar is achieved at 2.2% nano-silica and 0.75% superplasticizer by weight of binders. Increasing nano-silica beyond this amount at the fixed SP dose is accompanied by decreasing slump flow and subsequent decrease in compressive strength and increase in porosity. Contrary to this, the compressive strength of the mortars increased with increasing SP within this range at the fixed nano-silica amount. The increasing slump flow aided proper packing density of mortar mixtures. Application of multiobjective optimisation is suitable for obtaining a solution of variables and objective functions that meet the requirements of all the variables involved in the mixture.

Acknowledgement

The authors acknowledge the financial support by the University Teknologi PETRONAS.

References

[1] M. Sonebi, E. García-Taengua, K. M. A. Hossain, J. Khatib, and M. Lachemi, "Effect of nanosilica addition on the fresh properties and shrinkage of mortars with fly ash and superplasticizer," *Constr. Build. Mater.*, vol. 84, pp. 269–276, 2015.

[2] H. Kim, T. Koh, and S. Pyo, "Enhancing flowability and sustainability of ultra high performance concrete incorporating high replacement levels of industrial slags," *Constr. Build. Mater.*, vol. 123, pp. 153–160, 2016.

[3] E. García-Taengua, M. Sonebi, K. M. A. Hossain, M. Lachemi, and J. Khatib, "Effects of the addition of nanosilica on the rheology, hydration and development of the compressive strength of cement mortars," *Compos. Part B Eng.*, Jul. 2015.

[4] B. S. Mohammed, A. B. Awang, W. San San, and C. P. Nhavene, "Properties of Nano Silica Modified Rubbercrete," *J. Clean. Prod.*, vol. 119, pp. 66–75, Feb. 2016.

[5] S. A. El, M. Heikal, and W. M. Morsi, "Hydration characteristic , thermal expansion and microstructure of cement containing nano-silica," *Constr. Build. Mater.*, vol. 59, pp. 151–160, 2014.

[6] L. P. Singh, S. R. Karade, S. K. Bhattacharyya, M. M. Yousuf, and S. Ahalawat, "Beneficial role of nanosilica in cement based materials – A review," *Constr. Build. Mater.*, vol. 47, pp. 1069–1077, Oct. 2013.

[7] I. Panas, A. Martinelli, and A. Matic, "Accelerating effects of colloidal nano-silica for beneficial calcium – silicate – hydrate formation in cement," vol. 392, pp. 242–248, 2004.

[8] H. Madani, A. Bagheri, and T. Parhizkar, "The pozzolanic reactivity of monodispersed nanosilica hydrosols and their influence on the hydration characteristics of Portland cement," *Cem. Concr. Res.*, vol. 42, no. 12, pp. 1563–1570, 2012.

[9] P. Hou, S. Kawashima, D. Kong, D. J. Corr, J. Qian, and S. P. Shah, "Modification effects of colloidal nanoSiO2 on cement hydration and its gel property," *Compos. Part B Eng.*, vol. 45, no. 1, pp. 440–448, Feb. 2013.

[10] H. Biricik and N. Sarier, "Comparative study of the characteristics of nano silica-, silica fume-

and fly ash-incorporated cement mortars," *Mater. Res.*, vol. 17, no. 3, pp. 570–582, 2014.

[11] L. Senff, J. A. Labrincha, V. M. Ferreira, D. Hotza, and W. L. Repette, "Effect of nano-silica on rheology and fresh properties of cement pastes and mortars," *Constr. Build. Mater.*, vol. 23, no. 7, pp. 2487–2491, 2009.

[12] A. Naji Givi, S. Abdul Rashid, F. N. A. Aziz, and M. A. M. Salleh, "Experimental investigation of the size effects of SiO2 nano-particles on the mechanical properties of binary blended concrete," *Compos. Part B Eng.*, vol. 41, no. 8, pp. 673–677, 2010.

[13] W. Li, Z. Wang, Y. Sun, L. Chen, L. Han, and Y. Zheng, "Application of response surface methodology to optimise ultrasonic-assisted extraction of four chromones in Radix Saposhnikoviae," *Phytochem. Anal.*, vol. 22, no. 4, pp. 313–321, 2011.

IOP Conf. Series: Materials Science and Engineering **201** (2017) 012014 doi:10.1088/1757-899X/201/1/012014

Formulation of geopolymer cement using mixture of slag and class f fly ash for oil well cementing

[1,a*]Dinesh Kanesan , [1,b] Syahrir Ridha and Prasath Rao [1,c]

[1]Petroleum Engineering Department Universiti Teknologi Petronas, 31750 Tronoh, Perak, Malaysia

[a]dinesh.kanesan@utp.edu.my, [b]syahrir.ridha@utp.edu.my, [c]prasathrao66@gmail.com

Abstract. The increase in greenhouse gas emissions has been a factor for the increase in global temperature. Geopolymer cement has been intensively studied to replace conventional ordinary Portland cement, however the focus is limited to civil purposes under atmospheric conditions. This research focuses on the formulation of geopolymer cement to be used in oil well cementing application by taking account the effect of sodium hydroxide (NaoH) molarity, ratio of alkali binder and fly ash, amount of dispersant for oilwell operation under temperature ranging of 80°C and 90C° and pressure of 1000 and 3000psi. The formulated composition is tested for fluid loss where the standard has been from 60 to 80 ml. The cement slurry is cured in a 50mm x 50mm x 50mm mold for period of 24 hours. Four manipulating variables were set in formulating the cement slurry namely, the ratio between fly ash and slag to alkali binder, ratio of sodium hydroxide (NaoH) to sodium silicate, molarity of NaoH and amount of dispersant added. After running a set of 16 experiment, sample (12) was found to possess the best rheological properties and fluid loss according to API RP10B. It was found that as the curing temperature and pressure increase, the compressive strength of the formulated geopolymer cement also increased.

1. Introduction

Cementing is a fundamental and important process in drilling a well. Oil well cementing provides zonal isolation and support to the casing in the wellbore. Cementing is done after the casing is placed in the wellbore, whereby cement slurry is pumped through the drill pipe into the annulus to fill the empty space between the casing and the wall of the formation. Cement slurry design is very much dependent on the depth and environment of the well placement. The design of cementing job for wells located deep in the reservoir is very much a challenge due to high temperature, pressure and the presence of corrosive fluids.

Typically, Ordinary Portland Cement (OPC) is used to form the cement slurry around the globe, but studies have shown that the OPC tends to degrade over the time due to chemical reaction which takes place in acidic wellbore environment [1][2]. Besides that, OPC also shows the trend of decreasing yield strength in application in high pressure and temperature wells [3]. It was found that, approximately one ton of CO_2 is released to the atmosphere for the production of one ton of OPC [4]. Due to the increasing awareness in addressing this issue, viable replacement for the conventional Portland cement is currently being reviewed and studied in detail. Geopolymer cement is one of the emerging alternative to replace the OPC in all construction and industrial processes. Increased strength and durability in shorter curing time is the key attribute of Geopolymer cement compared to OPC [5].

Geopolymer cement can be formulated using various cementitious component such as fly ash, slag, rock and ferro-sialate based materials, making it very much viable to form instead of relying throughout on calcium carbonate rocks as OPC. Geopolymer cement also shows superior qualities of being more resistant to acidic wellbore environment compared to OPC due to its low calcium carbonate content [6].

2. Methodology

The methodology was divided into four parts. The first part of this experiment was the formulation of geopolymer cement. The variables in formulating the cement slurry were the ratio of slag and fly ash to alkali binder, the ratio between Sodium Hydroxide and Sodium Silicate, the molarity of Sodium Hydroxide and the amount of dispersant added to cement slurry.

2.1. Geopolymer cement formulation

The ratio between Sodium Hydroxide and Sodium Silicate was set to constant of 1:1 throughout the experiment. Sodium Hydroxide and Sodium Silicate was used as alkali binder for this experiment, the ratio between Sodium Hydroxide and Sodium Silicate changed and analysed, molarity of sodium hydroxide was also varied. The molarity used for this experiment was 10, 12 and 14 mol/L. Other factors analysed was the amount of dispersant added to the cement slurry whereby the amount was varied between 3,6,10 and 12ml. Fly ash, slag, Sodium Hydroxide and Sodium Silicate is then mixed using the speed mixer at 1200 rpm for duration of one minute with the addition of dispersant. To increase the thickening time of the cement slurry, water (20% of total weight of the mixture) was also added to the mixture. In accordance to API standard, the cement slurry was designed according to parameter listed in Table 1.

Table1: API Cement Slurry Rheology

Test Parameter	Industry standard
Density	12.5 – 16 ppg
Viscometer	PV> Mud Cp and YP>5
Compressive strength	>500psi

2.2. Fluid Rheology and density test

The formulated mixture was tested for its fluid rheological properties using the viscometer to determine if the mixture adheres to the API standard. The test is carried out using Fann300 viscometer. Reading were taken after the viscometer is run for 5 seconds. The stabilised reading at 600rpm and 300rpm was taken respectively. The readings were used to calculate the Plastic Viscosity (PV) and Yield Point (YP) of the cement slurry.

2.3. Fluid Loss Test procedure

Fluid loss test was performed to identify the amount of fluid diffused out of the cement slurry to adjacent formation at high pressure and temperature. This test was performed using static fluid loss tester at pressure of 1000psi and temperature of 260 °C. Measuring cylinder is placed below the equipment after the pressure is applied to cement slurry. The time also observed using stopwatch. After the period of 30 minutes, the pressure was removed and the total amount of water loss was recorded.

2.4. Compressive Strength test

Cement cubes were obtained from the cement mold after the curing process. The dimensions of the cement cubic was measured, and the surface area of the cubic was calculated. The cement cubic was then loaded in the compressive strength tester and the load was increased uniformly. The force where the cement cube fails was noted and recorded as 'Fi'. The compressive strength of the cement cube was calculated using Equation (1).

$$\sigma i = \frac{Fi}{Ai} \ldots\ldots\ldots\ldots\ldots\ldots\ldots \text{Equation (1)}$$

Where;
σi= compressive strength, N/$mm2$
Fi= maximum load, N
Ai= cross-section area

3. Results and Discussion

3.1. Cement Slurry Rheology and density

After the formulation of each cement slurry, it was tested for its rheological properties and density. The formulation that matches the API standards was further tested with fluid loss test and compressive strength test, meanwhile other formulation were considered as failed. After formulating 12 geopolymer cement slurries, sample (12) was chosen as the best cement slurry which met the API standard requirements. The composition and rheological test of sample (12) is listed in Table 2 and 3.

Table 2: Composition of sample (12)

Total weight (g)	600
H20 (g)	120
FA(g)	150
Slag(g)	150
Sodium Hydroxide (g)	60
Sodium Silicate (g)	120
Dispersant (g)	10

Table 3 : Density and Rheology test result sample (12)

Density	13.4
Plastic viscosity	31
Yield point	6

The results show that composition of slag and fly ash mixture to alkali binder affects the fluid rheological properties. In the second study, the ratio between sodium hydroxide and sodium silicate was varied. It was identified that the ratio of 1:05 produces the best results in fluid rheological test. Addition of Sodium Silicate increases the viscosity of the cement slurry. However above the optimum amount of Sodium Silicate, further addition leads to gelling effect of the cement which eventually fails in the PV and YP test. For the molarity of the Sodium Hydroxide solution, the molarity of 12 mol/litre^3 gave better results compared to the 14 mol/litre^3 and 10 mol/litre^3. When running the experiment with 10 mol/litre^3 and 14 mol/litre^3 the samples failed in PV and YP test. The effect of dispersant was also analysed, whereby it was found that the addition of dispersants reduces the gelling effect of the cement slurry. Above the optimum amount, further addition causes the failure of cement slurry in PV and YP test meanwhile below the optimum amount it leads to the gelling of cement slurry.

3.2. Compressive Strength test

Cement mold of sample (12) was cured at different temperatures and pressures in a 50mm x 50mm x 50mm cement mold. The compressive strength test was carried out using the destructive compressive strength tester to investigate the mechanical behaviour under high temperature and pressure. The results of compressive strength of the cement molds are listed in Table 4.

Table 4: Test 1 and Test 2 results

Test 1	
Reading of compressive strength tester	50.8 kN
Curing pressure	1000 psi
Curing temperature	80° Celsius
Compressive strength	2946.4psi
Test 2	
Reading of compressive strength tester	79.7 kN
Curing pressure	3000 psi
Curing temperature	90° Celsius
Compressive strength	4622.6 psi

The increase in curing temperature and pressure resulted in an increase in the compressive strength of the formulated geopolymer cement. This is in contrast with OPC where the compressive strength decreases as curing temperature and pressure increases [7].

4. Conclusion

In this research on geopolymer cement formulated using mixture of Class C Fly Ash and slag, it was found that geopolymer cement could be a better replacement to the currently used OPC cement for oil well cementing applications in high pressure and temperature wellbore environment. It's also shown in this study that the compressive strength of geopolymer cement does not show a steep decrease compared to OPC in high pressure and temperature, instead it increases as the curing temperature and pressure is increased. In addition, this study has also identified the factors which determines the rheological properties of cement slurry to match the standard required by industry. However, to be applied in the oil well cementing industrial, further studies on the additives should be conducted to give an ideal cement slurry with improved compressive strength.

References

[1] T.Bakharev. Resistance of Geopolymer material to acid attack. Cement and Concrete Research 35 (2005) 658-670. August 2005.

[2] C.M. Nasvi, P.G. Ranjith, J. Sanjayan, Geopolymer as Well Cement and Variation of its Mechanical Properties Under Different Curing Temperature and Curing Mediums. Greenhouse Gases: Science and Technology. 2, (2011), 46-58.

[3] M.C.M Nasvi, P.G. Ranjith, J. Sanjayan 'A numerical study of triaxial mechanical behaviour of Geopolymer at different curing temperature: An application for geological sequestration wells'. Journal of Natural Gas Science and Engineering. August 2015.

[4] D.W. Law, A.A. Adam, T.K. Molyneaux, I. Patnaikuni, A.Wardhono, Long term durability properties of class F fly ash geopolymer concrete. Materials and Structures. (2014). 48:721-731

[5] F.A.Diego, B. Marilda.'Durability and cementing efficiency of fly ash in concretes'. Contruction and building materials. January 2012.

[6] Mucteba Uysal, Veysel Akyuncu 'Durability performance of concrete incorporating class F and class C fly ashes'.Contruction and building materials. February 2012

[7] F.N Okoye, J. Durgaprasad, N.B Singh 'Mechanical properties of alkali activated flyash/Kaolin based geopolymer concrete'. Contruction and building materials. September 2015

Design and manufacture a coconut milk squeezer

I Wayan Surata, Tjokorda Gde Tirta Nindhia, D. Budyanto, and A. E. Yulianto

Department of Mechanical Engineering, Faculty of Engineering, Udayana University, Jimbaran, Bali, 80361, Indonesia

iwasura@gmail.com, nindhia@yahoo.com

Abstract. The process of cooking oil production generally is started by grating the ripe coconut meat, then pressing the grated meat to obtain coconut milk, and finally heating the coconut milk to obtain the cooking oil. Pressing mechanism to obtain coconut milk is a very important step and decisive in the process of producing cooking oil. The amount of milk produced depends on the pressure applied at the time of pressing grated coconut. The higher the pressure, the more milk is obtained. Some commercial mechanical pressing tools that available in the market are not efficient due to the working steps too much and take long time per cycle of work. The aims of this study was to design and manufacture a power screw squeezer for the collection of coconut milk. Power screw produces a compressive force in the cylinder to push and press the grated coconut until the end of the cylinder while the coconut milk and coconut dregs flow out simultaneously. Screw press was designed using straight shaft configuration with square profile. Performance test was done to investigate the actual capacity and yield of milk produced. The results showed that squeezer of grated coconut worked well with capacity an average of 13,63 kg/h and coconut milk yield of 58%.

1. Introduction

Cooking oil included in the group of nine basic commodities that generally is needed by Indonesian community, therefore cooking oil must be available and be sold freely in the market every day. The production of cooking oil from coconut as raw materials, in addition to the main results in the form of coconut milk and cooking oil, also produced a byproduct in the form of coconut dregs or the solid part which is a valuable source of protein for livestock feeds. The cooking oil production process from coconut generally starting from grated the mature coconut meat, then pressing the grated meat to obtain coconut milk, and finally heating the coconut milk to obtain the cooking oil. Currently coconut grater tool has equipped an engine that driven by electrical power. While squeezing process is still done manually by hand using filters that made from the part of coconut tree or cloth. The local practice of squeezing the grated coconut is still done traditionally, because the equipment is simple, it is also cost-effective and does not require any particular skill [1], [2]. Although in the market has also been available a mechanical squeezer of screw-piston, and screw-hydraulic, but such tool is not efficient due to many steps that must be done in one working cycle. The working mechanism of this tool is very long and difficult for the operator, besides that working posture and dimension of the tools are not ergonomics, cause fatigue and musculoskeletal disorders [3] – [5]. The aims of this research were to design and manufacture a mechanical tool for squeezing grated coconut using a power screw mechanism to produce coconut milk continuously. Power screw unit is a tool that is very commonly used for oil extraction, because the working steps are very simple, continuous, and easy to operate. The main data taken into consideration for prototype design are: the variable auger rotational speed, the variable pressure in the barrel, the diameter of barrel, and the diameter of the auger shaft [6]. Generally, the rotational speed of a power screw for pressing process ranges between 15 - 40 rpm [6] – [9]. The next variable is the working pressure inside the barrel. Reference [10] shows the final pressure in a squeezing process of jatropha oil plants (jatropha seeds) minimum 4 MPa. Because the grated coconut is not a grain, but very soft material then the pressure is much lower. In this study assumed working pressure of 2 MPa.

2. Experimental Method

Power screw configuration with a straight shaft is the most common mechanism used for press tools, because the manufacturing process is easy. The distance of the pitch and diameter of the screw base is constant as depicted in Fig. 1.

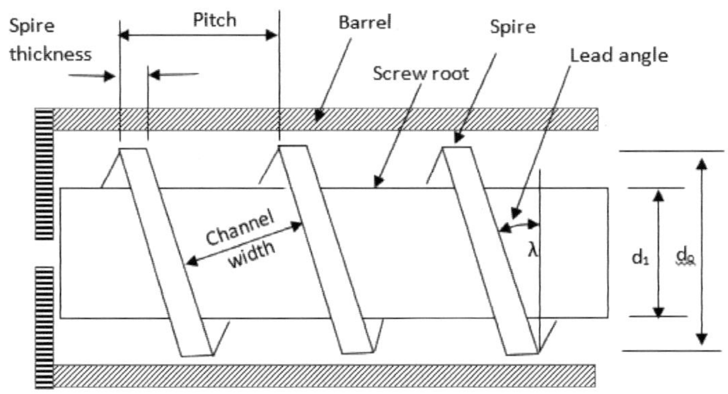

Figure 1. Geometry of power screw with straight shaft.

Press volume flow rate can be evaluated by using the relation as in [6]:

$$Q_v = V_{te}\,(1 - \varepsilon)\; n\; k\; 60\; (m^3/h) \tag{1}$$

where:

V_{te} - is the theoretical volume of the material displaced by the auger spire during a complete rotation, in the exhaust area (m^3)

n - auger rotation speed (rpm)

k - coefficient taking into account the material flowing back through the spire extremities, as well as the incomplete feed with material (k = 0,2 – 0,35)

ε - pressure ratio, calculated using equation $\varepsilon = (v_i - v_f)/v_i$, with v_i = initial volume of the material (m^3), and v_f = final volume (m^3).

The theoretical volume of the material displaced by the auger spire is calculated using the following equation:

$$V_{te} = \frac{\pi}{4}(d_2^{\,2} - d_1^{\,2})\,(p - \delta)\;\;(m^3) \tag{2}$$

where:

p - the auger spire pitch (m)

δ - thickness of the auger spire (m)

d_2 - outer diameter of the auger spire (m)

d_1 - inner diameter of the auger spire (m)

By replacing in equation (1) the expression of the theoretical volume given by equation (2), it results the expression of the press volume flow rate:

IOP Conf. Series: Materials Science and Engineering **201** (2017) 012015 doi:10.1088/1757-899X/201/1/012015

$$Q_v = \frac{\pi}{4}(d_2^2 - d_1^2)\ (p - \delta)\ (1 - \varepsilon)\ n\ k\ 60\ (m^3/h) \quad (3)$$

The yield of coconut milk is calculated by using the equation:

$$\text{Milk yield} = \frac{Wo - Wf}{Wo} \text{ x } 100\% \quad (4)$$

where:

w_o - initial weight of grated coconut (kg)

w_f - final weight of grated coconut (kg)

In this study, rotation of the screw shaft was designed for 15 - 30 rpm which rotated by electric motor. Working pressure assumed 2 Mpa. Furthermore selected barrel diameter do = 72 mm, the outer diameter of the screw d2 = 71 mm, and a screw shaft diameter d1 = 22 mm. Power screw length L = 300 mm, pitch distance p = 42.8 mm, and thread thickness δ = 1.2 mm. Grated coconut is smooth and slick material, therefore the friction coefficient between grated coconut and surface of the barrel is smaller. The coefficient of friction μ = 0.2, pressure ratio ε = 0.5, backflow coefficient k = 0.35, and based on the experiment volume of 1 m3 grated coconut equal with weight 352 kg [11]. Illustration prototype design of power screw for squeezing grated coconut and their components are shown in Fig. 2.

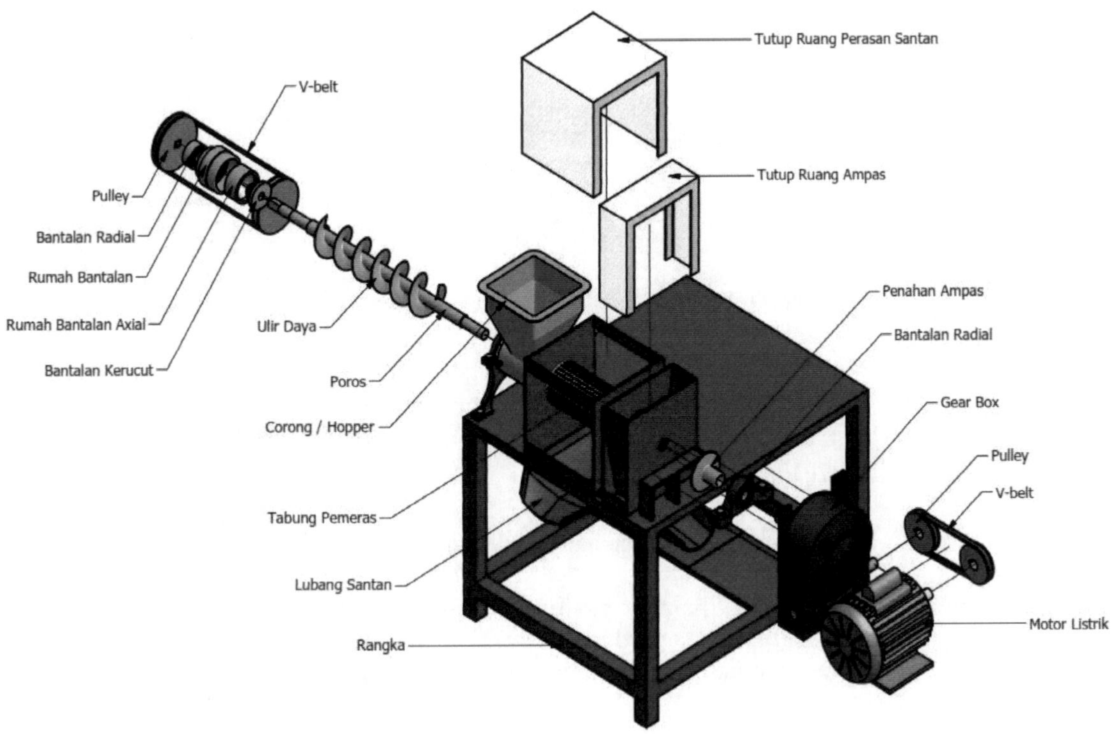

Figure 2. Design of power screw squeezer.

3. Results and Discussion

A complete assembly of grated coconut squeezer has been tested with variables of screw shaft rotation varied from 15, 25, and 30 rpm. Performance test included actual production capacity and yield of milk produced. The theoretical capacity was calculated refer to (3), and milk yield refer to (4). The results of the tests were presented in Table I.

Table 1. Performance test results.

Parameter	Unit	Analysis of result			Mean
		1	2	3	
Shaft rotation	rpm	15	25	30	23.33
Grated coconut weight	kg	1.00	1.00	1.00	1.00
Dregs weight	kg	0.41	0.42	0.42	0.42
Time for pressing	min	4.58	4.40	4.43	4.44
Milk yield	%	59.00	58.00	58.00	58.33
Actual capacity (experiment)	kg/h	13.10	13.63	13.85	13.52
Theoretical capacity	kg/h	8.25	13.75	16.50	12.83

The experiment results in Table I shown that the average capacity of the squeezer about 13.52 kg/h. This result was very satisfactory because the experimental results were in the range of theoretical calculations. The optimum condition occurred at rotation of 25 rpm, where the theoretical capacity equal to the experimental capacity about 13,63 kg/h, as shown in Fig. 3.

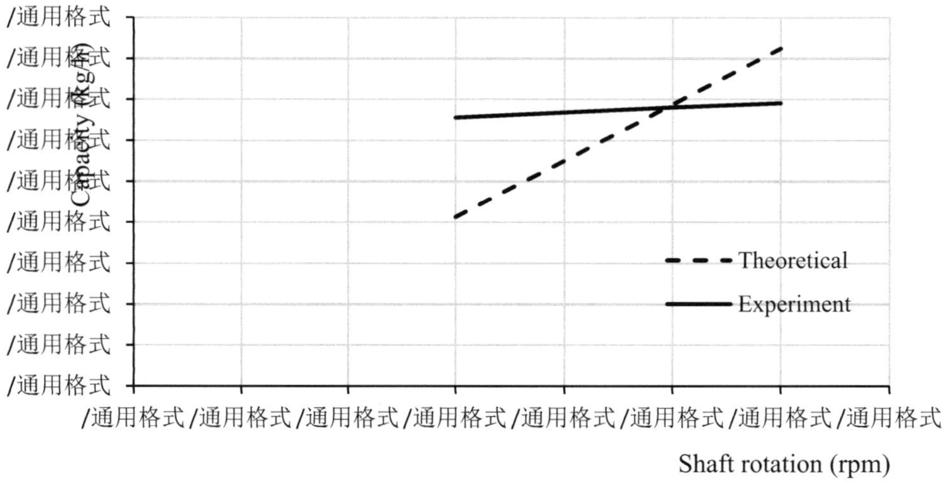

Figure 3. Capacity of the squeezer.

Based on the analysis in Table I, the milk yield was obtained an average of 58%. This result was higher when compared to hand squeezing which the milk yield about 52.9% [12]. Likewise, when compared to use a mechanical piston screw the milk yield about 51.25% [13].

4. Conclusion

The power screw squeezer for pressing grated coconut run successfully as designed. The performance of this pressing unit was better than screw piston type. The actual capacity about 13,63kg/h, and the milk yield produced an average of 58%. This squeezer can be used and suitable to support small industries in the process of making traditional cooking oils.

Acknowledgment

Authors are highly thankful to Udayana University for providing financial support to carry out the research work, under the scheme of Research Group grant (*Skim Penelitian Grup Riset*).

References

[1] R. Palungkun, *Aneka Produk Olahan Kelapa*. Bogor: Penebar Swadaya, 1992.

[2] B. Rindengan, H. Novarianto, *Pembuatan dan Pemanfaatan Minyak Kelapa Murni,* Jakarta: Penebar Swadaya, 2006.

[3] I. Ghazali, M. Tambunan, and Nazlina "Perancangan Alat Pemeras Kelapa Parut Menjadi Santan dengan Cara Pengepresan Manual yang Ergonomis". *J. Teknik Industri FT.USU,* vol. 2, No. 2, 2013

[4] A. Manuaba, "Ergonomi, Kesehatan, dan Keselamatan Kerja", in *Proc. Seminar Nasional Ergonomi*. Surabaya: Guna Widya, 2000.

[5] E. Grandjean, *Fitting the Task to the Man, a Textbook of Occupational Ergonomics*, Piladelphia: Taylor & Francis, 1998.

[6] B. Sorin-Stefan, M. Ionescu, G. Voicu, N. Ungureanu and V. Vladut, *Calculus Elements for Mechanical Presses in Oil Industry*. Politehnica University of Bucharest, Romania, 2013.

[7] K. C. R. Martins, P. H. C. Silva, V. L. Galvao, L. S. P. Martins, S. L. E. Matias, and R. J. R. Pereira, "Development of Prototype of Screw Press for The Collection of Oil from Seeds of Castor and Jatropha". *Brazilian Congress of Thermal Sciences and Engineering*, 2010.

[8] P. Sari, "Preliminary Design and Construction of A Prototype Calona Seed Oil Extraction Machine," MS. thesis, the Graduate School of Natural and Applied Sciences of Middle East Technical University, 2006.

[9] Y. Zheng, "Screw Pressing of Whole and DehulledFlaxeed for Organic Oil Rich in Omega-3 Fatty Acids," PhD. dissertation, North Dakota State University of Agriculture and Applied Science, 2003.

[10] P. Beerens, "Screw-pressing of Jatropha seeds for fuelling puposes in less developed countries," MS. thesis, Dept. of Sustainable Energy Technology, Eindhopen University of Technology, 2007.

[11] Rumus Hitung (May 2013). Tabel Massa Jenis dan Berat Jenis Berbagai Zat di Sekitar Kita [online] Available: http://rumushitung.com/2013/05 /31/tabel- massa-jenis-dan-berat-jenis/.

[12] F. G. Winarno, *Kelapa Pohon Kehidupan*, Jakarta: Gramedia Pustaka Utama, 2014

[13] I W. Surata, "Penggunaan Roda Tangan Berhendel pada Alat Pres Parutan Kelapa Mengurangi Keluhan Sistem Muskuloskeletal dan Meningkatkan Produktivitas Kerja Pembuat Minyak Kelapa Tradisional di Desa Ped Nusa Penida," MS. thesis, Program Pascasarjana Universitas Udayana, 2001.

IOP Conf. Series: Materials Science and Engineering **201** (2017) 012016 doi:10.1088/1757-899X/201/1/012016

Simulation and Characterisation of Planar Spring Based on PCB-FR4 in Electromechanical System for Energy Harvesting

A N Fuadah[1], N F Maulanisa[1], A Ismardi[1], G Sugandi[2]

[1] Engineering Physics Department, Telkom University, Bandung-Indonesia
[2] Research Center for Electronics and Telecommunication-Indonesia Institute of Science, PPET-LIPI, Bandung-Indonesia

corresponding author's e-mail : abrarselah@telkomuniversity.ac.id

Abstract. This paper presents comparison study of simulation and fabrication characterized two type planar springs at micro–fabricated electromagnetic power generator for an ambient vibration energy harvesting system. The power generator utilized a LASER–machined FR4-PCB planar spring, a copper coil, and NdFeB magnet. In order to change resonant frequency, we developed a gimbal suspension structure for the fabrication of spring. The NdFeB permanent magnet was applied as inertial mass. The system was specially designed to harvest low ambient vibrations from 20 to several hundred hertz and low acceleration. The dimension of fabricated energy harvester had 2.5 x 2.5 cm^2 in size. In this study we present two different design of cantilever, which is has two and four cantilever, respectively. The different designed given different resonance frequency to the system. The result of simulation giving resonance frequency of two cantilever membrane 22.6 Hz and four cantilever membrane 110.3 Hz. The measurements result has generated 0.135 V with resonance frequency 39 Hz of two cantilever membrane appropriate for human motions, four cantilever membrane has generated 0.174 V with resonance frequency106 Hz appropriate for machine industries.

Keyword: planar springs, ambient vibration, energy harvesting system, cantilever.

1. Introduction

Energy harvesting is capturing free environmental energy and converts it into useful electrical energy which can be used to power a wireless sensor node. Free environmental energy can be found in form of solar, thermal, wind and mechanical vibrations which are converted into electrical energy. A lot of research has been done in recent years on using ambient vibrations as a power source. The vibration energy harvester is a kinetic energy generator that converts mechanical energy in the form of vibrations present in the application environment into electrical energy. Kinetic energy is typically converted into electrical energy using electromagnetic (G Hatipoglu&Rey 2010), piezoelectric (Anton S. R.&A. 2008; Renaud M. et al. 2009) or electrostatic (Mitcheson P. D. et al. 2004) transduction mechanisms.

In the literature, there are many study of vibration energy harvesting using electrodynamics conversion energy mechanism either permanent magnet or also coil as moving part generator. In a resonant vibration energy generator, maximum power can be achieved when the environmental frequency matches with the resonant frequency of the mechanical resonator structure, but environmental vibrations are typically below 100 Hz and have broadband characteristics (Byung-Chul Lee&Chung 2013). Therefore, there are many researcher investigated PCB-FR4 as mechanical resonator material for generate very low resonant frequency because the low Young's modulus of FR-

IOP Conf. Series: Materials Science and Engineering **201** (2017) 012016 doi:10.1088/1757-899X/201/1/012016

4 material. Beside this, due to simple and low cost fabrication PCB process, the possibility efforts of integration of miniature PCB based vibration energy harvesting integrated with wireless sensor node system on a single PCB can be realized (Byung-Chul Lee&Chung 2013; Neil N. H. Ching et al.).

In this paper, two different configuration resonator of PCB-based planar spring for vibration energy harvesting have been compared to meet its frequency resonant with refer to maximum output voltage in the coil. The fabrication and characterization of both configuration of mechanical resonator are presented.

2. Design and Fabrication

Figure 1a shows a schematic of the electrodynamics vibration energy harvester (EVEH) system and the implementation of EVEH system in miniature as shown in fig 1b. The energy harvester composed of a thick PCB-FR4 as housing and fixed frame of generator, copper coil placed on the base of frame, and extra thin PCB-FR4 planar spring which has four cantilever beam and a platform for attachment permanent magnet NdFeB at the center of platform. The PCB planar spring and permanent magnet are used as movable part, mean while a coil attached and the frame are used as stator part of the generator. In principle work of the harvester to generated electric energy, when the harvester is subject to an excitation vibration, the proof mass on which a permanent magnet of induction is attached responds with deflection. According to Faraday's law of electromagnetic induction, the time rate of change of deflection induces a voltage in output coil terminal.

The Electrodynamics Vibration Energy Harvesters (EVEH) converts kinetic energy into electrical energy by electromagnetic transduction mechanisms. A general spring-mass model was built to describe the harvester system by C.B. Williams and R.B. Yates [6], as show in Figure 1a. This system can be modeled analytically (equation 1) as a mass-spring-damper consisting of a damping coefficient, and c_t, spring constant, k, and the seismic mass, m.

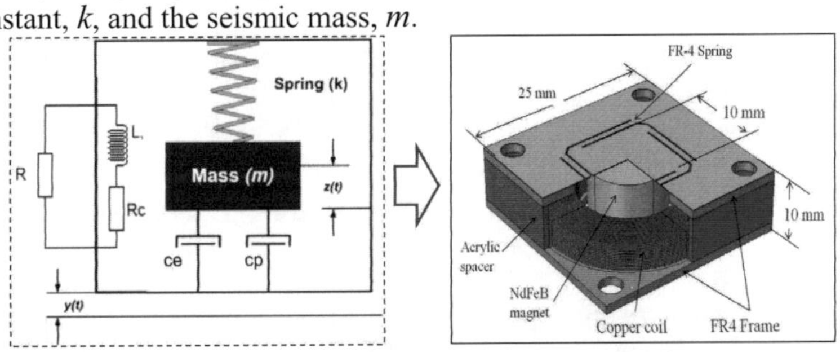

(a) (b)

Figure 1. Schematic of EVEH system (a), and Implemented of miniature EVEH system

The voltage generated in the wire coil caused by the presence of magnetic induction is moving relative to the coil, *z* as input. When the module system is vibrated by the displacement, $y(t)$, mass transfer, $z(t)$, produced by the relative mass attached to a spring. Energy losses in the system (consisting of the parasitic losses, *cp*, and the reduction of electrical energy generated by the mechanical transduction, *ce*), represented by the total attenuation coefficient, *ct.* From these analytic modeling, the equations of motion can be derived as follows

$$m\frac{d^2z}{dt^2} + c_t\frac{dz}{dt} + kz(t) + F_{mn}[z(t)] = F_d(t) \qquad (1)$$

Where F_{mn} is the magnetic force acting on the coil and the external driving force F_d is due to harmonic vibrations. The maximum energy that can be harvested by this system occurs when the external frequency is equal to that of the resonance frequency of the system, $\omega_n^2 = k/m$. Williams and Yates showed that the power dissipated in the damper (i.e, extracted by both transduction mechanisms and parasitic damping mechanism) is given by:

$$P_{av} = \frac{m\xi_t Y^2 \left(\frac{\omega}{\omega n}\right)^3 \omega}{\left[1-\left(\frac{\omega}{\omega n}\right)^2\right]^2 + \left[2\xi_t\left(\frac{\omega}{\omega n}\right)\right]^2} \quad (2)$$

where m is the mass of moving parts, Y and ω is the amplitude and angular frequency of the motion harmonic, respectively, ω_n is the angular frequency resonance of the mass-spring oscillator, $\xi_t = c_t / 2m\omega_n$ is total damping ratio, and ct is a total damping factor.

Mechanical resonator is a very important component of the EVEH as previously described. Mechanical resonators are used to increase the amplitude of mechanical vibrations that are transferred come from the surrounding environment. They are designed to match the frequency characteristics of the application environment. A mechanical resonator characterized by a resonant frequency.

$$f_n = \frac{1}{2\pi}\sqrt{\frac{n \cdot E \cdot w \cdot t^3}{4 \cdot L^3 \cdot (m_t + 0.24 m_c)}} \quad (3)$$

Where E is Young`s modulus of the cantilever material; n, w, t, and L are the numbers of cantilever beam ,width, thickness and length of the cantilever, respectively.

Figure 2 show the shape of planar spring type that has two (Fig.2a) and four (Fig.2b) cantilever beam and a platform in the center of planar spring. The width of planar spring cantilever beam 1.0 mm and length 21.5 mm (a), 10 mm (b) respectively. The planar spring was fabricated on double side PCB-FR4 with size 2.5 mm x 2.5 mm and thickness of FR4 substrate 0.1 mm and 0.015 mm thick copper layer for each side.

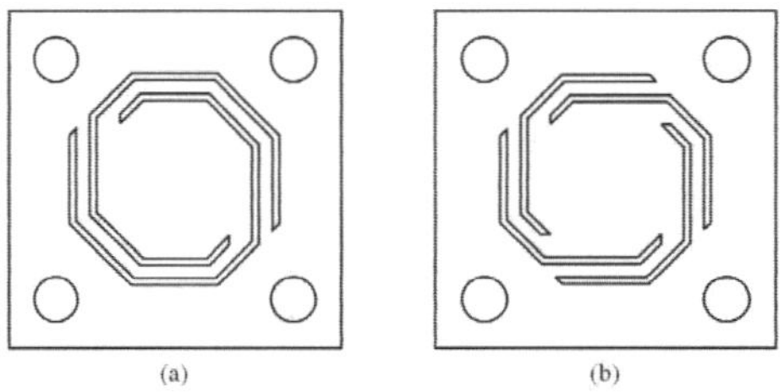

Figure 2. Design of planar spring for vibration energy harvesting with (a) two cantilever, and(b) four cantilever beam.

Two 3-Dimension structure of planar spring are considered for simulation. The designs of planar spring (Fig. 2) are simulated using finite element analysis by Comsol 4.2. Some parameters, such as width, length, and thickness are chosen to produce resonance frequency that is near to real value (experiment result). A permanent magnetic source was then applied to the system, with the specifications are densities 7.17 g/cm3, Mass 0.36 g, Area 0.39 cm^2 and High 3.08 mm.

After obtaining the simulation result, the membrane will be fabricated and then characterised for confirmation the result is matched with the simulation. The PCB planar spring is fabricated using laser cutting for released cantilever beam and magnet platform. The membrane will then characterised by using a loudspeaker as vibration source by changing input frequency that is controlled by audio signal generator. The resonance frequency will generate maximum voltage from the system.

3. Result and Analysis

3.1. Simulation Results

The result of simulation FR4 membranes coated with Cu using two cantilevers has obtained a resonant frequency of 39.93 Hz. The highest displacement of the membrane was also shown in the Figure 3, as it is indicated as red zone, while the blue one indicated frame of the membrane. The maximum displacement is leads to higher output power.

Following the same step for two cantilever design, the four cantilevers FR4 membranes coated with Cu was also simulated by using Comsol. The result showed that it has a resonance frequency of 110.73 Hz. The maximum voltage was also obtained at that point, as it is described in the Figure 4.

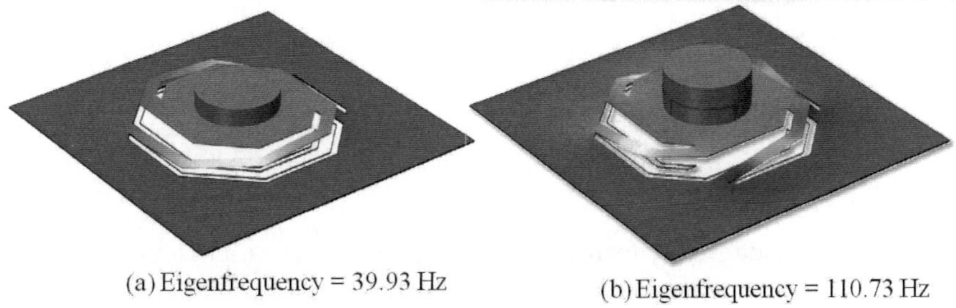

Figure 3. Eigen frequency response of two type coated Cu FR4 planar spring designed with two cantilever (a), and four cantilevers beam (b).

3.2. Experiment Results

Simple fabrication process for the mechanical resonator components and assembling components into one system energy harvester is shown in Figure 4. The process steps include design layout of the planar spring (Fig. 4a), then FR4 PCB substrate cleaning process and proceed with the process of cutting with a laser cutter in accordance design layout (Fig. 4b and 4c) and cleaning FR4 substrate. After the mechanical resonator fabrication, the next step is to characterize the resonance frequency of the resonator components, the components are assembled into a vibration energy harvesting system as shown in figure 4d.

The experiment was carried out using an audio signal generator, vibrator (loudspeaker), accelerometer and multi meter. The input frequency from the signal generator software converted by loudspeaker which will produce external mechanical vibration. When the external mechanical vibration or forces is enough to vibrate the membrane and then the mass component of the energy harvesting devices would move or oscillate. The electromagnetically induction process will be converted to electrical energy, and will be captured by a multimeter.

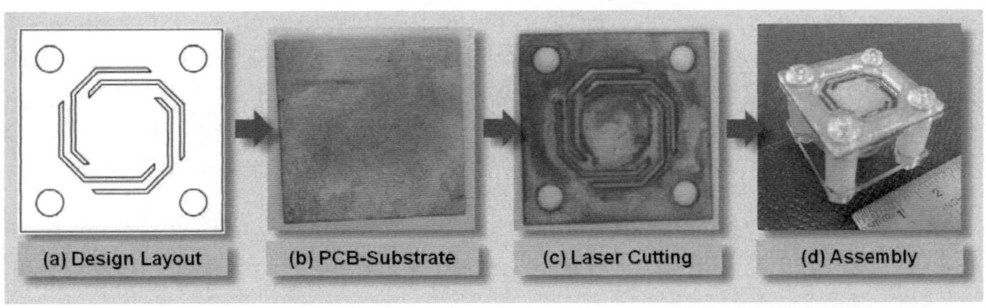

Figure 4. Step fabrication of planar spring (a-c) and assembling into EVEH system (d)

The characterisation of measurement for the membrane design with two cantilevers is shown in the Figure 5, as it is giving the input frequency of 32 – 40 Hz. The maximum voltage generated at frequency 37 Hz which has 0.13 V at external acceleration 0.5g, while for external acceleration 1g the frequency resonant was shifted to 39 Hz with open circuit voltage 0.15 V. The Figure 5 shows response frequency of planar spring with four cantilever beam, as it is giving the input frequency of 100 – 125 Hz. The maximum voltage generated at frequency 108 Hz which has 0.17 V at external acceleration 0.5g, meanwhile for external acceleration 1g the frequency resonant was shifted to 106 Hz with open circuit voltage 0.225 V. At the peak point, the graph down significantly into the lowest voltage, we analyse that the highest voltage shows resonance frequency of planar spring with two cantilever beam.

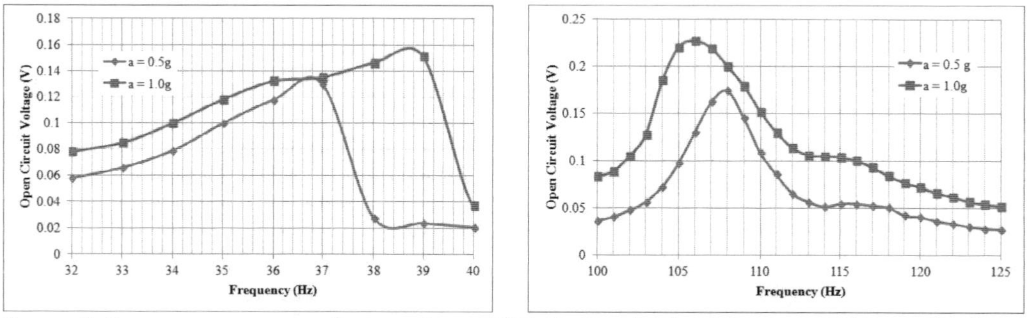

Figure 5. Frequency response for membrane FR4 designed (a) with two cantilevers, (b) four cantilevers beam versus external acceleration 0.5 g and 1.0 g.

4. Conclusion

This paper presents the design, fabrication and characterization of planar springs made from FR4-PCB for electrodynamics vibration energy harvesting applications (EVEH). The use of very thin material PCB-FR4 expected the energy harvesting systems can be integrated with electronic drive components on one chip PCB. Based on the results of measurements, the FR4 was used as the planar spring material is very suitable for the sources of energy vibrating with the acceleration is very low, below 1.0g with a vibration frequencies below and around 100 Hz. The different designed given different resonance frequency to the system. The result of simulation giving resonance frequency of two cantilever membrane 39.93 Hz and four cantilever membrane 110.73 Hz. The measurements result has generated 0.135 V with resonance frequency 39 Hz of two cantilever membrane appropriate for human motions, four cantilever membrane has generated 0.174 V with resonance frequency106 Hz appropriate for machine industries.

References

[1] G Hatipoglu & Rey, H. U. 2010. Fr4-Based Electromagnetic Energy Harvester for Wireless Sensor Nodes. IOP Publishing 19 (Smart Materials and Structures): 11.

[2] Byung-Chul Lee & Chung, G.-S. 2013. Design and Fabrication of Low Frequency Driven Energy Harvester Using Electromagnetic Conversion. Transaction on Electrical and Electronic Materials 14(3): 5.

[3] Mitcheson P. D., Miao P., Stark B. H., Yeatman E. M., Holmes A. S. & C., G. T. 2004. Mems Electrostatic Micro-Power Generator for Low Frequency Operation Sensors Actuators A 115(7).

[4] Neil N H Ching, Gordon M. H. Chan, Wen J. Li, Hiu Yung Wong & Leong, P. H. W. 2000. Pcb Integrated Micro-Generator for Wireless Systems. Symp on smart structures and Microsystems.

[5] Renaud M., Fiorini P., Shaijk R. & C., H. 2009. Harvesting Energy from the Motion of Human Limbs: The Design and Analysis of an Impact-Based Piezoelectric Generator Smart Mater. Struct. 18(16.

[6] S P Beeby, R N Torah & Tudor, M. J. 2008. Kinetic Energy Harvesting. ACT Workshop on Innovative Concepts.

[7] S gandi, M A Grace. 2015. Development of Miniature Planar Spring for Electrofynamics Vibation Energy Harvesting Using Extra Thin Printed Circuit Board. Proceedings of the International Conference on ECBA-2015.

[8] Bilen S G, Terman K Mc, Gilchrist B E and Hoyt R P. 2010. Electrodynamic Tethers for Energy Harvesting and Propulsion on Space Platforms. (California: AIAA Space 2010 -8844)

IOP Conf. Series: Materials Science and Engineering **201** (2017) 012017 doi:10.1088/1757-899X/201/1/012017

Validation of Finite Element Model used to Analyze Sheet Metal Punching Process in Automotive Part Manufacturing

N Chantarapanich[1,2], A Siripanya[1], S Sucharitpwatskul[3], S Wanchat[1,2]

[1] Department of Mechanical Engineering, Faculty of Engineering at Sriracha, Kasetsart University, Chonburi, Thailand 20230
[2] Industrial Production Technology Research Group, Faculty of Engineering at Sriracha, Kasetsart University, Chonburi, Thailand 20230
[3] National Metal and Materials Technology Center, Pathumthai, Thailand 12120

E-mail: nattapon@eng.src.ku.ac.th

Abstract. Punching process is an operation that a scrap is separated from a metal sheet by a punch. Improper setting of punching conditions may lead to excessive of material deformation around edge region (burr), which may weakening the strength of produced part. Analysis of punching mechanics would be beneficial reducing defective part. One of effective analysing tool for this application is Finite Element (FE) method. The aim of this study is to develop reliable FE model for analysis of punching process. The FE model was developed based on 2D. FE result was validated with experimental testing result by comparing burr height. It was found that FE result is -1.79% difference compared to experimental result. Good agreement between FE and experimental result was obtained.

1. Introduction

Most of automotive body panels, for instant, chassis frame, door, and fender [1,2], were built based on various sheet metal forming processes. Sheet metal punching is one of main manufacturing process for automotive body panel productions. Principle of punching is to form the sheet metal into desired shape by applying force on punch to separate scrap from a sheet metal through die [3], as shown in Figure 1(a).

The sheet metal punching process usually creates the localized problem around edge regions. One main observed problem is an excessive material deformation, called burr, as shown in Figure 1(b). Burr can cause stress concentration at edge of holes which may decrease in strength of produced sheet metal. Understanding the mechanics of materials in punching process is necessary to prevent burr formation. Analyzing mechanics of materials requires effective tool such as finite element (FE) method, which have became increasingly importance to analyze many manufacturing problem [4,5]. Since FE analysis is based on numerical method, at least one of FE model is required to be validated with experimental test to assure that FE model will best describe the reality. Typically, FE model is acceptable, if the difference between FE and experimental results is not over 20 % [6].

This paper is the first part of two parts paper belonging to burr prevention project for punching process study in one of automotive part manufacturer. The first part presents the development of FE model for sheet metal punching analysis whereas the second part will be later presented the developed measure to prevent the burr produced from punching process.

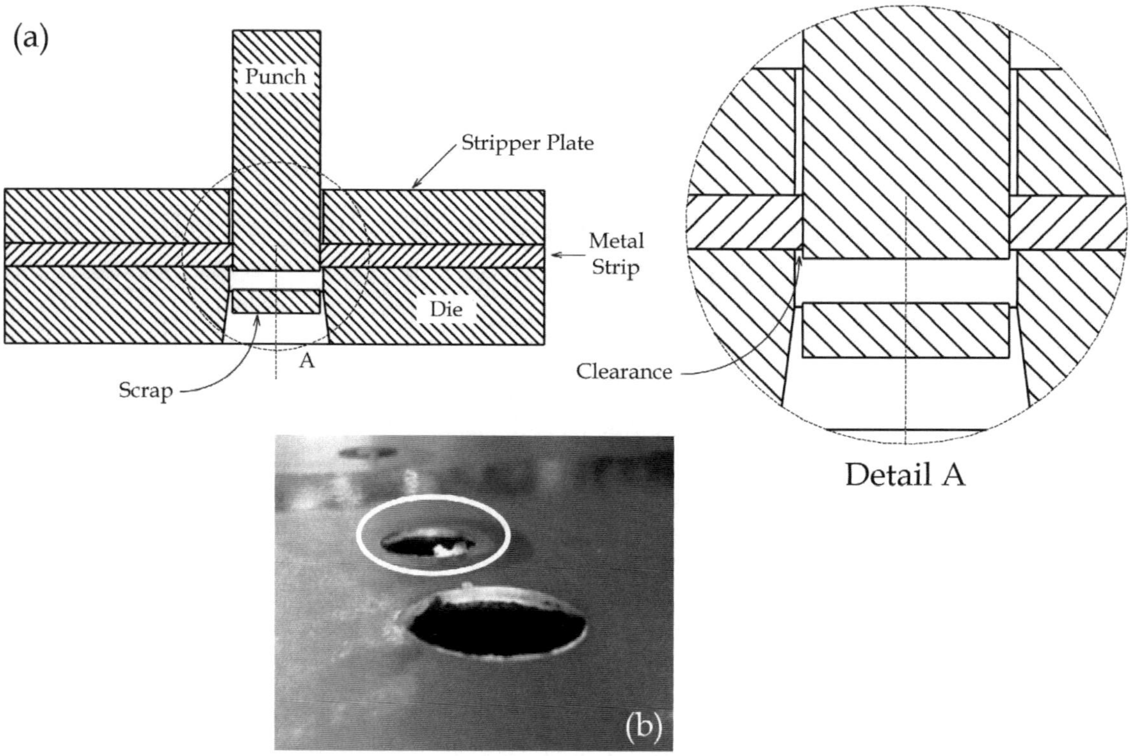

Figure 1. (a) Sheet metal punching process, (b) Burr.

2. Methodology

Development of FE model started from selection of one set of punch and die set for validation. In this case, circular shape as shown in Figure 2a was selected. In order to develop reliable FE model, it was necessary to acquire accurate data of (1) geometry, (2) material properties, and (3) boundary conditions [7].

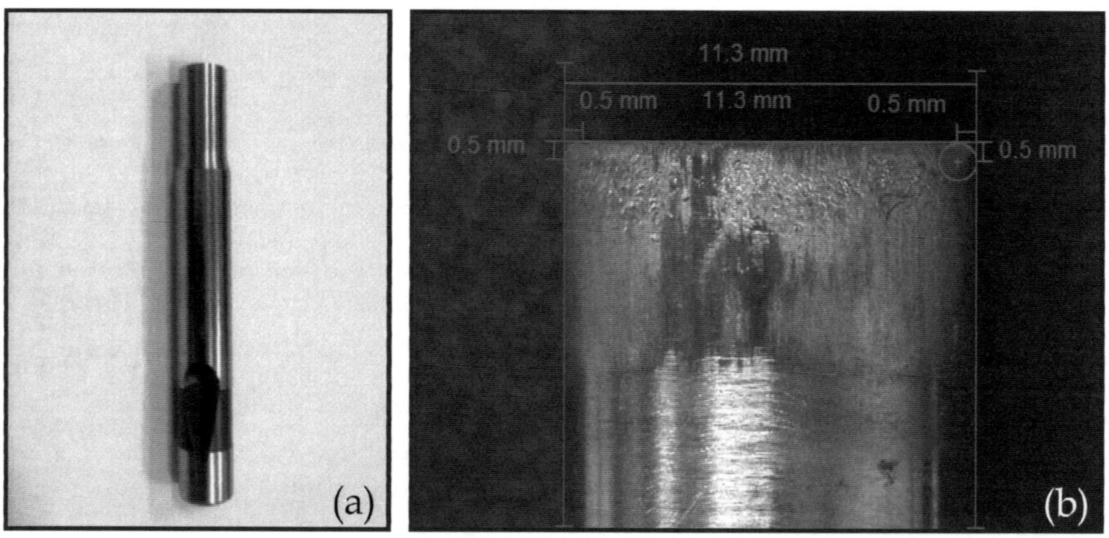

Figure 2. (a) Punch used in this study, (b) Microscope examination of punch edge.

For geometric model, since shape of punch and die process is symmetric, 2D model was developed instead of 3D model. Advantages of 2D geometric model included smaller number of element and nodes employed in analysis, thus less computational resource is required. 2D model was generated from CAD catalog from supplier by cross-section the mid plane of geometry. However, the 2D model derived from the CAD catalog cannot create exact shape at punch edge. This is because punch edge is usually deformed after the certain period of use. As a result, punch was then measured its shape dimension (radius) using microscope, as shown in Figure 2b. 2D geometric model was modified its edge shape according to measured dimension. In addition to punch geometry, clearance between punch and die, and sheet metal thickness were also measured to provide accurate information for CAD geometry modeling. From the measurement, clearance between punch and die was 1.13-mm and metal sheet thickness was 2.9-mm. Punch and sheet metal in CAD model were created as 2D surface models whereas punch holder, retainer, and die were modeled as wireframe.

Material properties assigned to different regions of FE model were from literature and tensile testing data. Punch is made of SKH51, its linear material property was referenced from [8]. The sheet metal used in this study was TIS 1999-2543. The material property of TIS 1999-2543 was acquired from the tensile test described in previous work of the authors [9]. To simulate separation of scrap from from metal sheet, Cockroft-Latham damage criteria with coefficient of 0.75 was applied. Boundary conditions applied to FE model was shown in Figure 3. Displacement during punching process was applied in –y direction at punch holder.

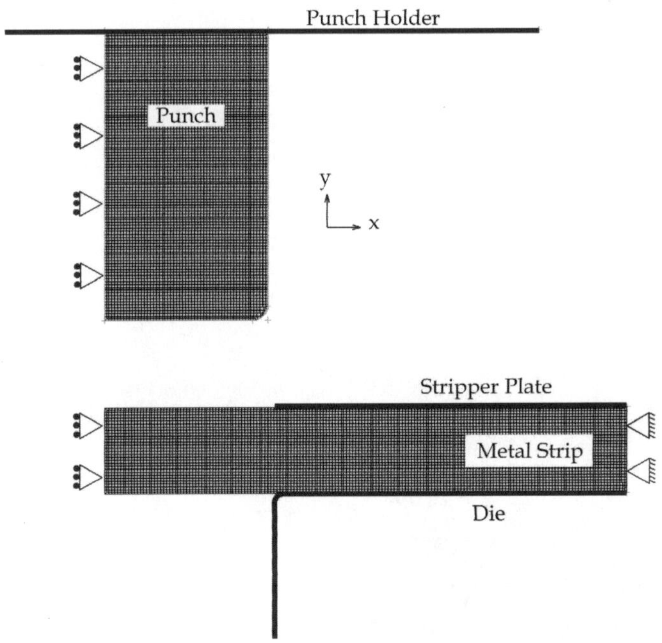

Figure 3. FE model and its boundary conditions.

FE analysis was performed using MSC Marc Mentat software package. Four-node quad element for FE analysis was generated using automatic mesh generation technique. Number of node and element employed in FE analysis were 11347 and 10987, respectively. Burr height obtained from FE analysis was compared to the burr height measured from sheet metal after punching process.

3. Results

Figure 4 shows FE result of metal sheet deformation, burr was observed around edge of hole near die surface. On the opposite surface, near stripper plate, surface around edge of hole was also sunk. Peak of burr height from FE analysis was 0.440 mm, whereas average burr height measured from finished product (n=10) was 0.448 mm (S.D. 0.06 mm).

Percentage difference between FE and experimental results is calculated using equation (1).

$$\%Difference = \frac{Result_{FE} - Result_{Exp}}{Result_{Exp}} \times 100 \tag{1}$$

From the calculation, percentage difference was -1.79%.

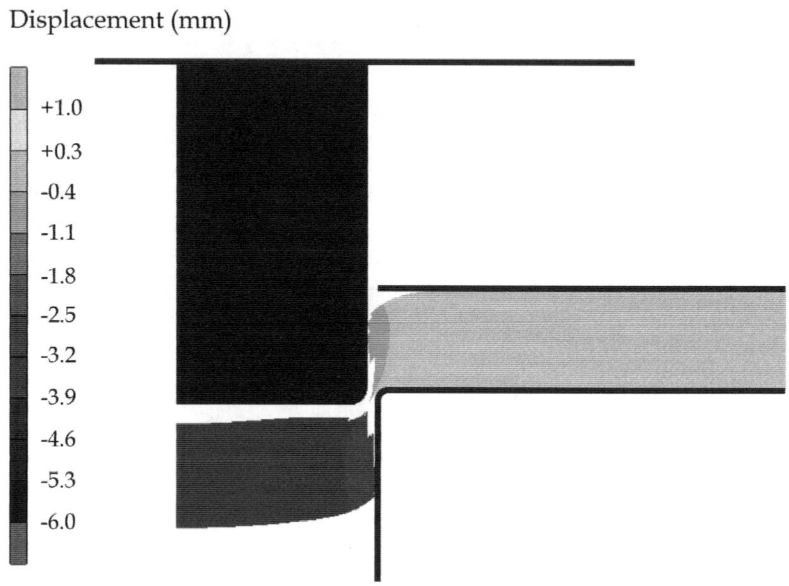

Figure 4. FE result shows deformation of material.

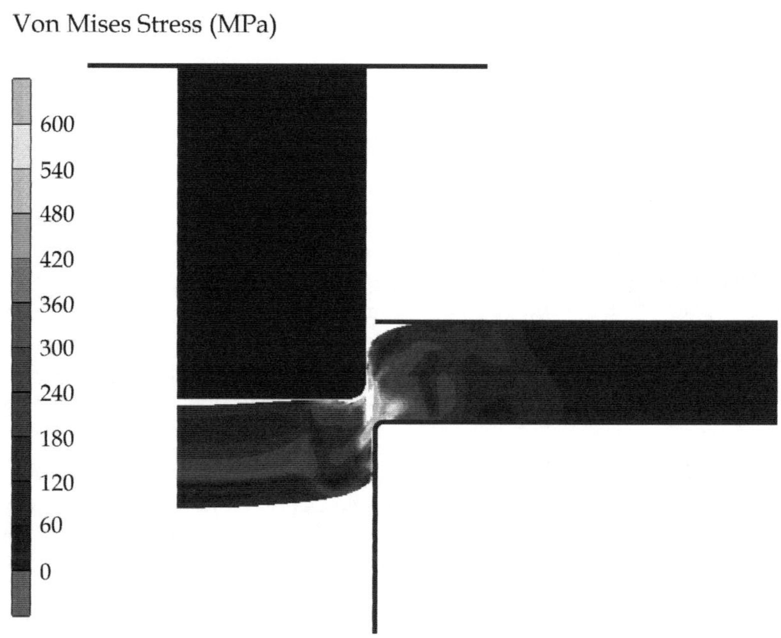

Figure 5. FE result shows stress during punching.

In addition to the results of material deformation used for validation, FE analysis allows to observe the stress distribution exhibited in material during punching process. It can be seen that high stress region of metal sheet occurs around the punch and die clearance, as shown in Figure 5. The value of maximum stress is beyond the failure stress of sheet metal material. Average stress on punch contact surface occurs reached its maximum values of 1400 MPa during initial contact to metal sheet.

4. Discussion and Conclusion

FE method is an effective tool to analyze various manufacturing problem [3,10]. Nevertheless, FE method is a numerical tool which its result requires to be validated with experimental result. To develop reliable FE model, three factors have to be accurately input for analysis i.e. geometry, material properties, and boundary conditions [7]. In this study, 2D FE model was developed for analyze punching process. FE Model was developed by (1) extracting actual shape of punch, die and sheet metal, (2) examining material property of punch and metal sheet, and (3) setting boundary condition as according to punching process. The obtained FE result was only -1.79% difference compared to the experimental results. Generally, the difference between FE and experimental results should be below 20% [6]. With this small difference in magnitude, good agreement between FE and experimental result was obtained.

References

[1] Kaneko S, Hiramoto J, Ishikawa T 2003 Development of Hot-Rolled Sheet Steel with Significant Increase in Tensile Strength Induced by Strain Age Hardening *Kawasaki Steel Technical Report* **48** 53–59

[2] Ounpanich D, Mutoh J, Yoshii K 2008 Fatigue behavior of hot-rolled steel plate at various stages of fabrication *Materials Science and Engineering A* **477** 386–394

[3] Brokken D, Brekelmans WAM, Baaijens FPT 2000 Predicting the shape of blanked products: a Finite element approach *Journal of Materials Processing Technology* **103** 51–56

[4] Umbrello D 2008 Finite element simulation of conventional and high speed machining of Ti6Al4V alloy *Journal of Materials Processing Technology* **196** 79–87

[5] García-Domínguez A, Claver J, Camacho AM, Sebastián MA 2015 Comparative Analysis of Extrusion Processes by Finite Element Analysis *Procedia Engineering* **100** 74–83

[6] Sivarao S, Leong ST, Yusof Y, Tan CF 2015 An experimental and numerical investigation of tensile properties of stone wool fiber reinforced polymer composites *Advanced Materials Letters* **6** 888–94

[7] Chantarapanich N, Nanakorn P, Chernchujit B, Sitthiseripratip K 2009 A finite element study of stress distributions in normal and osteoarthritic knee joints *Journal of the Medical Association of Thailand* **92** S97-S103

[8] China Steel Supplier 2013 Data Table for: Steel Grades: Tool Steel: SKH 51 *www.steelgr.com*

[9] Siripanya A, Chantarapanich N, Wanchat S 2015 Mechanical properties of a hot rolled stainless steel sheet *KKU Engineering Journal* **43** 207–209

[10] Jin X, Altintas Y 2012 Prediction of micro-milling forces with finite element method *Journal of Materials Processing Technology* **212** 542–552

IOP Conf. Series: Materials Science and Engineering **201** (2017) 012018 doi:10.1088/1757-899X/201/1/012018

Analysis of flexible layered shallow shells on elastic foundation

L Stupishin[1], **A Kolesnikov**[1] **and T Tolmacheva**[1]

[1] Urban, road Construction and Structural Mechanics Department, South-West State University, 305040, 50-let Octyabrya st., 94, Kursk, Russian Federation

lusgsh@yandex.ru

Abstract. This paper contains numerical analysis of a layered geometric nonlinear flexible shallow shell based on an elastic foundation. Rise of arch in the center of the shell, width, length and type of support are given. The design variable is taken to be the thickness of the shallow shell, the form of the middle surface forming and the characteristic of elastic foundations. Critical force coefficient and stress of shells are calculated by Bubnov-Galerkin. Stress, characteristic of elastic foundations – thickness dependence are presented.

1. Introduction

Layered shallow shell on elastic foundation is promising structures as both at the macro and micro level of consideration. Calculation methodologies can be used in building and nano-structured materials systems. Therefore, the development of methods of their analysis is an urgent task.

Often designers solve problems not only to calculate such a structure, but also to find the optimal parameters. Currently, most of these structures are designed with software based on the finite element method and method of sensitivity analysis of structures [1]. It is convenient model for engineering design gives good results only for linear problems. In the case of non-linear problems, you can get results, the accuracy of which is difficult to assess.

Therefore, the development of numerical methods for the design of structures operating in the nonlinear stage is necessary.

2. Equations for the determination of critical force coefficient and stress for layered shallow shells on an elastic foundation

Equations for the determination of critical force coefficient and stress for layered shallow shells on an elastic foundation given in [2, 3]:

$$\begin{cases} \frac{1}{Eh}\nabla^2\nabla^2\varphi + k_y\frac{\partial^2 w}{\partial x^2} + k_x\frac{\partial^2 w}{\partial y^2} - 2k_{xy}\frac{\partial^2 w}{\partial x\partial y} + \frac{\partial^2 w}{\partial x^2}\frac{\partial^2 w}{\partial y^2} - \left(\frac{\partial^2 w}{\partial x\partial y}\right)^2 = 0; \\ D\nabla^2\nabla^2 w - \frac{\partial^2\varphi}{\partial y^2}\left(k_x + \frac{\partial^2 w}{\partial x^2}\right) - \frac{\partial^2\varphi}{\partial x^2}\left(k_y + \frac{\partial^2 w}{\partial y^2}\right) + 2\frac{\partial^2\varphi}{\partial x\partial y}\left(k_{xy} + \frac{\partial^2 w}{\partial x\partial y}\right) = Z + 2t\nabla^2 w - kw. \end{cases} \quad (1)$$

Where

φ – stress function, w – deflection function.

k_x, k_y – curvature of the middle surface of the shell,

$$k_{xy} \approx \frac{\partial^2 F}{\partial x \partial y}, \tag{2}$$

F=$F(x,y)$ - the equation of the middle surface of the shell,
Z – loading function,
k – modulus of elastic foundation reaction,
t - soil shear parameter,

$$k = E_0(1-\nu_0)^{-1}\int_0^H \Psi'^2 dz, \tag{3}$$

$$t = E_0(4(1-\nu_0))^{-1}\int_0^H \Psi^2 dz. \tag{4}$$

Where

$$E_0 = E_s(1-\nu_s)^{-1}, \tag{5}$$

$$\nu_0 = \nu_s(1-\nu_s)^{-1}, \tag{6}$$

E_s и ν_s - elastic module and Poisson's ratio of the soil,
H – depth of the soil stratum,
$\Psi(z)$ - function of cross distribution displacements.
Function of cross distribution displacements:

$$\Psi(z) = \left(sh\gamma\frac{H-z}{a}\right)\left(sh\gamma\frac{H}{a}\right)^{-1}. \tag{7}$$

Where
γ - parameter that characterizes the vertical deformation profile within the soil continuum.

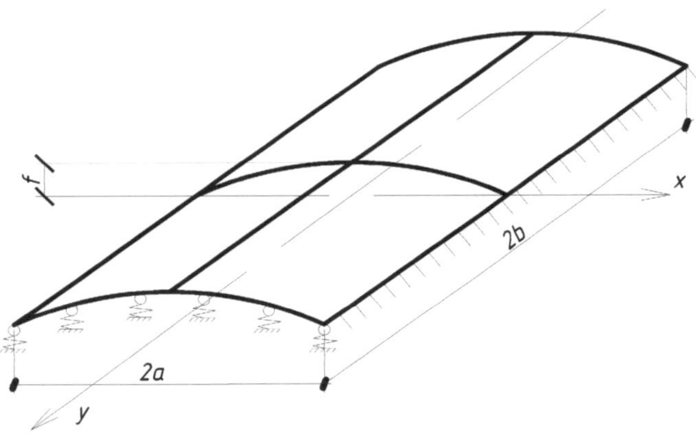

Figure1. Shallow shell on an elastic foundation.

Equivalent stresses in the shell are found by the fourth stress hypothesis. Stress coefficient can be described by the equation [4, 5]:

$$\sigma = \left(\frac{1}{2}\left[\left(\overline{\sigma_1} - \overline{\sigma_2}\right)^2 + \left(\overline{\sigma_3} - \overline{\sigma_2}\right)^2 + \left(\overline{\sigma_2} - \overline{\sigma_1}\right)^2\right]\right)^{1/2}, \tag{8}$$

where
$\overline{\sigma_1}$, $\overline{\sigma_2}$, $\overline{\sigma_3}$, – dimensionless stress coefficients:

$$\overline{\sigma_1} = \left|\frac{6}{t}\overline{DB}\left(\frac{\partial^2}{\partial x^2}Z_x Z_y + \nu\frac{\partial^2}{\partial y^2}Z_y Z_x\right)\right| + \left|\overline{A}\frac{\partial^2}{\partial y^2}Z_y Z_x t\right| + \left|\frac{g}{t^2}\overline{DB}\left(\frac{\partial^3}{\partial x^3}Z_x Z_y + \frac{\partial}{\partial x}Z_x\frac{\partial^2}{\partial y^2}Z_y\right)\right|, \tag{9}$$

$$\overline{\sigma_2} = \left|\frac{6}{t}\overline{DB}\left(\nu\frac{\partial^2}{\partial x^2}Z_x Z_y + \frac{\partial^2}{\partial y^2}Z_y Z_x\right)\right| + \left|\overline{A}\frac{\partial^2}{\partial x^2}Z_x Z_y t\right| + \left|\frac{g}{t^2}\overline{DB}\left(\frac{\partial^3}{\partial y^3}Z_y Z_x + \frac{\partial}{\partial y}Z_y\frac{\partial^2}{\partial x^2}Z_x\right)\right|, \tag{10}$$

$$\overline{\sigma_3} = \left|\frac{6}{t}(1-\nu)\overline{DB}\frac{\partial}{\partial x}Z_x\frac{\partial}{\partial y}Z_y\right| + \left|\overline{A}\frac{\partial}{\partial x}Z_x\frac{\partial}{\partial н}Z_y t\right|. \tag{11}$$

$\overline{D}$ - dimensionless stiffness coefficient of shell:

$$\overline{D} = D\left(Eh^3\right)^{-1}, \tag{12}$$

$$D = Eh^3\left(12\left(1-\nu^2\right)\right)^{-1}. \tag{13}$$

$$\overline{A} = A\cdot a^2\left(Ef_0^5\right)^{-1}, \tag{14}$$

$$\overline{B} = B\cdot a^2\left(f_0^3\right)^{-1}, \tag{15}$$

$$g = f_0(a)^{-1}. \tag{16}$$

A, B – unknowns of Bubnov-Galerkin method:

$$A = -(J_1)^{-1}\left(BJ_2 + B^2 J_3\right), \tag{17}$$

$$\begin{aligned} B = {} & (6C_1)^{-1}\big(36C_3C_2C_1 + 108qC_1^2 - 8C_2^3 + 12\sqrt{3}(4C_3^3C_1 - C_3^2C_2^2 + 18C_3C_2C_1q + \\ & + 27q^2C_1^2 - 4qC_2^3)^{1/2}C_1\big)^{1/3} - 0{,}67\left(3C_3C_1 - C_2^2\right)\big(C_1(36C_3C_2C_1 + 108qC_1^2 - 8C_2^3 + \\ & + 12\sqrt{3}(4C_3^3C_1 - C_3^2C_2^2 + 18C_3C_2C_1q + 27q^2C_1^2 - 4qC_2^3)^{1/2}C_1\big)^{1/3} - C_2(3C_3)^{-1}\big)^{-1}. \end{aligned} \tag{18}$$

$$C_1 = 2EhJ_3^2\left(J_1J_4\right)^{-1}, \tag{19}$$

$$C_2 = 3EJ_3J_2\left(J_1J_4\right)^{-1}, \tag{20}$$

$$C_3 = DJ_1^2\left(J_1J_4\right)^{-1} + EhJ_2^2\left(J_1J_5\right)^{-1}. \tag{21}$$

$J_1 - J_5$ - elements of the equation (1):

$$J_1 = \int_{-a}^{a}\int_{-b}^{b}\left(\nabla^2\nabla^2\overline{w}\right)\overline{w}dxdy, \tag{22}$$

$$J_2 = \int_{-a}^{a}\int_{-b}^{b}\left(\frac{\partial^2 F}{\partial y^2}\frac{\partial^2\overline{w}}{\partial x^2} + \frac{\partial^2 F}{\partial x^2}\frac{\partial^2\overline{w}}{\partial y^2} - 2\frac{\partial^2 F}{\partial x\partial y}\frac{\partial^2\overline{w}}{\partial x\partial y}\right)\overline{w}dxd, \tag{23}$$

$$J_3 = \int_{-a}^{a}\int_{-b}^{b}\left(\Delta\overline{w}\right)\overline{w}dxdy, \tag{24}$$

$$J_4 = \int_{-a}^{a}\int_{-b}^{b}Z\overline{w}dxdy, \tag{25}$$

$$J_5 = \int_{-a}^{a}\int_{-b}^{b}\left(2t\nabla^2\overline{w} - k\overline{w}\right)\overline{w}dxdy. \tag{26}$$

h – shell thickness, *E* - elasticity module, *ν* - Poisson's ratio, *Zx, Zy* – beam function.

Critical force coefficient for layered shallow shells on an elastic foundation can be described by the equation:

$$p_{cr} = 2\left(27C_1^2\right)^{-1}\left[\left(C_1^2 - 3C_1C_3\right)^{3/2} + C_2\left(C_2^2 - 4{,}5C_1C_3\right)\right] \tag{27}$$

For layered shells on elastic foundation of most interest is the research of stress.

Shells on an elastic foundation often makes layered. It can be considered representation of modules of elasticity of several layers to one.

The distance between the upper surface and the neutral of the layered shell can be determined by the equation:

$$z_0 = \left(E_1 h_1^2 + 2E_2 h_1 h_2 + E_2 h_2^2\right)\left(2(E_1 h_1 + E_2 h_2)\right)^{-1}, \tag{28}$$

where E_1, E_2 - elastic module of the upper and lower layer of shells,

h_1, h_2 - thickness of the upper and lower layer of shells.

Rigidity of the two-layer shallow shells can be described by the equation:

$$B = \left(E_1 h_1 + E_2 h_2\right)\left(1 - \nu_1 \nu_2\right)^{-1}; \tag{29}$$

$$D = \left(E_1 (z_0^3 - (z_0 - h_1)^3) + E_2 (h - z_0)^3 + (z_0 - h_1)^3\right)\left(3(1 - \nu_1 \nu_2)\right)^{-1}. \tag{30}$$

Where h_1, h_2 - thickness of the upper and lower layer of shells.

Main elasticity relations that bind the stress and strain state of the two-layer shell:

$$N_1 = B(\varepsilon_1 + \nu\varepsilon_2); \tag{31}$$

$$N_2 = B(\varepsilon_2 + \nu\varepsilon_1); \tag{32}$$

$$S = B(1 - \nu)\varepsilon_3; \tag{33}$$

$$M_1 = -D(\chi_1 + \nu\chi_2); \tag{34}$$

$$M_2 = -D(\chi_2 + \nu\chi_1); \tag{35}$$

$$T = -D(1 - \nu)\chi_3. \tag{36}$$

The equations enable us to determine and analyze the stress-strain state of the shell at any point.

3. Analysis of stress in layered shallow shells on elastic foundations

Common software systems based on the finite element method using does not allow to study of stress shallow shells on elastic foundation at the required level.

The analytical method allows analyzing the effect of various parameters of the shallow shells and choosing the algorithm to optimize its shape.

The changes of shallow shells on an elastic foundation depending on the variation form of the shells and characteristic of an elastic foundation are considered [6, 7].

Figure 2 depicts the dependence of the stress on the thickness of shell (Fig. 2, a) and on the parameter that characterizes the vertical deformation pro-file within the soil continuum (Fig. 2, b) for simple support and fixed support.

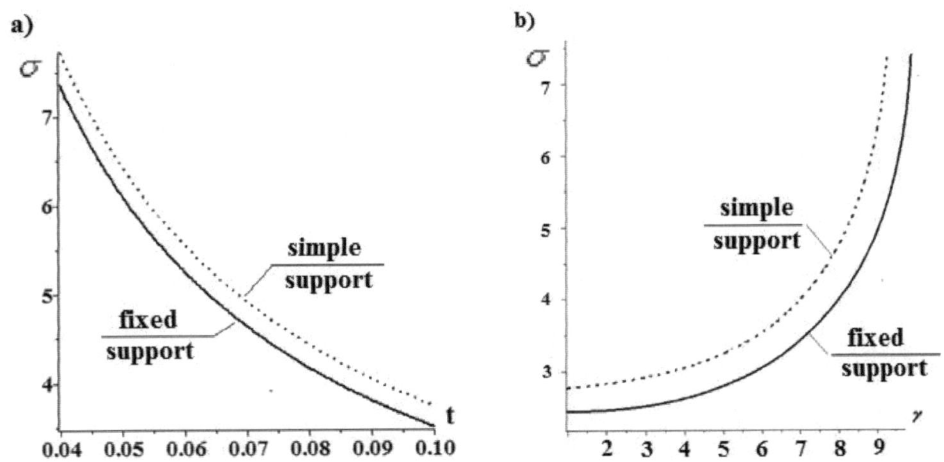

Figure 2. Illustration of the dependence of the stress: a) on the thickness of shell; b) on the parameter that characterizes the vertical deformation profile within the soil continuum.

The graphs show that the stress functions are smooth and unimodal. This will simplify the optimum forms of geometric nonlinear shallow shells an elastic foundations analysis.

The stress increases when the values of the parameter that characterizes the vertical deformation profile within the soil continuum grow up and the values of the thickness decrease. Type of support has an effect on the value of the stress.

Figure 3 depicts the dependence of the stress on the thickness of layered shell and on the parameter that characterizes the vertical deformation profile within the soil continuum for fixed support.

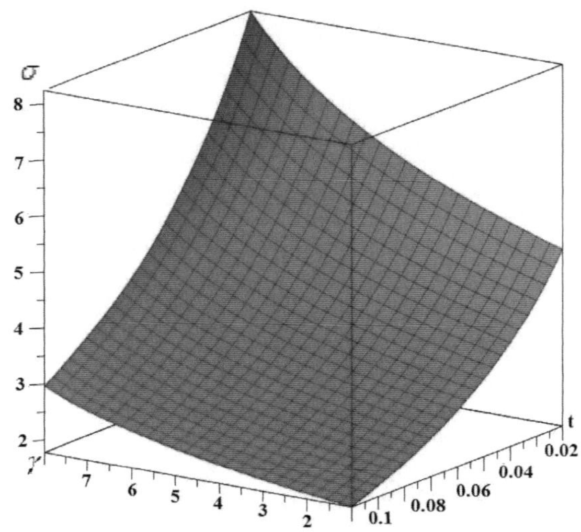

Figure 3. Illustration of the dependence of the stress on the thickness of shell and on the parameter that characterizes the vertical deformation profile within the soil continuum for fixed support.

Figure 3 shows that there are various combinations of the values of shell thickness and the parameter that characterizes the vertical deformation profile within the soil continuum at which the stress in shell is the same. Similar relationships were found for the critical load. Therefore, it is possible to find the optimal combination of these parameters with limiting the value of the stress, critical load the thickness (volume) of the shell or the characteristics of the elastic foundation [7-9].

4. Conclusions

This methodology can be applied to determine stresses and critical loads for geometrically nonlinear layered shallow shells on an elastic foundation of variable shape of the middle surface, thickness of shell and the characteristics of the elastic foundation. The graphs show the patterns of change of the stress of shells when changing the characteristics of the shell and the foundations. This can be used for designing and reconstructing of buildings and designing of nano-structured materials systems

This work was supported by the grant of the President of the Russian Federation № MK-9203.2016.8.

References

[1] Haug E J, Choi K K and Komkov V 1986 *Design sensitivity analysis of structural systems* (New York: Academic press, inc.) p 428
[2] Stupishin L and Kolesnikov A 2014 *Adv. Mat. Res.* **919-921**, 144-147
[3] Stupishin L and Kolesnikov A 2014 *Appl. Mech. & Mat.* **501-504**, 766-769
[4] Stupishin L and Nikitin K 2014 *Appl. Mech. & Mat.* **501-504**, 514-517.
[5] Stupishin L and Kolesnikov A 2014 *Appl. Mech. & Mat.* **580-583**, 3062-3065
[6] Andreev V *OPTI XII* 2012 189-201
[7] Andreev V *Appl. Mech. & Mat.* **166-169**, 354-358

Numerical research orthotropic geometrically nonlinear shell stability using the mixed finite element method

L Stupishin[1], K Nikitin[1] and A Kolesnikov[1]

[1] Urban, road Construction and Structural Mechanics Department, South-West State University, 305040, 50-let Octyabrya st., 94, Kursk, Russian Federation

lusgsh@yandex.ru

Abstract. A methodology for shell stability research and determining buckling load, based on the mixed finite element method are proposed. Axisymmetric geometrically nonlinear shallow shells made of orthotropic material are considered. The results of numerical research of stability by changing the shape of shells, ratio of elastic modulus of the material and parameters of the support contour are presented.

1. Introduction
The developed methodology is based on the Galerkin method and finite element method in a mixed formulation. This approach is described in the works [1, 2, 3], has several advantages over the classical formulation of the finite element method. Solution of the nonlinear problem is carried out using one of the algorithm continuation methods. The technique can be applied in various fields, including the analysis of nano-scale objects.

2. Description of the methodology
We used an algorithm continuation method [4], is as follows:

$$\xi_k^0 = t \cdot dX_{k-1},\ X_k^0 = X_{k-1} + \xi_k^0,$$
$$\begin{cases} J(X_k^i)\Delta X_k^{i+1} = -F(X_k^i) \\ (\xi_k^i, \Delta X_k^{i+1}) = \frac{1}{2}(t^2 - (\xi_k^i, \xi_k^i)) \end{cases},$$
$$X_k^{i+1} = X_{k-1} + \Delta X_k^{i+1},\ \xi_k^{i+1} = X_k^{i+1} - X_{k-1}, \tag{1}$$

where, $F(X)$ - nonlinear system of n equations with n+1 unknown variables what is solved; X - vector of unknown variables. In our problem it includes a system of nodal degrees of freedom and the parameter of loading q. The system of equations $F(X)$ has a set of solutions that can be represented by curve solutions; k - 0, 1, 2 ... - number of step along the curve solutions; i - 0, 1, 2 ... - the number of refinement iteration at ach step k; $F(X^i_k)$ and X^i_k – respectively is system of equations and vector of its solutions at step k and refinement iteration i; ξ^0_k - initial direction vector for search a solution X^i_k at step k; t - the step value; dX_k - unit vector tangent to the curve solutions at current step k; dX_{k-1} - unit vector tangent to the curve solutions at previous step; X^0_k - initial vector of solution of system equation $F(X^i_k)$ at step k; X_{k-1} - last refined vector of solution of system equation $F(X^i_k)$ at previous step; $J(X^i_k)$ - Jacobi matrix system of equations $F(X^i_k)$ at step k and refinement iteration i; ξ^i_k - direction vector for search a solution X^i_k at step k and refinement iteration i; ΔX^i_k - vector of iterative refinement of solution vector X^i_k at step k and refinement iteration i; X^{i+1}_k – refined vector of solution of system equation $F(X^i_k)$ at step k and refinement iteration i; ξ^i_k - refined direction vector for search a solution X^i_k at step k and refinement iteration i.

Refinement is being until the following conditions hold:

$$\left\|\Delta X_k^{i+1}\right\| > \varepsilon, \tag{2}$$

where, ε - constant that specifies the search accuracy solutions.

The value of dX_k is determined by solving the system of equations of the form:

$$\begin{cases} J(X_k)dX_k = 0 \\ (dX_k, dX_k) = 1 \end{cases}. \tag{3}$$

To determine the stress-strain state of the shell used finite element method, matrices and vectors which are derived using Galerkin method. An expression describing the state of the j-th finite element is of the form:

$$F_j(X^i_{k,j}) = \{r_1 \quad r_2 \quad r_3 \quad r_4\}^T, \tag{4}$$

where,

$$r_1 = \left(\frac{\eta-\chi}{3}\overline{C}_2 + \chi\overline{C}_{12} - e_1\overline{C}\right)\Phi_1 + e_3\theta_1 + \left(e_1\overline{C}_1 + \frac{\eta-\chi}{6}\overline{C}_2\right)\Phi_2 + e_4\theta_2 +$$
$$+\left(\eta^2 + 3\chi\eta - 4\chi^2\right)\frac{\theta_1^2}{40} + \left(2\eta^2 + \chi\eta - 3\chi^2\right)\frac{\theta_1\theta_2}{60} + \left(3\eta^2 - \chi\eta - 2\chi^2\right)\frac{\theta_2^2}{120}, \tag{5}$$

$$r_2 = -e_3\Phi_1 + \left(\frac{\eta-\chi}{3}\overline{D}_2 - \chi\overline{D}_{12} - e_1\overline{D}_1\right)\theta_1 + e_4\Phi_2 + \left(e_1\overline{D}_1 + \frac{\eta-\chi}{6}\overline{D}_2\right)\theta_2 -$$
$$-\frac{qhC\sqrt{\mu}}{120}\left(-8\chi^4 ++ 7\eta^2\chi^2 + 7\eta\chi^3 - 3\eta^3\chi - 3\eta^4\right) - \frac{s_i}{6}\left(\eta^2 + \eta\chi - 2\chi^2\right) +$$
$$+\left(4\chi^2 - 3\chi\eta - \eta^2\right)\frac{\theta_1\Phi_1}{20} + \left(3\chi^2 - \chi\eta - 2\eta^2\right)\frac{\theta_2\Phi_1}{60} + \left(3\chi^2 - \chi\eta - 2\eta^2\right)\frac{\theta_1\Phi_2}{60} +$$
$$+\left(2\chi^2 + \chi\eta - 3\eta^2\right)\frac{\theta_2\Phi_2}{60}, \tag{6}$$

$$r_3 = \left(e_2\overline{C}_1 + \frac{\eta-\chi}{6}\overline{C}_2\right)\Phi_1 + e_4\theta_1 + \left(\frac{\eta-\chi}{3}\overline{C}_2 + \eta\overline{C}_{12} - e_2\overline{C}_1\right)\Phi_2 - e_3\theta_2 +$$
$$+\left(4\eta^2 - 3\chi\eta - \chi^2\right)\frac{\theta_2^2}{40} + \left(3\eta^2 - \chi\eta - 2\chi^2\right)\frac{\theta_1\theta_2}{60} + \left(2\eta^2 + \chi\eta - 3\chi^2\right)\frac{\theta_1^2}{120}, \tag{7}$$

$$r_4 = e_4\Phi_1 + \left(e_2\overline{D}_1 + \frac{\eta-\chi}{6}\overline{D}_2\right)\theta_1 + e_3\Phi_2 + \left(\frac{\eta-\chi}{3}\overline{D}_2 + \eta\overline{D}_{12} - e_2\overline{D}_1\right)\theta_2 -$$

$$\frac{qhC\sqrt{\mu}}{120}\left(-7\chi^4+23\eta^2\chi^2-7\eta\chi^3-7\eta^3\chi-12\eta^4\right)-\frac{s_i}{6}\left(2\eta^2-\eta\chi-\chi^2\right)+$$
$$+\left(\chi^2+3\chi\eta-4\eta^2\right)\frac{\theta_2\Phi_2}{20}+\left(2\chi^2+\chi\eta-3\eta^2\right)\frac{\theta_1\Phi_2}{60}+\left(3\chi^2-\chi\eta-2\eta^2\right)\frac{\theta_1\Phi_1}{60}+$$
$$+\left(2\chi^2+\chi\eta-3\eta^2\right)\frac{\theta_2\Phi_1}{60}, \quad (8)$$

$$e_1=\frac{\eta^2+\eta\chi+4\chi^2}{6(\chi-\eta)},\quad e_2=\frac{4\eta^2+\eta\chi+\chi^2}{6(\chi-\eta)},\quad e_3=\frac{K}{12}\left(\eta^2+2\eta\chi-3\chi^2\right),\quad e_4=\frac{K}{12}\left(\eta^2-\chi^2\right), \quad (9)$$

$$X^i_{k,j}=\{\Phi_1,\theta_1,\Phi_2,\theta_2,q\}^T, \quad (10)$$

χ & η - respectively radial coordinate the start and end node of the j-th axisymmetric finite element of shell, $X^i_{k,j}$ - vector, which includes the of all the degrees of freedom nodes of a finite element.

Other details relating to used of the mixed finite element including assembly algorithm generalized matrix given in [1 - 3].

Designed calculation algorithm determines the stress-strain state of the shell at each step of the method of continuation on parameter. In the process of calculating the deformation analysis of its changes and fixes their extreme values. According to it determines the values of the upper and lower buckling loads corresponding to the upper and lower extreme point on the graph of the deformation of the load (i.e., the upper and lower at the turning point).

3. Checking the results

The results of calculations were compared for the values of the upper and lower buckling loads at [5]. The difference in the values is not more than 1.2%.

To establish the convergence of the algorithm was carried out test calculations for different finite element mesh density in the radial direction. The results show convergence forces, moments and displacements with increasing density of the grid. Acceptable for practical calculations results are obtained by partitioning the shell at least 20 elements in the radial direction.

4. Investigation of stability

The developed methodology was used to study the influence on the values of the upper and lower buckling loads of the following parameters:
- parameter that specifies the shape of the shell (parameter Z);
- parameters of elastic fastening support contour the shell (parameters n and m);
- the ratio of the elastic modulus of the shell material in the radial (E_1) and circumferential direction (E_2) (parameter Θ).

Shape generatrix a shell of revolution is defined by the function parameter Z:

$$f(\overline{\rho})=f_{\max}\rho^Z, \quad (11)$$

where, f_{max} - the rise in the center of the shell, ρ - dimensionless coordinate, which changes from 0 in the center of the shell to the value 1 on the support contour.

It is assumed that the shell is based on the elastic support contour, which specifies the parameters of the flexibility n and m. Parameter n describes flexibility of the support contour in the radial direction, and may range from -0.3 (when displacement in the radial direction are prevented) to infinity (if not restricted displacement in a radial direction). The parameter m describes flexibility of the support contour in its torsion and may range from +0.3 (if the

IOP Conf. Series: Materials Science and Engineering **201** (2017) 012019 doi:10.1088/1757-899X/201/1/012019

reference contour is a hinge and rotation not restricted) to infinity (if the rotation of the support contour are prevented).

In carrying out investigations, the parameters varied the shell in the following ranges: Z - from 1 to 2.5; n - from -0.3 to 100; m - from 0.3 to 100. Considered shallow shell of revolution with base radius a = 1 [m] and height f_{max} = 1 [m]; thickness h = 4 [mm], Poisson's ratio of the shell material ν1 = ν2 = 0.3. The shell is loaded uniformly distributed load of intensity *p*.

Some results are shown in Fig. 1 and Fig. 4, which presents graphs of the dimensionless deflection depending on the values of the load on the shell and the values of variable parameters. Fig. 2, Fig. 3, Fig. 5 and Fig. 6 are graphs of the upper and lower buckling loads, depending on the values of the variable parameters of the shell.

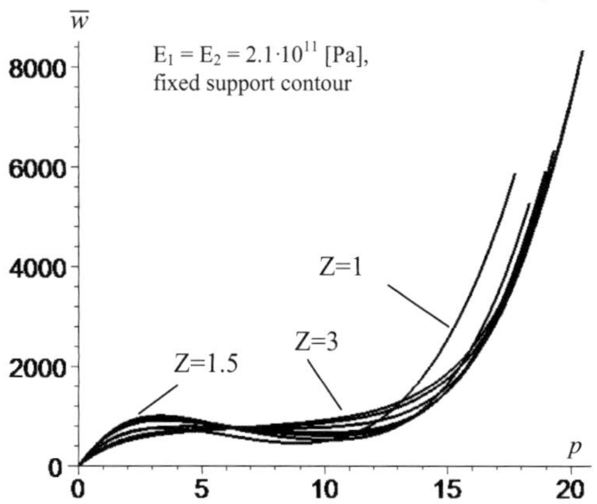

Figure 1. The graph change dimensionless displacements in the center of the shell when the value of the load p for different values of Z.

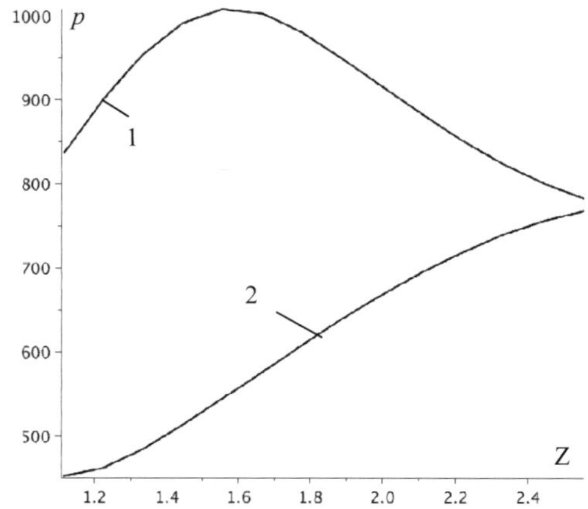

Figure 2. The graphs change the values of buckling loads p, depending on the value of Z (1 - upper buckling load 2 - lower buckling load).

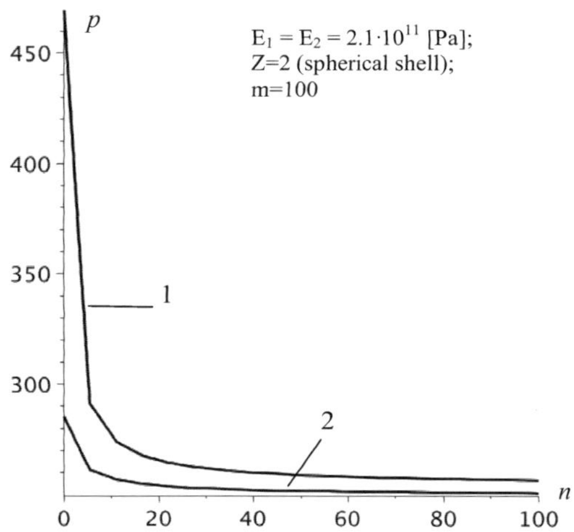

Figure 3. The graphs change the values of buckling loads, depending on the value of n (1 - upper buckling load 2 - lower buckling load).

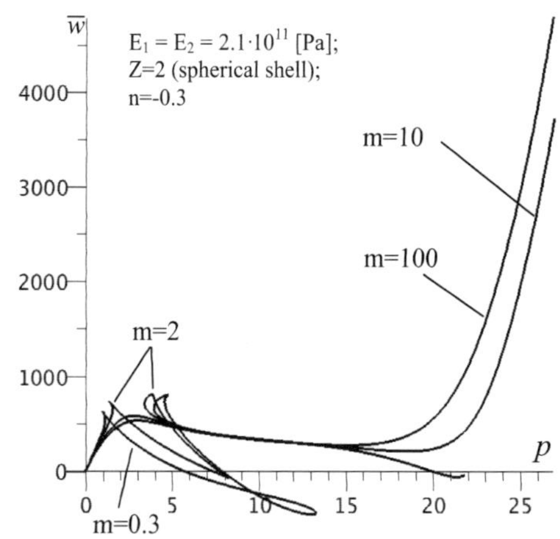

Figure 4. The graphs change dimensionless displacements in the center of the shell when the value of the load p for different values of m

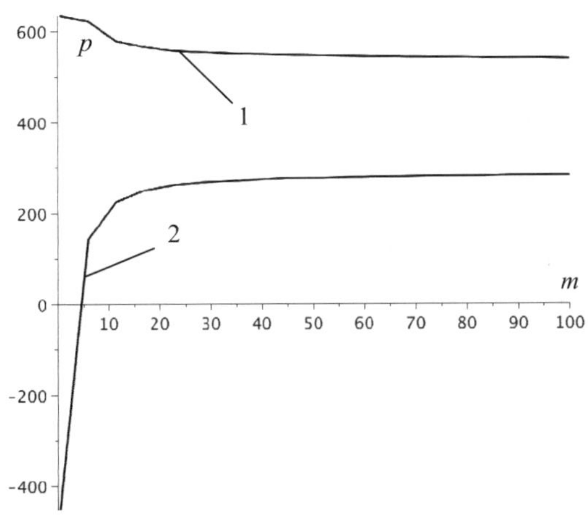

Figure 5. The graphs change the values of buckling loads, depending on the value of m (1 - upper buckling load 2 - lower buckling load).

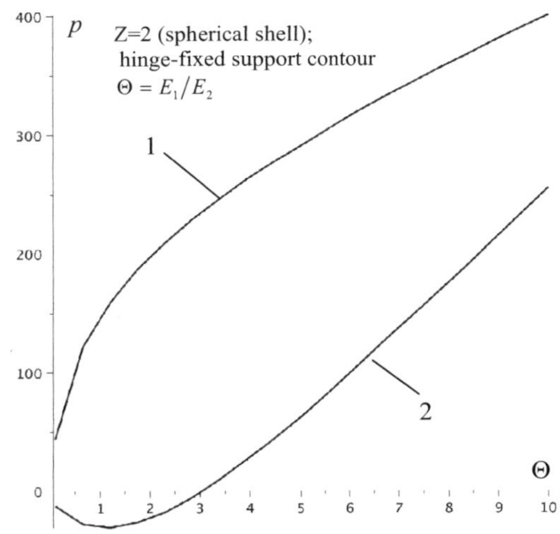

Figure 6. The graphs change the values of buckling loads, depending on the value of Θ (1 - upper buckling load 2 - lower buckling load).

When you change the Z observed extreme values of the upper buckling load which is located near Z = 1.6 (see Fig. 2). We can assume that the shell of such forms have the highest safety factor than other forms of the shell under the influence of uniformly distributed loads. At the same time, the extremum for the lower buckling load is absent. By increasing the shape parameter Z is observed only increase the values of the lower buckling load (Fig. 2). The largest gap between the values of the upper and lower buckling loads occurs precisely at the point of extremum upper buckling load.

Change the parameters n and m shows the growth of buckling load with an increase in the stiffness of the support contour the radial direction and increase flexibility in respect of the rotation (Fig. 3, Fig. 5). The highest values of buckling loads correspond to the hinge-fixed the support contour.

Increasing the parameter Θ results in an increase the upper buckling load (Fig. 6). At the same time, with Θ = 1 there is an extreme value of the lower buckling load.

5. Conclusions

Developed methodology based on mixed finite element method shows good accuracy and convergence of the results. It is more effective in comparison with finite elements in other formulations. Applying the methodology of investigating the shell's stability allows you to find more optimal combination of size, shape for the shell and the parameters of it's of the support contour. Presented methodology may be used for building and analysis of nano-scale objects in matereals.

This work was supported by the grant of the President of the Russian Federation #MK-9203.2016.8

References

[1] Stupishin L and Nikitin K 2014 *Adv. Mat. Res.* **919-921**, pp 1299-1302
[2] Stupishin L and Kolesnikov A 2014 *Appl. Mech. & Mat.* **501-504**, pp 766-769
[3] Stupishin L and Nikitin K 2014 *Appl. Mech. & Mat.* **501-504**, pp 514-517
[4] Grigolyuk E. and Shalashilin V. 1988 *Problems of nonlinear deformation* (Moscow: Nauka Publ)
[5] Valishvili N *1976 Methods of analyzing shells of revolution on electronic digital computers* (Moscow: Mashinostroenie Publ)

Comparative analysis of buckling criteria for engineering structures. Single-degree-of-freedom systems

L. Stupishin[1]

[1] Urban, road Construction and Structural Mechanics Department, South-West State University, 305040, 50-let Octyabrya st., 94, Kursk, Russian Federation

lusgsh@yandex.ru

Abstract. The criteria of the buckling of engineering structures in terms of the systems with lumped parameters are discussed in the article. The examples of setting buckling problems and their solutions using Timoshenko and Bryan criteria's, and the criterion of the critical levels of energy are given. The analysis of the compared approaches, their strengths and shortcomings in terms of the single-degree-of-freedom systems are presented.

1. Introduction

The buckling is one of the most important limit states, which received much attention over the longer time. However, there is still no consensus on the causes of its manifestations and, accordingly, its criteria for characterizing. The most popular among the engineers and researchers of energy criteria are two generally accepted formulations: the buckling criterion in the form of Timoshenko, and the buckling criterion in the form of Bryan [1, 2]. But it is difficult to argue that these approaches allow us to investigate the entire spectrum of tasks caused of structures buckling. One alternative could be a variation criterion of the critical energy levels proposed in [3]. Below is a comparative analysis of these approaches to the buckling of engineering structures investigations with a simple example of a system with one degree of freedom.

Consider elastically deformable system exposed to contraction shown in Figure 1. The proposed schema differs from the classical Timoshenko's, that it describes the initial stage of deformation - contraction of the rod, at which the accumulation of strain energy. It is necessary to describe the behavior of the system, according to the ideas of Brian. Therefore, the system formally has two degrees of freedom: the longitudinal movement and rotation due to the loss of initial state. However, to describe the buckling will be used by the one degree of freedom that characterizes the deviation of the bar from the initial position at the buckling.

2. Timoshenko criterion

Let us consider an elastically deformable system subjected to direct compression shown in Figure 1. Following the author of the criteria, we neglect the phase of the energy accumulation by the system. We consider the moment of the system initial deformation shape loss, i.e. a deviation from the rectilinear state.

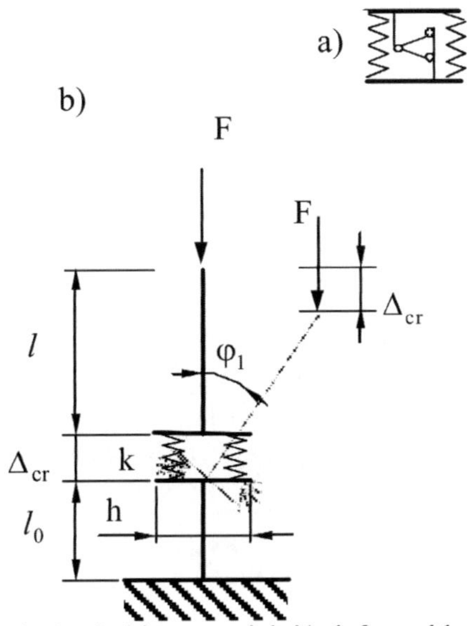

Figure 1 a) elastic hinge model; b) deformable system model

The external force Fc will do possible work when a system transits to a new deformed state.

$$W_{ex} = F_c\Delta, \tag{1}$$

where the displacement is defined through an infinitesimal angle of the rod displacement φ, as follows:

$$\Delta = l\varphi^2/2. \tag{2}$$

The springs performed the real work of internal forces to put the system into a new equilibrium state:

$$W_{in} = \int Md\varphi \tag{3}$$

The moment generated by the reactive forces in the springs due to small displacements is defined as follows:

$$M = rh^2\varphi/2. \tag{4}$$

The total energy of the system can be written as follows:

$$\Pi = F_c l\varphi^2/2 - rh^2\varphi^2/4. \tag{5}$$

The condition of the system total energy minimum for equilibrium state is as follows:

$$\partial\Pi/\partial\varphi = 0 \tag{6}$$

It gives the critical force value:

$$F_c = rh^2/2l. \tag{7}$$

The critical force value is the same as the value obtained for the classical model given in [1, 2]. The proposed model of the system (Figure 1), which takes into account the history of deformation until the loss-of-stability moment, allows finding critical displacements.

$$2r\Delta^2/2 = F\Delta/2. \tag{8}$$

From which we obtain:

$$F_c = 2r\Delta_c. \tag{9}$$

Finally, taking into account (6), we get:

$$\Delta_c = h^2/4l. \tag{10}$$

Note that the displacements were obtained as a result of dividing deformation process into stages, rather than applying Timoshenko criterion.

3. Bryan criterion

In this case, it is important to take into account the stage of deformation internal energy accumulation by elastic body when the system is being compressed.

The critical value of the longitudinal force exerted by the springs to the point of stability loss as in previous section (9) will be

$$N_c = 2r\Delta_c. \quad (11)$$

The bending moment is determined at the displacement finite values as follows:

$$\mathrm{M} = \mathrm{rh}^2\sin(\varphi/2). \quad (12)$$

After the substitution of the expression (11) for (3) and integrating, we obtain an expression of the internal forces work that caused the deviation of the system from the original equilibrium position:

$$W_{in} = 2rh^2(\cos(\varphi/2) - 1). \quad (13)$$

The internal deformation energy, accumulated as a result of possible work of the load constant in magnitude, has the same form as in the previous section:

$$\mathrm{W_{ex}} = 2\mathrm{r}\Delta_c l(1-\cos\varphi). \quad (14)$$

The internal energy of the system in a new equilibrium state is as follows:

$$\mathrm{U} = 2\mathrm{r}\Delta_c l(1-\cos\varphi) - 2\mathrm{rh}^2(1-\cos(\varphi/2)). \quad (15)$$

The condition of stationary energy can be presented as follows:

$$\partial \mathrm{U}/\partial\varphi = 0 \quad (16)$$

It leads to the following expression:

$$\Delta_c = \mathrm{h}^2\sec(\varphi/2)/(4l). \quad (17)$$

Minimum replacement value when the loss of stability takes place is written as follows:

$$\Delta_c^{\min} = \mathrm{h}^2/(4l), \quad (18)$$

which coincides with the solution obtained in the previous section.

Critical force can be determined by formula (8)

$$\mathrm{N_c} = \mathrm{rh}^2\sec(\varphi/2)/(2l). \quad (19)$$

Its minimum value is as follows:

$$\mathrm{N_c^{\min}} = \mathrm{rh}^2/(2l). \quad (20)$$

Figure 2 shows a curve of the changes in the system critical displacements depending on the angle of deviation, which is periodic in its nature. This feature becomes evident only when the finite displacements in Brian criterion are taken into account.

The critical force value is identical to the critical load obtained in [1, 2], despite the differences in the physical models of the rods. This is because Brian criterion investigates the system close to the critical equilibrium position in the same way as Timoshenko criterion does.

4. The criterion of the critical levels of energy

The condition of self-equilibrated state of the system can be represented as the difference of the work of internal forces maintaining this state in the system:

$$\mathrm{W_{in}} = \mathrm{W_{in}^{N}} - \mathrm{W_{in}^{\varphi}} \to \min, \quad (21)$$

on condition that the displacements are normalized (which is necessary when accounting finite displacement values):

$$\overline{\Delta}_c{}^2 + \varphi_c^2 = 1 \quad (22)$$

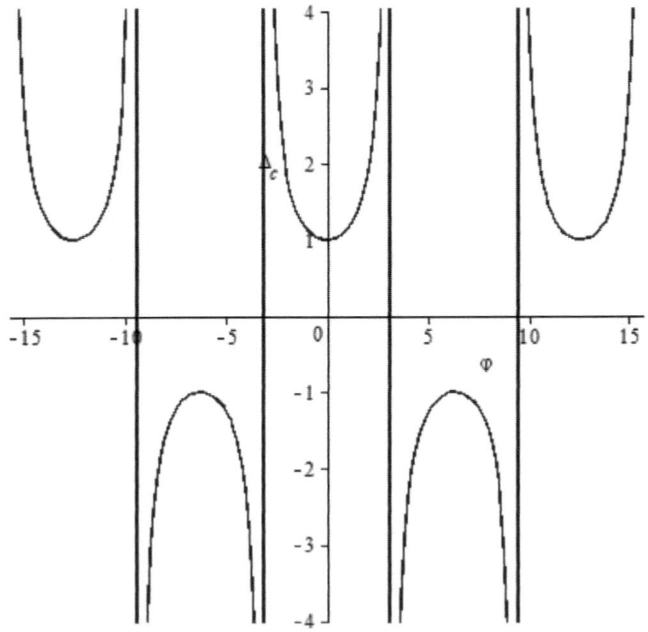

Figure 2 Critical displacement curve

Let us introduce dimensionless variables:

$$\overline{\Delta}_k = \frac{l\Delta_k}{h^2},\ \overline{\Delta}_c = \frac{l\Delta_c}{h^2},\ \overline{W}_{in}^{\varphi} = \frac{lW_{in}^{\varphi}}{rh^2},\ \overline{W}_{in}^{N} = \frac{lW_{in}^{N}}{rh^2}. \quad (23)$$

Then Lagrange's equation for the set task will have the form:

$$L = 2(\cos(\varphi_c/2) - 1) - \overline{\Delta}_c{}^2 + \lambda(\overline{\Delta}_c{}^2 + \varphi_c{}^2 - 1) \quad (24)$$

From the condition of stationary Lagrangian for the variable

$$\frac{\partial L}{\partial \overline{\Delta}_c} = 0,\quad \frac{\partial L}{\partial \varphi_c} = 0, \quad (25)$$

we obtain

$$\lambda_{\Delta_c} = 1,\ \lambda_{\varphi_c} = \frac{\sin(\varphi_c/2)}{2\varphi_c} \quad (26)$$

For small critical angles, it is possible to estimate the displacement value and the inner critical compressive force:

$$\Delta_c = \overline{\Delta}_c \lambda_{\varphi_c} = \frac{h^2 \sin(\varphi_c/2)}{2l\varphi_c} \approx \frac{h^2}{4l} \quad (27)$$

$$N_c = 2r\Delta_c = \frac{2rh^2 \sin(\varphi_c/2)}{2l\varphi_c} \approx \frac{rh^2}{2l} \quad (28)$$

The results are identical to the results obtained in the previous sections for small and finite values of the angles (7), (10), (16), (18).

From the normalization condition, it is possible to estimate the value of the longitudinal displacements up to the moment of stability loss:

$$\overline{\Delta}_k = -\sin(\varphi/2) \pm \sqrt{1-\varphi^2}. \tag{28}$$

This function has the minimum value when

$$\frac{d\overline{\Delta}_k}{d\varphi} = -\cos(\varphi/2)/2 \pm \varphi\Big/\sqrt{1-\varphi^2} = 0. \tag{29}$$

This equation can be solved when the angle $\varphi = 0{,}685611$.

5. Conclusions

Tymoshenko criterion examines the system close to the initial equilibrium state without taking into account deformation history when the deviation from the initial state is infinitesimal. As a result, it only allows evaluating the value of the critical load.

Bryan criterion examines the internal forces work when finite displacements are small and close to the initial equilibrium state. Due to taking into account energy accumulation at the previous stages of deformation, it also makes it possible to obtain the values of displacements before critical state. However, no other equilibrium states, except initial vertical equilibrium state, were not revealed, despite the fact that actually the deformed design scheme was studied. The assumption concerning the finiteness of displacements allows revealing a qualitative regularity of the displacements and loads changes close to the initial equilibrium state, i.e. their periodical character.

The criterion of the critical levels of energy makes it possible to determine not only the values of critical loads and displacements, but also to obtain evaluations of the angles of the system displacements to a new equilibrium state. Although the system was studied close to the equilibrium state, the normalization condition makes it possible not to be limited by the small values of deviations.

6. References

[1] Alfutov N.A. 1991 *Osnovy rascheta na ustoichivost uprugih sistem* (Moscow, Mashinostroenie) p 336
[2] Rzhanizin A.R. 1955 *Ustoichivost ravnovesiya uprugih sistem* (Moscow, Gostehizdat) p 475
[3] Stupishin L 2014 *Appl. Mech. & Mat.* **579-578**, 1584-1587
[4] Stupishin L and Kolesnikov A 2014 *Adv. Mat. Res.* **919-921**, 144-147
[5] Stupishin L and Kolesnikov A 2014 *Appl. Mech. & Mat.* **501-504**, 766-769
[6] Stupishin L and Nikitin K 2014 *Appl. Mech. & Mat.* **501-504**, 514-517.
[7] Stupishin L and Kolesnikov A 2014 *Appl. Mech. & Mat.* **580-583**, 3062-3065

The effect of biogas desulfurization to acidity of lubricant oil of the biogas fuelled engine

Tjokorda Gde Tirta Nindhia[1*], I Wayan Surata[1], Ari Wardana[1]

[1]Deparment of Mechanical Engineering, Engineering Faculty, Udayana university, Jimbaran, bali, Indonesia, 80361

E-mail: nindhia@yahoo.com/tirta.nindhia@me.unud.ac.id

Abstract This research is established for the purpose of the understanding the stability of the acidity of lubricant oil in biogas fuelled engine due to the absence of hydrogen sulfide (H_2S). As was recognized that other than Methane (CH_4) , there are also other gas impurities in the biogas such as carbon dioxide (CO_2), hydrogen sulfide (H_2S), moisture (H_2O) and ammonia (NH_3). Due to H_2S contents in the biogas fuel, the engine was found failure. This is caused by corrosion in the combustion chamber due to increase of lubricant acidity. To overcome this problem in practical, the lubricant is increased the pH to basic level with the hope will be decrease to normal value after several time use. Other method is by installing pH measurement sensor in the engine lubricant so that when lubricant is known turn to be acid, then lubricant replacement should be done. In this research, the effect of biogas desulfurization down to zero level to the acidity of lubricant oil in the four stroke engine was carried out with the hope that neutral lubrication oil to be available during running the engine. The result indicates that by eliminating H_2S due desulfurization process, effect on stability and neutrality of pH lubricant. By this method the engine safety can be obtained without often replacement the lubricant oil.

1. Introduction

Biogas is promoted to be a fuel for stationary engine. It is about not efficient to storage in the form of compressed biogas in the steel container as compression natural gas (CNG). The biogas is directly flowed from the digester to the temporally storage and continued to be injected as a fuel of the four stroke engine on site to run an electric generator. Pure methane is not possible to be obtained directly from anaerobic digestion process. The gas that is obtained (biogas) generally consist amount of trace such as water (H_2O), hydrogen sulfide (H_2S), and ammonia (NH_3). The main composition of biogas are methane (CH4) and carbon dioxide (CO_2) [1,2].

The function of a lubricant in the engine or machine is to create a film as a barrier between moving mechanical parts. The function of this barrier is to reduce friction and wear. The lubricant also acts as a coolant, to prevent corrosion and avoiding harmful formation of deposit. The base oil that is used for lubricant can not cover all of these requirement. The additives are added to the lubricant formulation which include anti oxidant, anti wear, corrosion inhibiter, detergent and viscosity modifier [3]. The additive usually neutralize acidic components taht is generated during running the engine, and give effect on reducing corrosion. Additives also improve overall efficiency in the engine. To extend the lifetime of the lubricant is also become concern for the reason of efficiency [3]. The lubricant oil must free from components which corrode metal parts of an engine or a machine. High sulfur fuels such as biogas will generates by-products of acidic sulfate. The increase of oxidation and sulfating by-products results in an increase in acidity levels and oil viscositys. These by-products increase the production of sludge and varnishes which significantly degrade lubrication performance[3]. Recently for practical purpose and simplicity, The company that produce biogas engine is completed the engine with special oil lubricant which has high basic number (BN) with the hope will act as reserve alkalinity which is to become neutral for several hour of operation [4]. But still the instability of oil viscosity need to be investigates if this technique will be a choice of the user. Other alternative that is develop to avoid biogas engine failure due to H_2S in the biogas fuel without any concern for elimination H_2S contents in biogas is by monitoring the oil acidification by using a chemical sensor. An early work regarding this method is developed by using copper film that was deposit on titanium [5]. This method was improved by using a sacrificial layer (600 nm of lead metal film) that is immersed into the oil. There will be a material loss due to corrosion. This material loss is monitored by measuring the capacitive coupling between sacrificial layer and readout electrodes[6]. Due to irregularity concentration of H_2S in the biogas, the acidity of lubricant will also have fluctuation and this condition can be monitored by using this method. The lubricant should be changed when the acidity reach its critical value [6]. This method is not addressed to prolong the life time of the lubricant.

Biogas desulfurization previously is not attracting attention in utilization of biogas as fuel of an engine. This condition was caused by limitation of appropriate technology for biogas desulfurizer. In the early work of utilization biogas as an engine fuel in china for example, the H_2S is ignored and let it burn in the combustion chamber. The system is simple by compressed in the cylinder thank, passing trough mixer and mixed with diesel fuel [7]. It is sure then the engine will failure in the very soon.Recently, desulfurizer is known can be prepared even from waste material. Example for this idea is by using iron chips from manufacturing waste [8]. The chips just annealed and compacted, then biogas is flowed in to this filter. It was reported that this type of desulfurizer is possible for regenerative used [2]. Other example of using waste material for desulfurizer is by using zinc case from used zinc-carbon battery [9].This research is investigate and effect of eliminating H_2S from the biogas to the acidity of the lubricant of the biogas fuelled engine.

2. Experiment

The engine that was used for this purpose was non contact transistor ignition (TCI), 4-stroke biogas fuelled engine, inclined single cylinder, compression ratio 9:1, with displacement 196 cc. The engine was coupled with electric generator for capacity 1000 watt . The engine need about 0.5 litters lubricant oil during operation. The lubricant that was used for this purpose was SAE 10W30, with viscosity index 141, the colour was brown, density 0.8858 ^{o}Cg/cm3, and flash point 218. The schematic of the experiment can be seen in Fig. 1. The biogas was produced in an anaerobic digester (1) and was streamed to the biogas desulfurizer (2). The biogas desulfurizer was made from iron or steel chips that was previously annealed and compacted to become a billet for easy installation and handling inside the pipe as explained in the previous publication [2]. The biogas that was released from desulfurizer was ensured to be free from H_2S impurities by using H_2S gas detector (3). If the H_2S still detected in the stream, then the billet of the iron chips is added in the desulfurizer. Biogas bag was prepared for biogas storage (4) and from here biogas was pumped to the part of the mixer in the biogas fuelled engine (5) to be mixed with oxygen from the air. The engine was let running for 100 hours. The engine was stopped running for every 10 hour, and the pH was measured with 3 time repetition for validity. The result is plotted in a graph and analyze.

IOP Conf. Series: Materials Science and Engineering **201** (2017) 012021 doi:10.1088/1757-899X/201/1/012021

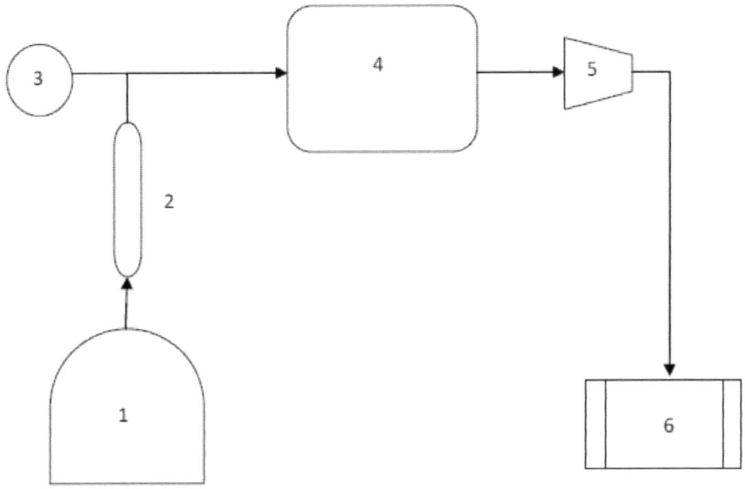

Figure. 1 Schematic of the experimental. 1. Anaerobic digester, 2. Desulfurizer, 3. Hidrogesn sulfide gas sensor , 4. Storage bag, 5. Biogas pump, 6. Biogas engine coupled with electric generator

3. Result and discussion

The engine is found still in good performance up to 100 hours operation and keep produce power up to its maximum capacity of 1000 watt. The desulfurizer working well and no H_2S impurity is detected in the biogas. The result of pH measurement for 100 hour operation can be seen in Fig.2. The fluctuation of the pH was about to found during early hour of operation until reach 50 running hour. The fluctuation range is from 7.6 the lowest and to reach 7.9 the highest. After passing 50 hour operation until final testing at 100 hours operation the pH is found stable around range 7.4 to 7.5. It can be understood that since the H_2S is eliminated from the biogas, the other impurities that active is ammonia (NH_3). With the moisture (H_2O) existence in the biogas, the reaction between NH_3 and H_2O possible to occur with resulting fluctuation to more base following reaction in Equation 1.

$$NH_3 + H_2O \leftrightarrow NH4^+ + OH^- \quad (1)$$

The NH_3 is polar molecules and have an ability to form hydrogen bonds. This give explanation related with high solubility of ammonia in water. The chemical reaction also occurs when ammonia dissolves in water. Ammonia acts as a base in aqueous solution, hydrogen ion is acquired from H_2O to yield hydroxide ions (OH^-) and ammonium ($NH4^+$). The production of hydroxide ions when ammonia dissolves in water gives aqueous solutions of ammonia which have basic (alkaline) properties. From this stand point the, the operation of biogas fuelled engine give the user 3 options related with avoiding failure of the engine due to H_2S corrosion. The 1st alternative is by using special oil lubricant that have performance as reserve alkalinity [10]. The price of the Lubricant may expensive since should be created for special purpose. The advantage of this alternative is, the lubricant replacement is not necessary to be conducted so often. The 2nd alternative is by mounting the chemical sensor of oil acidification [5,6]. This alternative gives the user an accurate time for oil replacement.

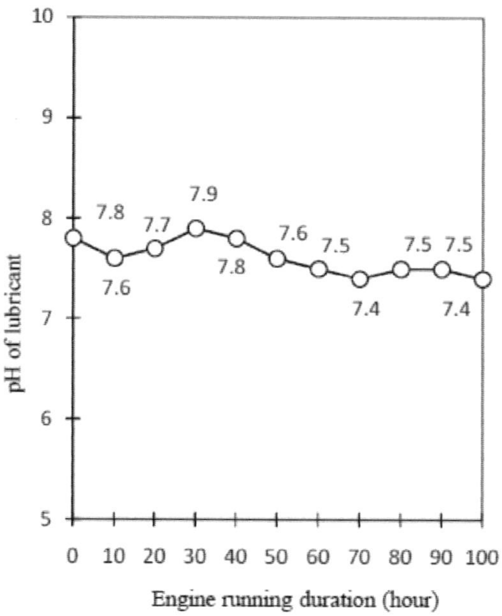

Figure. 2 The pH of oil Lubricant, measured every 10 hours for 100 hours running of biogas engine

The oil that will be used may be regular oil that available in the market with reasonable price, but this alternative come with disadvantage that the oil should be replacement quite often. The environmental issue related with waste of used oil, and the sensor that is consumable may increase the cost for maintenance. The 3rd alternative is by installation of desulfurizer as was explained in this paper. This alternative gives opportunity to the user to use regular oil that available in the market with reasonable price. The oil replacement is not so often. And by recent development in biogas desulfurization technology, which make possible by using waste material [2] for desulfurizer, this can reduce the cost of production desulfurizer. The desulfurizer is also possible to be used repetitively. Other benefit is the biogas installation is free from dangerous and toxic H_2S gas that may harmful if there is a leak in the installation system. It can be completed id this discussion that by progress in production of activated carbon [11], which is make availability of activated carbon with reasonable price, the system then can be improved with biogas purification from CO_2 impuritieas to increase calorific value of biogas.

4. Conclusion

The biogas desulfurization give advantages in application of biogas as a fuel of an engine by keeping the pH of the Engine lubricant in neutral condition so that corrosion and engine failure can be avoided. This method make possible for the use of regular oil that available in the market with reasonable price. The replacement of the oil lubricant can be performed not so often. The method have more advantage because recent biogas desulfurization technology can be realize by using waste material such as iron chips which is available abundantly and can be use repetitively

5. Acknowledgements

This research was supported financially by ministry of Research and Technology and Higher Education The republic of Indonesia (Ristekdikti) with also support from Fulbright Program from Government of The United State of America by giving opportunity to deepen the analyzes at the Anaerobic Digester Research and Education Canter (ADREC) Michigan State

University, East Lansing, Michigan, USA. Special thanks addressed to Prof Dana Kirk of ADREC for useful advice and Support .

References

[1] Ryckebosch, E, Drouillon, M., and Vervaeren, H., Techniques for transformation of biogas to biomethane, biomass and bioenergy, 35, 15 March (2011), pp.1633-1645

[2] Nindhia, T.G.T , Sucipta, I M., Surata, I W. , Atmika, I K.A., Negara, D.N.K.P.,and Negara, K.M.T., Processing of Steel Chips Waste for Regenerative Type of Biogas Desulfurizer, International Journal of Renewable Energy Research, Vol.3, No.1,(2013), pp. 84-87

[3] Mortier, R.M., Fox, M.F., and Orszulik, S.T., Chemistry and Technology of Lubricants, (2010),Springer Dordrecht Heidelberg, London, New York.

[4] Diedrich, D.C., Moreira, E. , Carlos, J. Zank, C., Analysis of The Lubricanting Oil's Life of Biogas Powered Generators, 22nd International Congress of Mechanical Engineering (COBEM 2013), November 3-7, 2013, Ribeirão Preto, SP, Brazil

[5] Agoston, A, Dorr, N., and Jakoby, B., Corrosion sensors for engine oils-laboratory evaluate.on and field tests, Sensors and Actuators B, 127 (2007), pp. 15-21

[6] Sen,S., Schneidhofer,C.,and Dörr,N., Online monitoring of the oil acidification using a chemical sensor measuring corrosiveness, Procedia Engineering 47 (2012), pp. 466 – 469.

[7] Qiu, J.C., Wei,L.T., and Li, Z.J., A Study on Compressed Biogas and its Application to the Compression Ignition Dual-Fuel Engine, *Biomass* 20 (1989), pp. 53-59.

[8] Surata, I W. , Nindhia, T.G.T , Atmika, I K.A., Negara, D.N.K.P. , and Putra, I W.E.P., Simple Conversion Method from Gasoline to Biogas Fueled Small Engine to Powered Electric Generator, Energy Procedia 52 (2014), pp. 626 – 632.

[9] Nindhia, T.G.T, Surata, I W.,Swastika, I D.G.P., and Widiana, P., Processing Zinc from Waste of Used Zinc-carbon Battery with Natrium Chloride (NaCl) for Biogas Desulfurizer, Key Engineering Materials, Vol. 705,(2016), pp. 368-373, 2016.

[10] Diedrich, D.C. , Moreira, E., Carlos, J., and Zank, C., Analysis of The Lubricanting Oil's Life of Biogas Powered Generators, 22nd International Congress of Mechanical Engineering (COBEM 2013), November 3-7, 2013, Ribeirão Preto, SP, Brazil.

[11] Negara, D.N.K.P., Nindhia, T.G.T., Surata, I W. and Sucipta, M., Development and Application of Bamboo Activated Carbons and Their Potency as Adsorbent Material for Adsorbed Natural Gas (ANG); An Overview, Key Engineering Materials, Vol. 705,(2016), pp. 126-130.

IOP Conf. Series: Materials Science and Engineering **201** (2017) 012022 doi:10.1088/1757-899X/201/1/012022

Creation of hydrophobic surfaces using a paint containing functionalized oxide particles

Paul Albert L. Sino[1], Marvin U. Herrera[1] and Mary Donnabelle L. Balela[2]

[1]Institute of Mathematical Sciences and Physics, University of the Philippines Los Baños, Laguna, Philippines 4031
[2]Sustainable Electronic Materials Group, Department of Mining, Metallurgical, and Materials Engineering, University of the Philippines, Diliman Quezon City, Philippines 1101

plsino@up.edu.ph, muherrera@up.edu.ph, mlbalela1@up.edu.ph

Abstract. Hydrophobic surfaces were created by coating various substrates (aluminum sheet, soda-lime glass, silicon carbide polishing paper, glass with double-sided adhesive) with paint containing functionalized oxide particles. The paint was created by functionalizing oxide particles (ground ZnO, TiO_2 nanoparticles, or TiO_2 microparticles) with fluorosilane molecules in absolute ethanol. Water contact angle of samples shows that the coated substrate becomes hydrophobic (water contact angle $\geq$ 90°). Among the oxides that were used, ground ZnO yielded contact angle exemplifying superhydrophobicity (water contact angle $\geq$ 150°). Scanning electron micrograph of paint-containing TiO_2 nanoparticles shows rough functionalized oxides structures which probably increase the hydrophobicity of the surface.

1. Introduction

Superhydrophobic surfaces (SHS) [1-5] have gained much attention in the scientific and industrial communities due to their wide range of possible applications such as self-cleaning [6], anti-icing [7], anti-corrosion [8] and anti-fouling [9]. SHS are surfaces characterized by their high contact angle to water, greater than 150°.

To achieve superhydrophobicity, suitable roughness in combination with low surface energy is necessary. However, some of the recent methods [10-14] that can be used to create superhydrophobic surfaces like lithography and plasma processing involves harsh conditions, toxic reagents and complex equipment, thus, limiting their application. Fabricating paint [15] is a facile approach to create SHS. Aside from avoidance to complexity and possible damage to the material, the use of paint can be easily adapted for large scale production and volumes. Furthermore, paint can be sprayed, dipped or extruded on both hard and soft surfaces thus, its flexibility.

In our previous study, we fabricated a paint that when applied to various substrates creates superhydrophobic surfaces [16]. The paint consists of ethanol, fluorosilane and dual-scale titania nanoparticles (~21 nm and ~450 nm). In this research, same methodology was utilized but the particles used are ZnO (ground), TiO_2 nanoparticles (~100 nm) and TiO_2 microparticles (~50 μm). Compared to dual-scale TiO_2 nanoparticles, ZnO and TiO_2 microparticles are cheaper.

Three sets of paint were created. The first two contains TiO_2 nanoparticles (~100 nm or ~50 μm). Meanwhile, the third paint contains ground ZnO (~ 15-100 nm) particles. The fabricated paints were coated to various substrates (aluminum sheet, soda-lime glass, silicon carbide polishing paper, glass with double double-sided adhesive) to create superhydrophobic surfaces. The paint was created by mixing oxide particles in ethanolic suspension containing perfluorooctyltriethoxysilane (fluorosilane).The oxide particles were used to enhance the roughness of the surface, enabling the entrapment of air molecules. Meanwhile, the silane molecules can decrease the surface energy. With the particles coated with silane, a hydrophobic surface can be created.

2. Methodology

2.1. Preparation of Paint

The paint is composed of functionalized oxide particles and ethanol. The oxide particles used are ZnO (ground), TiO_2 nanoparticles (~100 nm) and TiO_2 microparticles (~50 μm). The oxides were all obtained from Sigma-Aldrich. Three sets of paint were created. The functionalized oxide particles act as hydrophobic agent. Meanwhile, absolute ethanol serves as the volatile suspension medium which evaporates when painted on the substrate. It also acts as the medium wherein the functionalization of oxide particles occurs.

The paint was created using the following steps: (1) fluorosilane (Alfa Aesar 1H-1H,2H,2H-perfluorosilane, 97%) was mixed with ethanol, (2) the oxide particles (TiO_2 or ZnO) were added to the solution.

2.2. Coating of Substrate and Characterization

Coating was done by dipping the substrate in the paint. The substrates used are soda-lime glass, aluminum sheet, SiC polishing paper (150 grits) and soda-lime glass with double-sided adhesive. The painted substrates were allowed to dry in ambient air for at least 180 sec.

The static contact water angle was measured from the image of the water droplet on the treated surface. To do this, an image processing software, DropAnalysis was used. Also, the morphology of soda-lime glass coated with paint containing TiO_2 (~100 nm) was observed using a field emission scanning electron microscope (FESEM, FEI Helios Nanolab 600i). The sample was first coated with gold-palladium (Au/Pd) using physical vapor deposition (PVD) method before the morphological characterization.

3. Results and Discussion

Paint with functionalized oxide particles (ground ZnO, TiO_2 mono-scale nanoparticles, TiO_2 microparticles) were successfully fabricated and coated on different substrates (aluminum sheet, soda-lime glass, silicon carbide polishing paper, glass with double double-sided adhesive). All coatings produce hydrophobic surfaces at varying degrees.

Paint with ground ZnO produces water contact angle greater than or equal to 150° (Figure 1), indicating superhydrophobicity. Contact angle incur on coated aluminium is 159 ± 4° (Figure 1a) while that of soda-lime glass is 166 ± 2° (Figure 1b). Meanwhile, the droplet contact angle on coated SiC polishing paper and coated glass with double-sided adhesive are 155 ± 5° (Figure 1c) and 155 ± 3° (Figure 1d), respectively. All surfaces coated with functionalized ground ZnO-containing paint produces superhydrophobic surfaces. These were not observed on the other types of paint that were used.

The substrates coated with paint-containing mono-scale nano-TiO_2 (~100 nm) creates droplets with average water contact angle greater than 140° (Figure 2). Coated aluminium sheet and coated glass with double-sided adhesive incurs contact angles of 144 ± 5° (Figure 2a) and 149 ± 5° (Figure 2d), respectively. Meanwhile, the average contact angle of coated SiC polishing paper (Figure 2c) and coated soda-lime glass (Figure 2b) are greater than 150° (161 ± 2° and 150 ± 6°, respectively).

The paint containing mono-scale TiO_2 microparticles (50 μm) that was coated on various substrates produced relatively low water contact angle, but still hydrophobic (Figure 3). Based on the images, the contact angle of the water droplet on painted aluminum (Figure 3a) is 128 ± 6°. Meanwhile, for SiC polishing paper (Figure 3c) and glass with double-sided adhesive (Figure 3d), the water contact angle are 114 ± 8° and 114 ± 5°, respectively. The glass coated with paint (Figure 3b) has a contact angle 134 ± 3°, the highest among the set of substrates.

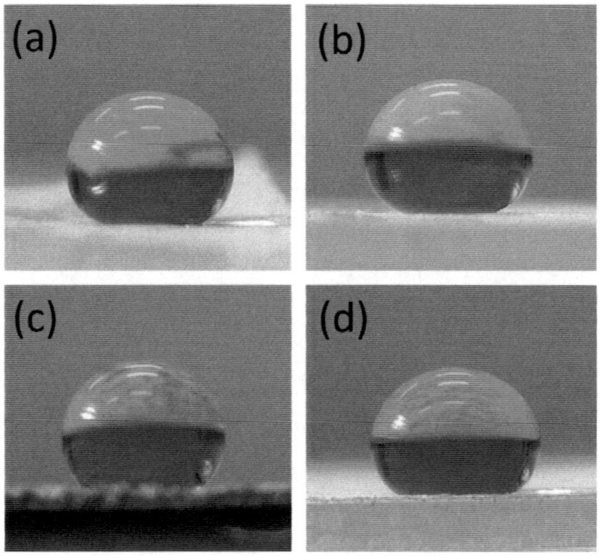

Figure 1. Droplets of water on the paint containing ground ZnO particles coated on (a) aluminum sheet, (b) soda-lime glass, (c) SiC polishing paper and (d) glass with double-sided adhesive.

Figure 2. Droplets of water on the paint containing TiO2 particles (~100 nm) coated on (a) aluminum sheet, (b) soda-lime glass, (c) SiC polishing paper and (d) glass with double-sided adhesive.

Air molecules which are trapped on pockets such as grooves, scratches and pores helps improve the hydrophobicity of the surface. Such air molecules pushes liquid droplets to form spherical structures, instead of flattening on the surface. Presence of structures that possibly act as air pockets were observed on the SEM micrograph (Figure 4). These structure includes (1) spaces between particles and (2) the protrusions on the functionalized particles.

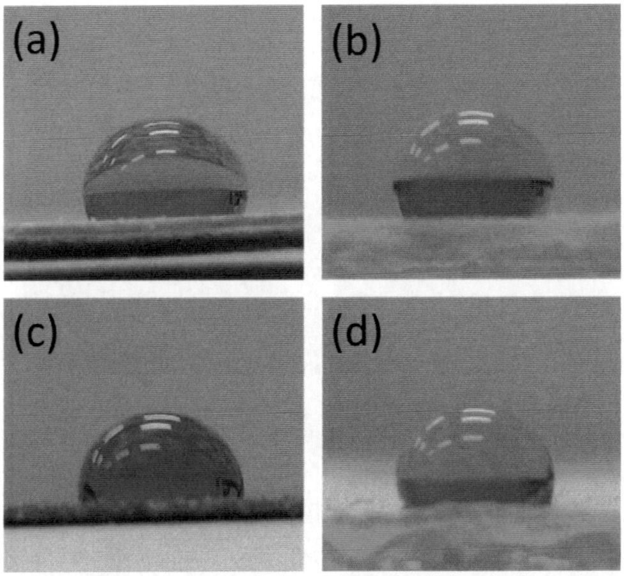

Figure 3. Droplets of water on the paint containing TiO2 particles (~50 µm) coated on (a) aluminum sheet, (b) soda-lime glass, (c) SiC polishing paper and (d) glass with double-sided adhesive.

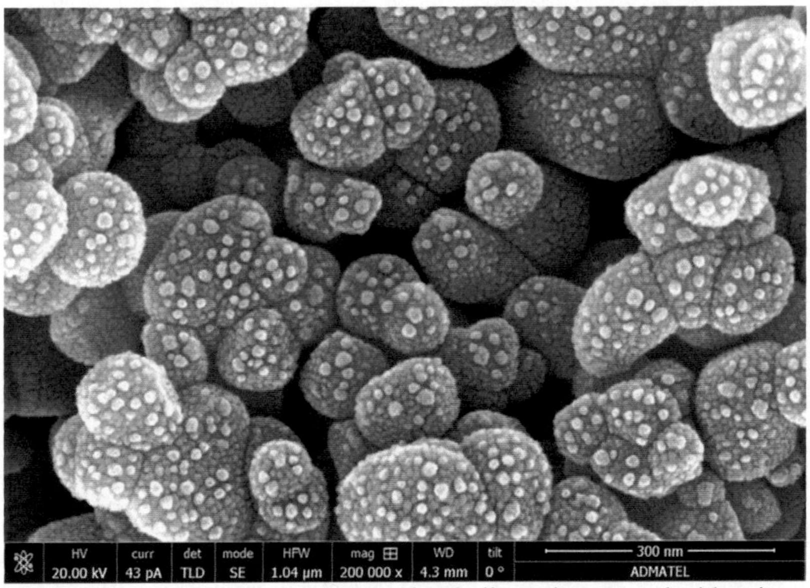

Figure 4. Scanning electron micrograph of soda-lime glass coated with paint containing functionalized-Titania nanoparticles.

4. Summary and Conclusion

In summary, the researchers have successfully fabricated three sets of paints that can be coated to various substrates to create hydrophobic surfaces. The paints were created by mixing oxide particles (ground ZnO, mono-scale TiO_2 nanoparticles or mono-scale TiO_2 microparticles) in an ethanolic suspension containing fluorosilane. The particles can roughen the surface while fluorosilane the surface energy. Based on the results, the substrates coated with paint containing ground ZnO or TiO_2 nanoparticles exhibited higher water contact angle compared to those coated with paint TiO_2 microparticles.

Compared to our previous study [16], the coated surfaces also exemplified hydrophobicity. Some surfaces are superhydrophobic while others have relatively low contact angle, but still hydrophobic. The hydrophobicity is attributed to the fluorosilane-functionalized oxide particles.

References

[1] Wang Z, Zhu H, Cao N, Du R, Liu Y and Zhao G 2017 *Mater. Lett.* **186** 274-8
[2] Abdulhussein A, Kannarpady G, Wright A, Ghosh A and Biris A 2016 *Appl. Surf. Sci.* **384** 311-32
[3] Li X, Zhang L, Ma X and Zhang H 2016 *Surf. Coat. Technol.* **307** 243-53
[4] Liu S, Liu X, Latthe S, Gao L, An S, Yoon S, Liu B and Xing R 2015 *Appl. Surf. Sci.* **351** 897–903.
[5] Hu W, Yu Y, Chen H, Lau K, Craig V, Brink F, Withera R and Liu Y 2015 *Appl. Surf. Sci.* **357** 2022–7.
[6] Kumar D, Li L and Chen Z 2016 *Prog. Org. Coat.* **101** 385-90
[7] Wang N, Xiong D, Li M, Deng Y, Shi Y and Wang K 2015 *Appl. Surf. Sci.* **355** 226-32
[8] Zhang H, Yang J, Chen B, Liu C, Zhang M and Li C 2015 *Appl. Surf. Sci.* **359** 905-10
[9] Arukalam I, Oguzie E and Li Y 2016 *J. Colloid Interface Sci.* **484** 220-8
[10] Cho SW, Kim JH, Lee HM, Chae H and Kim CK 2016 *Surf. Coat. Technol.* **306** 82-6
[11] Sharifi N, Pugh M, Moreau C and Dolatabadi A 2016 *Surf. Coat. Technol.* **289** 29-36
[12] Kothary P, Dou X, Fang Y, Gu Z, Leo SY and Jiang P 2017 *J. Colloid Interface Sci.* **487** 484-92
[13] Gao L, Zhou W, Wang Y, Wang S, Bai C, Li S, Liu B, Wang J and Li YL 2016 *Optik* **127** 5211-4
[14] Xie L, Tang Z, Jiang L, Breedveld V and Hess D 2015 *Surf. Coat. Technol.* **281** 125–32
[15] Zhou L, Xu S, Zhang G, Cai D and Wu Z 2016 *Appl Clay Sci.* **132-133** 290-5
[16] Sino PA, Herrera MU and Balela MD 2016 *Key Eng. Mater.***705** 268-72.

IOP Conf. Series: Materials Science and Engineering **201** (2017) 012023 doi:10.1088/1757-899X/201/1/012023

Experimental analysis of high oxygen concentration influences on horizontal flame spread over PA6 and epoxy

Ruichao Wei[1,2], Richard Yuen[2] and Jian Wang[1,a]

[1]State Key Laboratory of Fire Science, University of Science and Technology of China, Hefei, Anhui, 230026, P. R. China
[2]Department of Civil and Architectural Engineering, City University of Hong Kong, Hong Kong 999077, P. R. China

[a]wangj@ustc.edu.cn

Abstract. The combustion characteristic of material in high oxygen concentration is directly related to the safety of elevated oxygen environment similar to oxygen bombs and future spacecraft. In this study, flame front and flame spread rate in high oxygen concentration are discussed. Thermoplastic material polyamide (PA6) and thermosetting material epoxy are selected as comparison material. The variation of flame front over epoxy is more stable than PA6 which has dripping phenomenon during the flame spread process. The flame spread rate of both PA6 and epoxy have a power law with the increasing oxygen concentration.

1. Introduction

The Increasing attentions have been paid to the study of flame spread characteristics of polymer materials under different conditions. Zhou et al [1, 2] studied the flame spread characteristics of rigid polyurethane. The effect of external radiation, pressure and sample size is considered. The result showed that the flame spread rate and flame height decrease with the increase of sample width and flame spread rate in Hefei is higher than Lhasa. Besides, the flame spread rates of rigid polyurethane increases under the condition of increasing external radiation. Xiao Chen [3, 4] performed experiments on PMMA under the condition of different sample orientation and give a better understanding from the heat and mass transfer theory about the effect of orientation on flame spread. In addition, many scholars have studied the effect of melting and dripping phenomenon on the flame spread of thermoplastic materials [5-11]. Also, there are a few researches on combustion behavior of materials under high oxygen concentrations [12-14].

2. Experimental setup

The experiments are conducted in HC-2 oxygen index apparatus. Oxygen concentration is changed by changing the proportion of oxygen and nitrogen. Oxygen concentration is expressed by the percent of oxygen in the air. The diameter of the glass is 8cm and the distance between sample and underside is 25cm. Bottom of glass cover is filled with small glass ball and fine sand, thereby reducing the speed of airflow and making the gas move into the glass evenly. We improve the material upholder to make apparatus enable to measure parameters of horizontal flame spread. The ambient temperature and humidity of experiment were 15±2°C and 45%±5% respectively. The samples are 40mm in length,

4mm in diameter. A high speed Sony camera (HDR-XR160) is placed 1.5 m from the apparatus to record the flame spread process.

3. Result and discussion

Pyrolysis front is hard to measure but we can obtain the position of flame front from the video information recorded by the high definition DV at the side. So we define fire spread rate by the change of the position of flame front as

$$V_f = \frac{x_f(t+\Delta t) - x_f(t)}{\Delta t} \quad (1)$$

In the process of flame spread, the difference between thermoplastic material and thermosetting material is expressed as Figure 1.

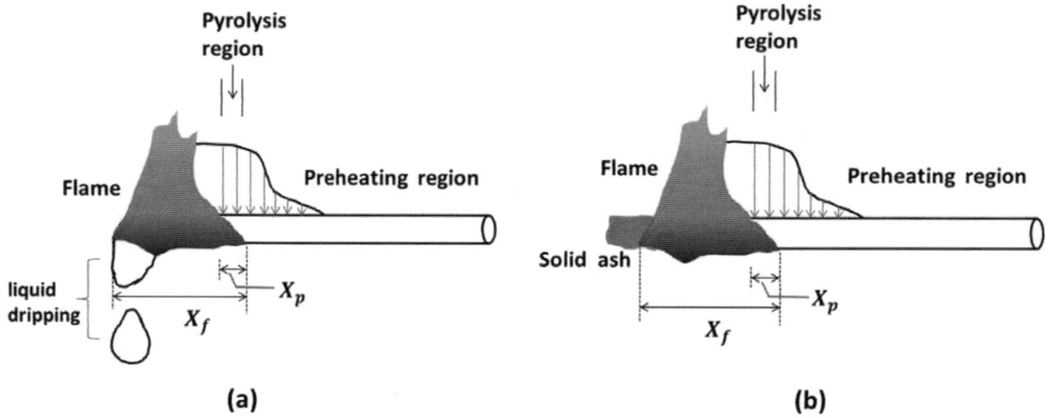

Figure 1. Illustrations of horizontal flame spreading: (a)thermoplastic sample PA6 (b)thermosetting sample Epoxy.

The local slope of the curve is defined as instantaneous flame spread rate. Different fluctuations can be observed for the samples of PA6 and epoxy. The flame front variation between PA6 and epoxy under 30% oxygen concentration is shown in Figure 2.

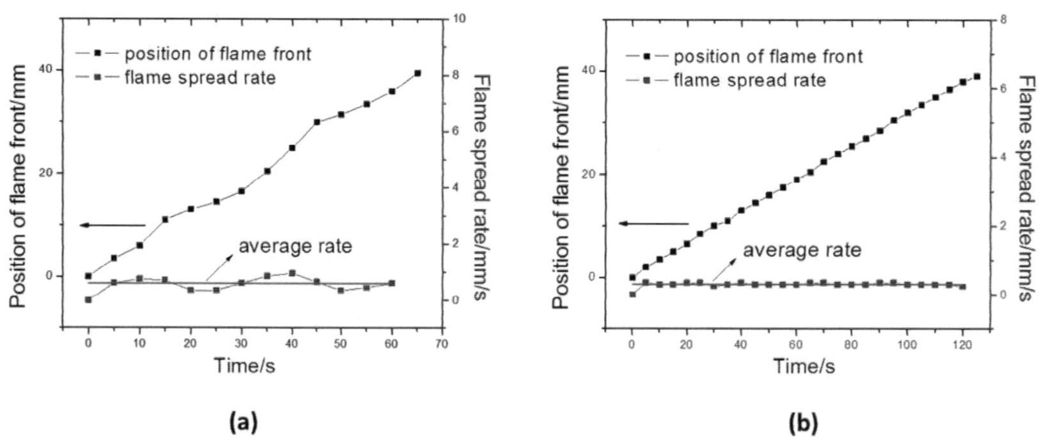

Figure 2. Flame front trajectories and time evolutions of the flame spread rate under 30% oxygen concentration: (a) PA6 and (b) epoxy.

With the increase of environmental oxygen concentration, gas phase chemical reaction was accelerated and the react between pyrolysis gas and oxygen become intense, increasing the heat release rate and flame temperature. High flame temperature increases the convictive flux and radiate flux, thus speeding up the horizontal flame spread. The variation of flame front about PA6 and epoxy from 25% to 50% of oxygen concentration is shown in Figure 3.

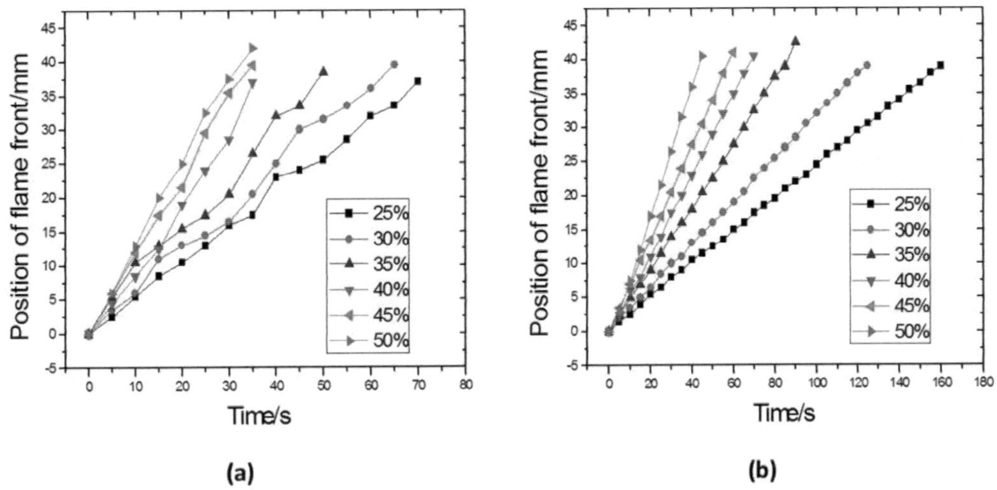

Figure 3. The variation of flame front under oxygen concentration from 25% to 50%: (a) PA6 and (b) epoxy.

Horizontal flame spread rate and heat flux from the flame for thermally thick material developed by Quintiere [15] proposed a formulation for flame spread rate of thermal thick material as

$$v_f \approx \frac{4(\dot{Q}_F^{"})^2 \delta_P}{\pi k \rho c (T_{ig} - T_\infty)^2} \tag{2}$$

$\dot{Q}_F$ is a part of combustion heat, so it can be written as

$$\dot{Q}_F = \zeta \dot{q}_f \tag{3}$$

According to oxygen consumption method [12], the relationship between flame combustion heat flux and oxygen concentration can be expressed as

$$\dot{q}_f = E\dot{V}_{298}\chi_{O_2}\left(\frac{\phi}{\phi(\alpha-1)+1}\right) \tag{4}$$

Combining Eqs. (2), (3) and (4), the relationship between flame spread rate and oxygen concentration can be expressed as

$$v_f \propto \chi_{O_2}{}^m \tag{5}$$

Figure 4. shows the fitting results of those two kinds of materials. The best-fit relationship between thermoplastic material PA6 and thermosetting are

$$v_{PA6} = 4.86\chi_{O_2}{}^{1.34}, with 25\% \le \chi_{O_2} \le 50\% \tag{6}$$

$$v_{Epoxy} = 3.12\chi_{O_2}{}^{1.87}, with 25\% \le \chi_{O_2} \le 80\% \tag{7}$$

The fitting results of those two kinds of materials shows that the horizontal flame spread rate over both thermoplastic materials and thermosetting materials increase according to a power low with the increasing oxygen concentration.

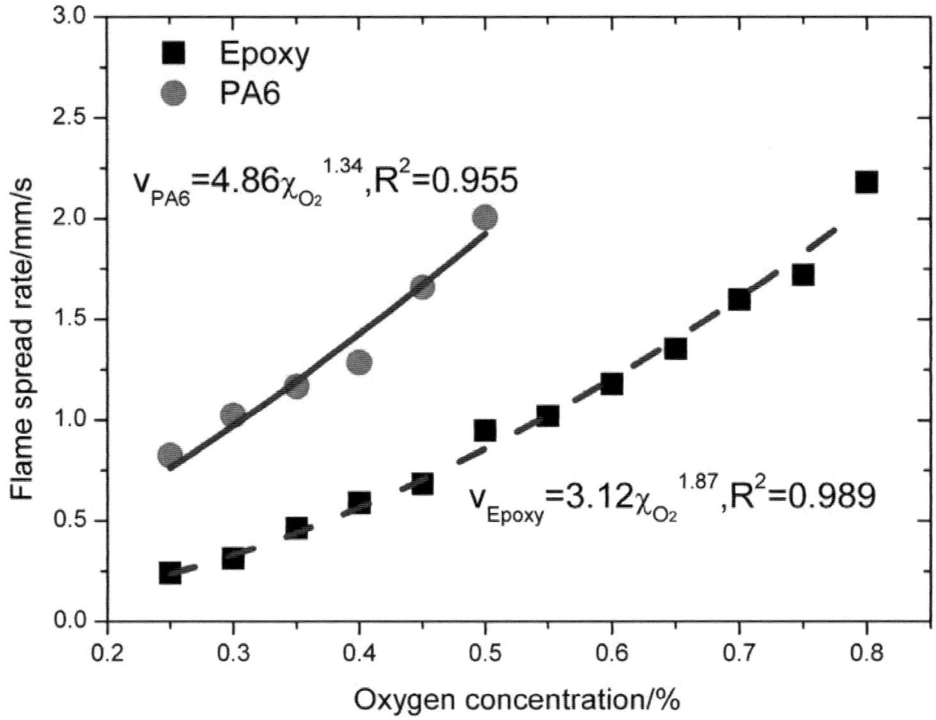

Figure 4. Flame spread rate over PA6 and Epoxy as a function of oxygen concentration. Points indicate the experimental date, and the dotted lines are the best fits.

4. Conclusions

In this work, a series of small size experiments of horizontal flame spread over PA6 and epoxy under different high oxygen concentration have been conducted to study the effect of ambient oxygen concentration flame front and flame spread rate. The difference of flame front between thermoplastic material and thermosetting material are also discussed in this paper. The main result of this work is shown as follows:

The movement flame front of PA6 is not stable changing with time under high oxygen concentration; Compared to thermoplastic material PA6, thermosetting material epoxy has stronger stability.

The horizontal flame spread rate increases according to a power law with the increase of oxygen concentration.

5. Acknowledgement

This research was supported by the National Natural Science Foundation of China (No. 51376172) and the grant from the Research Grant Council of the Hong Kong Special Administrative Region, China (contract grant number CityU 11301015).

References

[1] Zhou Y et al. Horizontal Flame Spread Characteristics of Rigid Polyurethane and Molded Polystyrene Foams Under Externally Applied Radiation at Two Different Altitudes. Fire

Technology 2015; 51(5): 1-22.

[2] Zhou Y et al. Experimental study of horizontal flame spread over rigid polyurethane foam on a plateau: effects of sample width and ambient pressure. Fire & Materials 2015; 39: 127-138.

[3] Chen X et al. Experimental and theoretical analysis on lateral flame spread over inclined PMMA surface. International Journal of Heat & Mass Transfer 2015; 91: 68-76.

[4] Chen X et al. Effects of sample orientation on pyrolysis and piloted ignition of wood. Journal of Fire Sciences 2014; 32(6): 483-497.

[5] Sherratt J, D Drysdale. The effect of the melt-flow process on the fire behaviour of thermoplastics. in InterFlam 2001; 9 th International Fire Science & Engineering Conference. 2001.

[6] Bockelie M J, T Qing. A computational model for fire growth and spread on thermoplastic objects. US Department of Commerce, National Institute of Standards and Technology. 2007.

[7] Ohlemiller T, J Shields. Aspects of the fire behavior of thermoplastic materials. NIST Technical Note 2008; 1493: 158.

[8] Butler K M, A model of melting and dripping thermoplastic objects in fire. Fire and materials 2009; 341-352.

[9] Xie Q, H Zhang, R Ye. Experimental study on melting and flowing behavior of thermoplastics combustion based on a new setup with a T-shape trough. Journal of hazardous materials 2009; 166(2): 1321-1325.

[10] Zhang J, T Shields, G Silcock. Effect of melting behaviour on upward flame spread of thermoplastics. Fire and Materials, 1997; 21(1): 1-6.

[11] Zhang J, Y Wang, X-d Lu. Study on melting behavior of polymers during burning. Fire Safety Science 2005; 8: 637-646.

[12] Mekki K et al. Wind-aided flame spread over charring and non-charrring solids: An experimental investigation. in Symposium (International) on Combustion 1991.

[13] CHAO Y H C, A C FERNANDEZ-PELLO. Concurrent Horizontal Flame Spread: The Combined Effect of Oxidizer Flow Velocity, Turbulence and Oxygen Concentration. Combustion Science & Technology 1995; 110-111(1): 19-51.

[14] Osorio A F et al. Limiting conditions for flame spread in fire resistant fabrics. Proceedings of the Combustion Institute 2013; 34(1): 2691-2697.

[15] DiNenno P J. SFPE handbook of fire protection engineering 2008; SFPE.

Effect of Substrate Orientation on the Growth of Germanium Oxide in Dry Oxygen Ambience

SK Sahari[1], NA Abdul Halim[1], K Muhammad[1], M Sawawi[1], AA Hamzah[2], B Yeop Majlis[2]

[1]Faculty of Engineering, Universiti Malaysia Sarawak, 94300 Kota Samarahan, Sarawak
[2]Institute of Microelectronics Engineering and Nanoelectronics, Universiti Kebangsaan Malaysia, Bangi, Selangor

[1]sskudnie@unimas.my

Abstract. The present investigation deals with the effect of substrate orientation effect on the growth of thermally oxidized Ge. The thermal oxidation was performed at temperature between 375 and 550°C in dry oxygen ambient under atmospheric pressure. The thickness of thermally oxidized Ge films was measured by spectroscopic ellipsometry and the chemical bonding structures were characterized by using x-ray photoelectron spectroscopy (XPS). No orientation dependence was observed for the oxidation at temperature of 375°C while for oxidation at 490 and 550°C, Ge oxidation and GeO desorption rate of (100) orientation yield higher rate than (111). The larger atomic space of (100) orientation explains the higher oxidation and desorption rate at Ge surface.

1. Introduction

Germanium (Ge) is one of the alternative channels for replacing Silicon (Si) for metal oxide semiconductor Field Effects Transistors (MOSFETs) because of higher carriers mobility than Si [1-3]. However, the thermodynamic instability at the high k/Ge interface is the most critical issue that must be overcome to realize a superior metal insulator semiconductor (MIS) [1-3]. On the other hand, an appropriate choice of substrate orientation and channel orientation for high performance Ge MOSFET are very important to determine the band structure of material and transport performance of the devices [4]. Several existing experimental reports showed unconformity with simulation results and inconsistency results among the group of researchers especially on the comparison between (100) and (110) surface orientation on the performance trend of Ge MOSFETs [5]. There was a report reported that Ge(110) shows the highest hole mobility for Ge MOSFETs application while another report showed the Ge(100) is the best substrate for MOSFETs application [6-7]. Moreover, Quan-Xin et al. reported that no orientation dependence between Ge (100) and (111) on the GeMOSFETs performance [8]. The independency between Ge (100) and Ge (111) on the performance of the GeMOSFETS is still in the open question.

Therefore, in this report, we try to clarify the independency between Ge(100) and Ge(111) by thermal oxidation. The substrate orientation dependence is investigated on the growth of Ge oxide at temperature between 375 and 550°C.

2. Experimental procedure

In this work, the p- type Ge(100) and Ge(111) substrates having resistivity 10-20 ohm were used. After wet chemical cleaning which employed the same method in the previous research [9-10] (Sahari et. al., 2013, Mishima et. al. 2013), thermal oxidation was performed at temperature between 375 and and 500 °C in dry oxygen ambience. The thickness of Ge oxide was measured by using spectroscopic ellipsometry (SE) with wavelength at 192-1000 nm and chemical bonding structure was characterized by high resolution x-ray photoelectron spectroscopy (XPS).

3. Results and discussion

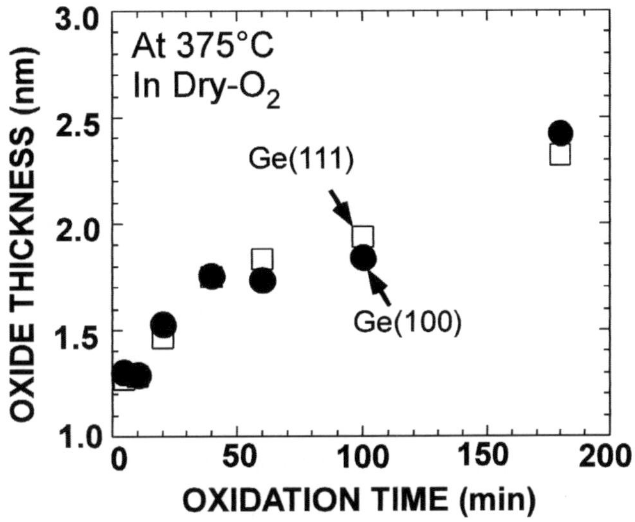

Figure 1. Ge oxide growth on Ge(100) and on Ge(111) by thermal oxidation at 375°C in pure dry-O_2 ambience.

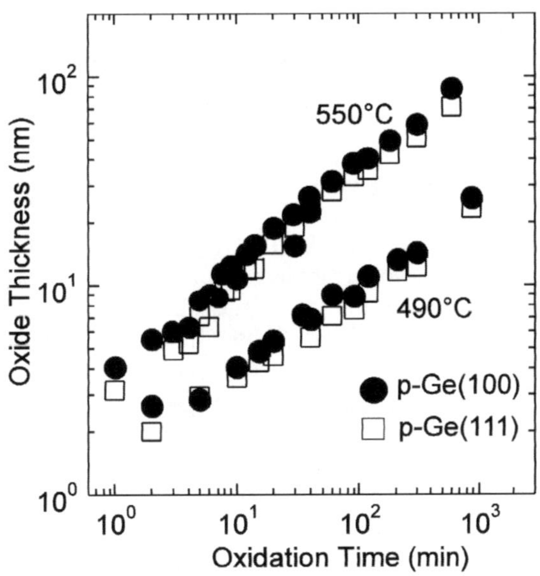

Figure 2. Ge oxide growth on Ge(100) and on Ge(111) by thermal oxidation at 490 and 550 °C in pure dry-O_2 ambience.

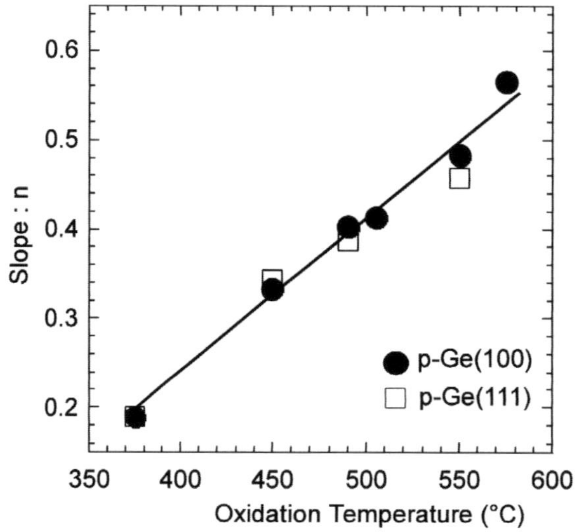

Figure 3. Slope versus oxidation temperature of different crystallographic orientation

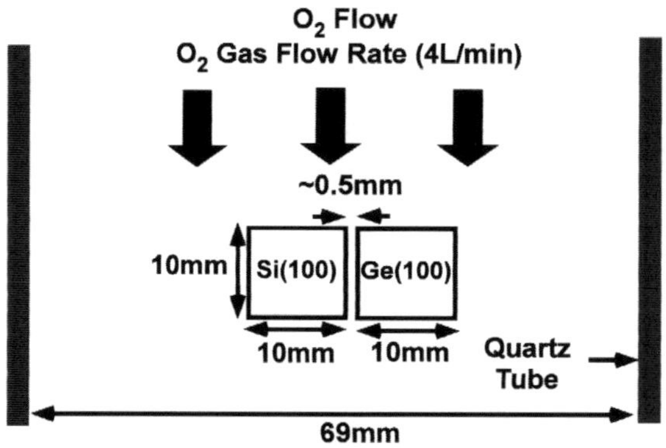

Figure 4. Experimental setup to confirm GeO desorption during Ge thermal oxidation.

Ge oxide growth on Ge(100) and Ge(111) at 375°C to 550°C with the oxidation time were evaluated from spectroscopic ellipsometry (SE) as shown in Figure. 1. While taking into account a measurement error of about 0.2 nm for the oxide thickness, the thermal oxidation rate on the Ge surface at 375°C was almost the same between the two different crystallographic orientations, which indicates no significant difference in Ge surface reaction rates of the O_2 molecules between the Ge(100) and (111) surfaces. In the oxidation time over 40 min, the oxide thickness for both surface orientations increased linearly with the oxidation time. Considering the Deal and Grove model explaining the thermal oxidation of Si surface for oxidation at temperature of 700° [11], this linear increase implies that oxidation at 375°C for both surface orientation is rate-limited by the surface reaction. On the other hand, the growth of Ge oxide on Ge(100) at over 490ºC is faster than that on (111) as shown in Figure 2. The faster oxidation rate of Ge(100) is presumably caused by the larger atomic atom space than oxygen molecule as discussed in previous report [12]. In addition, the Ge oxide thickness was exponentially increased with the oxidation time over 10 min, which confirmed there was a rectilinear relation in the log-log plot for both surface orientations. The slope n was extracted by using linear extrapolation, and the n values for the Ge(100) and (111) surfaces are summarized as in Figure 3. From the figure, the slope *n* of both surface crystallographic orientations, the slope at each oxidation temperatures almost remains unchanged irrespective of the orientations.

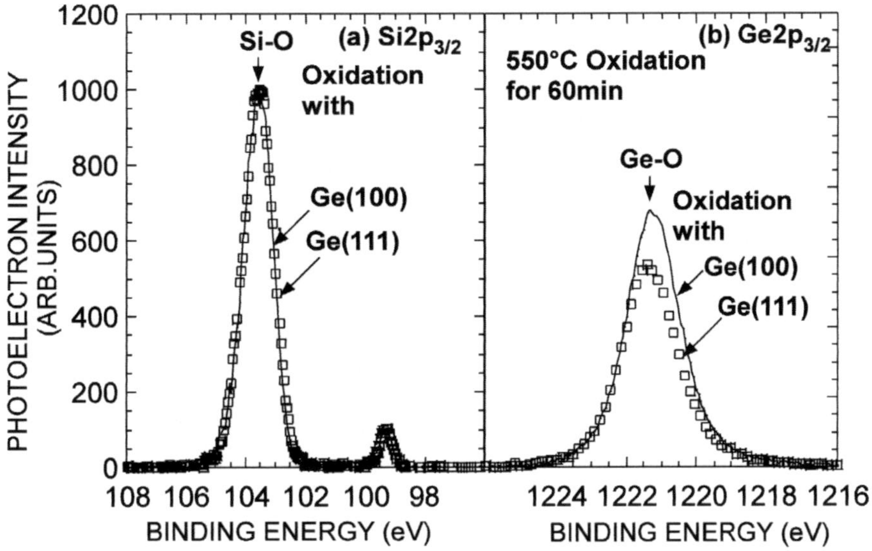

Figure 5. Si2p$_{3/2}$ and (b) Ge2p$_{3/2}$ spectra for 11 nm-thick SiO_2/Si set nearby Ge(100) or Ge(111) during oxidation at 550 °C for 60 min. Photoelectron take-off angle was set at 90°.

To get an insight into the GeO desorption during thermal oxidation, 11nm-thick SiO_2/Si was placed nearby Ge(100) or Ge(111) in the oxidation furnace and oxidized simultaneously as schematically illustrated in Figure. 4, and then XPS analysis of SiO_2 surface was carried out. Figure 5 shows the Si2p$_{3/2}$ and Ge2p$_{3/2}$ spectra taken for the SiO_2 surface after oxidation at 550 °C for 60 min. In each spectrum, the binding energy calibration and the photoelectron intensity normalization were made by Si2p$_{3/2}$ signals. From the result, no changes have been observed for Si2p$_{3/2}$ signal of 11.0nm-thick SiO_2 oxidized with Ge(100) and Ge(111) while the Ge2p$_{3/2}$ signal originating from attached GeO_x on SiO_2 was dependent on the crystallographic orientation. It was found that the amount of GeO desorption from Ge(100) surface was higher in comparison to the Ge(111) case as shown in Figure 5(b). Taking into account of the crystallographic orientation dependence of Ge-oxide growth, some of GeO units (or oxygen vacancies) generated at the oxide/Ge interface are likely to be re-oxidized (or annihilated) with O_2 flux through the oxide until reaching the oxide top surface, which results in enhanced oxide growth.

4. Conclusion

The thermal oxidation of the Ge(100) and Ge(111) surfaces were investigated. The slope of Ge oxide growth on Ge(100) at 375°C with oxidation time shows the same value as native oxide growth and this slope gradually decreases with increasing temperatures over 490°C for the oxide thickness less than 20 nm. Almost no crystallographic orientation dependence on the oxidation rate was observed for. In addition, the growth of Ge oxide on Ge(100) is faster than that on (111) at 490 and 550°C. With increasing oxidation temperature and in longer oxidation time, the slope of Ge oxide growth was significantly changed where the slope increases with increasing temperature. The GeO desorption during thermal oxidation was confirmed from the XPS analysis of Si substrate placed next to the Ge(100) during oxidation over 550°C.

Acknowledgement
The authors would like to thank the Ministry of Higher of Education in providing Research Acculturation Collaborative Effort (RACE) grant and Fundamental Research Grant Scheme (FRGS). The gratefully acknowledge Universiti Malaysia Sarawak for the support.

References

[1] Sze S M and Ng K K 2007 *Physics of Semiconductor Devices* (New York, Wiley 3rd ed) p.786
[2] Zhang LL, Gunji M, Thombare S, and McIntyre PC, 2013, *IEEE electron Device Lett.* 34, p. 732
[3] Zhang R, Taoka N, Huang PC, Takenaka M, and Takagi S, 2011, Technical Digest of the International Electron Devices Meeting, December 5-7, p.457
[4] Takagi S, Irisawa T, Tezuka T, Numata T, Nakaharai S, Hirashita N, Moriyama Y, Usuda K, Takenaka M, and Sugiyama N, 2008, IEEE Trans. Electron Devices 33, p. 22
[5] Zhang R, Huang P, Lin J, Takenaka M, and Takagi S, 2012 Technical Digest of International electron Devices Meeting, p.371
[6] Zhang R, Iwasaki T, Taoka N, Takenaka M, and Takagi S, 2011 Symposium on VLSI Technology, p.56
[7] Lee C H, Nishimura T, Tabata T, Wang SK, Nagashio K, Kita K, and Toriumi A, 2010 Technical Digest of International Electron Devices Meeting, p. 416
[8] Yun Q, Li M, An X, Lin M, Liu P, Li Z, Zhang B, Xia Y, Zhang H, Zhang X, Huang R, and Wang Y, 2014 Chin Phys. B 23, p. 118506
[9] Sahari S K, Mohamad Sutan N, Sapawi R, Yakub I, Awg Salleh, DNS, Lit A, 2014, Key Engineering Materials 594-595, p. 1069
[10] Mishima K., Murakami, H, Ohta, A, Sahari, SK, Fujioka, T, Higashi, S, Miyazaki, S, 2013 Journal of Physics: Conference Series 417
[11] Deal B and Grove AS, 1965, *Appl. Phys.*36(12), p. 3770
[12] Sahari S K, Murakami H, Fujioka T , Bando T, Ohta A, Makihara K, Higashi H, Miyazaki S, Jpn. J. Appl. Phys. 50, p. 04DA12

IOP Conf. Series: Materials Science and Engineering **201** (2017) 012025 doi:10.1088/1757-899X/201/1/012025

Electrical characteristics of zinc oxide-copper oxide heterojunction at different illumination intensities

Teresa Lean Baquiran, Maria Carmela T. Garcia, Marvin U. Herrera

Physics Division, Institute of Mathematical Sciences and Physics, University of the Philippines Los Baños, Philippines 4031

mtgarcia2@up.edu.ph

Abstract. The current-voltage characteristics of copper oxide (CuO) - zinc oxide (ZnO) heterojunctions under different illumination intensities were observed in this study. ZnO and CuO films were deposited on separate graphite electrodes via electrophoretic deposition (EPD). The films underwent heat treatment and were physically joined to form the p-n junction. The current-voltage characteristics of the junctions were obtained under forward and reverse bias under different illumination intensities. It was observed that the ZnO-CuO junctions exhibited diode-like behavior in the dark, indicated by the passing of current in forward bias and restricting in reverse bias. Under illumination, the junctions exhibited photodiode-like behavior, with increasing reverse current at greater illumination intensities.

1. Introduction

In the past years, metal oxide semiconductors have attracted interest due to ease of fabrication, low cost, and chemical stability. They have applications in the field of electronics [1-2], optoelectronics [3], gas sensing [4-5], biosensing [6-7], photocatalysis [8-10] and solar cells [11-12].

Joining an n-type semiconductor and a p-type semiconductor can create a p-n junction which has a potential barrier and depletion region. The existence of the potential barrier can lead to the preferential direction of current, which is a characteristic of a diode.

Diodes are operated using two modes - forward and reverse bias. Connecting the p-type region of the diode to the positive terminal of the voltage source and the n-type region to the negative terminal would allow the device to operate under forward bias. In this configuration, current flows in one direction around the circuit. Switching the connections would change the mode of operation of the diode to reverse bias. In this configuration, the depletion region widens and inhibits the diffusion of electrons and holes across the device, therefore restricting current flow around the circuit. A photodiode is a type of diode which is operated in the reverse bias configuration. The photodiode converts light into current when illuminated with light of sufficient energy.

In a p-n junction, the area near the interface (depletion regions) is a charged region. When illuminated, electron-hole pairs can separate and could participate in the conduction of charges across the junction. Such event is very much evident at the reverse bias where the depletion region is much wider.

Several literatures have shown that the electrical characteristics of oxide-based junctions exemplify diode-like behavior [13-23]. Zinc oxide is an intrinsic n-type semiconductor with wide band gap of 3.37 eV while copper oxide is a p-type semiconductor with narrow band gap of 1.2 eV. Joining the two semiconducting oxides can led to the creation of a p-n heterojunction. This study aims to determine the response ZnO-CuO heterojunctions under reverse bias with different illumination intensities.

2. Methods

Zinc oxide and copper oxide were deposited using electrophoretic deposition (EPD) on separate graphite electrodes. The zinc oxide and metal oxide powder underwent grinding to ensure homogeneity before being placed in separate beakers. They were then suspended in acetone and subjected to a power supply to deposit them on the graphite electrodes. The deposition of zinc oxide utilized a 500V power source while a 60V power source was used to deposit copper oxide. After deposition, the samples were air dried and placed in a furnace at 600°C for heat treatment to ensure uniformity. Afterwards, the ceramics were physically joined to form a p-n heterojunction.

The current-voltage characteristics of the samples were obtained using a variable voltage source and a picoammeter. Characterization was done under a UV-filtered incandescent lamp at different illumination intensities to determine the samples's photoresponse. IV curves of the samples were measured for both the forward bias and reverse bias configurations..

3. Results and Discussion

We have successful created a p-n heterojunction by physically joining n-type ZnO and p-type CuO. Figure 1 shows the current-voltage plot of zinc oxide-copper oxide junction at dark conditions. At forward bias, the current shows rectifying behavior while at reverse bias, the current is low. Such behavior is characteristic of a p-n junction in a diode.

The response of the sample at different illumination intensities are shown in Figures 2 and 3. Figure 2 plots current-voltage at different illumination intensities while Figure 3 shows the corresponding current-illumination intensity at different reverse bias voltages. Both figures show that the reverse bias current increases with illumination intensity. This trend is similar to the behavior of photodiodes under reverse bias.

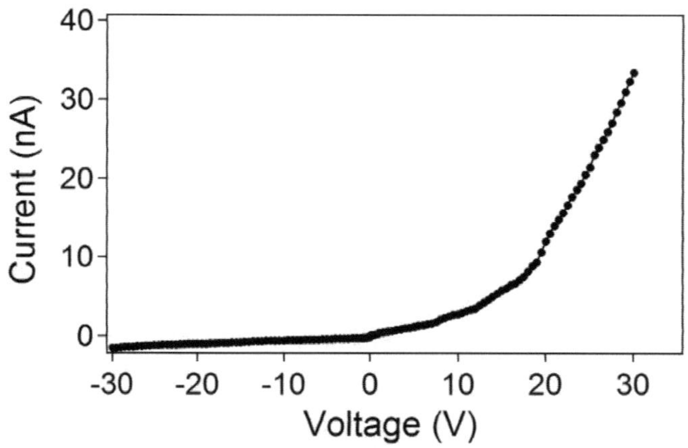

Figure 1.Current-voltage plot of zinc oxide-copper oxide junction under dark conditions

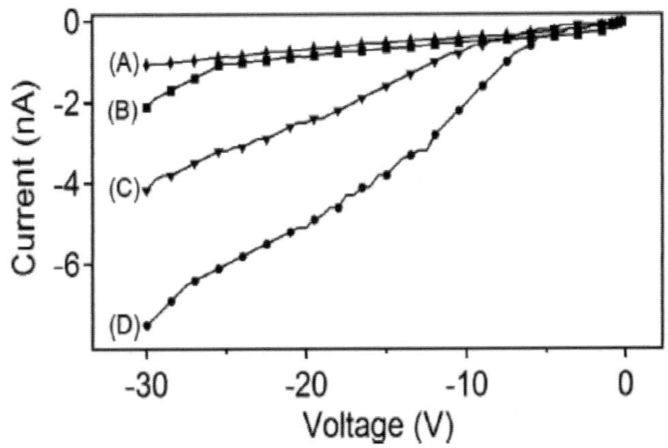

Figure 2.Current-voltage plot at reverse bias of zinc oxide-copper oxide junction at different illumination intensities: (A) 0, (B) 1000, (C) 2000 and (D) 3000 lux

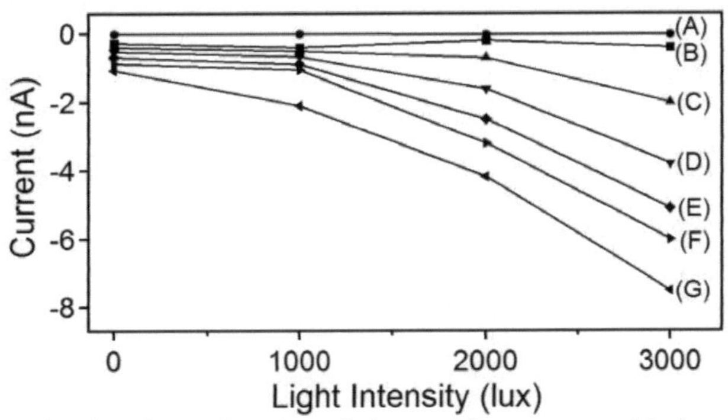

Figure 3.Current-illumination intensity plot of zinc oxide-copper oxide junction at different reverse bias voltages: (a) 0V; (b) -5V; (c) -10V; (d) -15V; (e) -20V; (f) -25V; (g) -30V

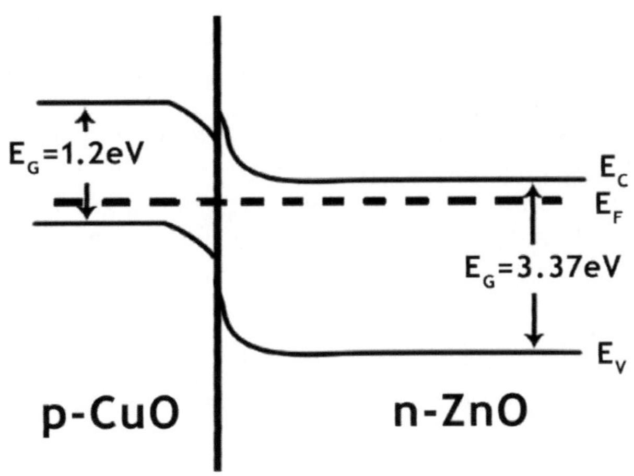

Figure 4.Energy band diagram of n-ZnO/p-CuO junction where E_F is the Fermi level, E_C is the conduction band energy, E_V is the valence band energy and E_G is the band gap

Similar to conventional p-n diode, the joining of n-type ZnO and p-type CuO creates potential barrier and depletion region that is responsible for the diode-like behavior of the sample. The energy band diagram [10, 24-27] of the zinc oxide-copper oxide junction is shown in Figure 4. When the two oxides are joined, charge transfer happens at the interfaces until it reaches equilibrium. At equilibrium, the Fermi levels (E_F) becomes equal. At the n-type semiconductor (ZnO) positive charges lumped near the interface, creating a positively charged depletion region. In addition, such lumping of charges creates upward bending of the valence (Ev) and conduction (Ec) bands which forms the potential barrier. On the other hand, negative charges lumped at the p-type semiconductor (CuO) side creates a negatively charged depletion region and downward bending of the energy bands.

When light with sufficient energy hits the semiconducting material, excitons (electron-hole pairs) are created. At the positively charged depletion region, the holes of excitons are separated and are repelled, producing the reverse bias current. Likewise, at the negatively charged depletion region, the electrons of excitons are separated and are repelled, once again resulting to a reverse bias current. (Note that the existences of depletion regions are more evident at reversed bias condition). The exciton density increases with illumination intensity thus it is expected that reverse bias current also increases with light intensity as shown in Figures 2 and 3.

4. Conclusion

Copper oxide (CuO) and zinc oxide (ZnO) heterojunctions were successfully formed by depositing them separately on graphite electrodes using electrophoretic deposition technique. The p-n heterojunctions were achieved by physical connection of the CuO-deposited graphite and the ZnO-deposited graphite. The heterojunctions exhibited rectifying behavior when forward biased and allowed low current to pass when reverse biased, indicating that a p-n junction was successfully formed. When exposed to increasing illumination intensities the reverse current in the p-n junction increases, similar to that of a photodiode.

Acknowledgements

The authors are grateful for the funding support given by the University of the Philippines Los Banos Basic Research Program.

References

[1] Ameen S, Akhtar MS, Ansari SG, Yang O and Shin H 2009 Electrophoretically deposited polyaniline/ZnO nanoparticles for pn heterostructure diodes *Superlattice. Microst.* **46** 872-880

[2] Nam G, Baek S and Park I 2014 Growth of ZnO nanorods on graphite substrate and its application for Schottky diode *J. Alloy Cmpd* **613** 37–41

[3] Antony A, Pramodini S, Poornesh P, Kityk IV, Fedorchuk AO and Sanjeev G 2016 Influence of electron beam irradiation on nonlinear optical properties of Al doped ZnO thin films for optoelectronic device applications in the cw laser regime *Opt. Mater.* **62** 64-71

[4] Chen Y, Xu P, Xu T, Zheng D and Li X 2017 ZnO-nanowire size effect induced ultra-high sensing response to ppb-level H_2S *Sensor. Actuat. B-Chem.* **240** 264-272

[5] Zhao M, Wang X, Ning L, Jia J, Li X and Cao L 2011 Electrospun Cu-doped ZnO nanofibers for H_2S sensing *Sensor. Actuat. B.* **156** 588– 592

[6] Bhat SS, Qurashi A and Khanday FA 2017 ZnO nanostructures based biosensors for cancer and infectious disease applications: Perspectives, prospects and promises *TrAC Trend Anal. Chem.* **86** 1-13

[7] Yang T, Chen M, Kong Q, Luo X and Jiao K 2017 Toward DNA electrochemical sensing by free-standing ZnO nanosheets grown on 2D thin-layered MoS_2 *Biosens. Bioelectron.* **89** 538-544

[8] Chang T, Li Z, Yun G, Jia Y and Yang H 2013 Enhanced Photocatalytic Activity of ZnO/CuO

[9] Liu H and Yang T 2003 Photocatalytic inactivation of Escherichia coli and Lactobacillus

helveticus by ZnO and TiO_2 activated with ultraviolet light *Process Biochem.* **39** 475-481

[10] Almanza DLV, Garcia MCT, Prodigalidad AD and Herrera MU 2016 Visible light photocatalytic activity of Zinc Oxide-Copper Oxide composites *Key Eng. Mater.***705** 385-389

[11] Marimuthu T, Anandhana N, Thangamuthu R, Mummoorthi M and Ravi G 2016 Synthesis of ZnO nanowire arrays on ZnO-TiO_2 mixed oxide seed layer for dye sensitized solar cell applications *J. Alloy Cmpd.* **677** 211-218

[12] AbdulAlmohsin S, AL-Mutoki SM and Li Z 2012 Fabrication and Characterization of Aluminum-doped ZnO/PANI Hybrid Solar Cells *J. Ark. Acad. Sc.* **66** 26-30

[13] Barış B 2014 Analysis of device parameters for Au/tin oxide/n-Si(1 0 0) metal–oxide–semiconductor (MOS) diodes *Physica B: Condens. Matter* **438** 65–69

[14] Chiou A, Wu S, Hsiao R, Hsu C 2016 TiO_2–silicon nanowire arrays for heterojunction diode applications *Thin Solid Films* **616** 116–121

[15] Lee B, Kim C, Lee Y, Lee S, Kim DY 2015 Dependence of photocurrent on UV wavelength in ZnO/Pt bottom-contact Schottky diode *Curr. Appl. Phys.***15** 29–33

[16] Liu Y, Yu J, Lai PT 2014 Investigation of WO_3/ZnO thin-film heterojunction-based Schottky diodes for H_2 gas sensing *Int. J. Hydrogen Energy* **39** 10313–19

[17] Mones ES, Gillado AV, Herrera MU 2016 Photoresponse of Zinc Oxide -Polyaniline Junction at Different Light Intensities *Key Eng. Mater.***705** 186-189

[18] Park JS, Kim JH, Kim JY, Kim DH, Kang D, Sung JS and Seong TY 2016 Hybrid indium tin oxide/Ag nanowire electrodes for improving the light output power of near ultraviolet AlGaN-based light-emitting diode *Curr. Appl. Phys.* **16** 545–548

[19] Sahu VK, Misra P, Ajimsha RS, Das AK and Singh B 2016 Effect of growth temperature on diode parameters of n-ZnO/p-Si heterojuction diodes grown by atomic layer deposition *Mater. Sci. Semicond. Process.***54** 1–5

[20] Suresh R, Ponnuswamy V, Sankar C, Manickam M and Mariappan R 2016 Influence of Co concentration on the structural, optical, morphological and photo-diode properties of cerium oxide thin films *Ceram. Int.* **42** 12715-25

[21] Turgut G, Duman S, Sonmez E, Ozcelik FS 2016 A study of Eu incorporated ZnO thin films: An application of Al/ZnO:Eu/p-Si heterojunction diode *Mater. Sci. Eng., B* **206** 9–16

[22] Yu J, Chen G, Li CX, Shafiei M, Ou JZ, du Plessis J, Kalantar-zadeh K, Lai PT and Wlodarski W 2011 Hydrogen gas sensing properties of Pt/Ta_2O_5 Schottky diodes based on Si and SiC substrates *Sensor. Actuat. A-Phys.* **172** 9–14

[23] Zhu R, Zhang X, Zhao J, Li R, Zhang W 2015 Influence of illumination intensity on the electrical characteristics and photoresponsivity of the Ag/ZnO Schottky diodes *J. Alloy Cmpd.* **631** 125–128

[24] Chabane L, Zebbar N, Lamri Zeggar M, Aida MS, Kechouane M and Trari M 2015 Effects of CuO film thickness on electrical properties of CuO/ZnO and CuO/ZnS hetero- *Mater. Sci. Semicond. Process.* **40** 840–847

[25] Katoch A, Choi SW, Kim JH, Lee JH, Lee JS and Kim SS 2015 Importance of the nanograin size on the H2S-sensing properties of ZnO–CuO composite nanofibers *Sensor. Actuat. B-Chem* **214** 111–116

[26] Kim JH, Katoch A, Kim SS 2016 Optimum shell thickness and underlying sensing mechanism in p–n CuO–ZnO core–shell nanowires *Sensor. Actuat. B-Chem* **222** 249–256

[27] Zhang Y, Yin J, Li L, Zhang L and Bie L 2014 Enhanced ethanol gas-sensing properties of flower-like p-CuO/n-ZnO heterojunction nanorods Sensors *Sensor. Actuat. B-Chem* **202** 500-507

IOP Conf. Series: Materials Science and Engineering **201** (2017) 012026 doi:10.1088/1757-899X/201/1/012026

Copper laminated composites reinforced with Al_2O_3 nanoparticles by suspension method and hot pressing

G.Roudini[1, a *], A.Rasti Ghahfarokhi[1] and A. Behzadmehr [2]

[1]Department of Engineering Materials, Faculty of Engineering, University of Sistan and Baluchestan, Zahedan, Iran
[2]Mechanical Engineering Department, Faculty of Engineering, University of Sistan and Baluchestan, Zahedan, Iran

[a *]ghodratollah.roudini@eng.usb.ac.ir

Abstract. Laminated composite of copper with alumina nanoparticle reinforcements has application such as electrical industries. Electrical junctions should have high electrical, thermal conductivity with good mechanical strength. In this research the layers interface of pure copper were reinforced by alumina nanoparticles. For the production of the composite, first alumina nanoparticles were dispersed on Cu layers by suspension method and then the layers were hot pressed at 950°C under 20 MPa pressure. After composite making, the microstructure, the tensile and impact of these composite were studied. The results showed that by increasing amount of alumina nanoparticles up to 0. 5 wt %, tensile and impact strength were increased and for the composites with more than 0.5 wt % because of agglomeration of nanoparticles, these properties were decreased.

1. Introduction

Metal laminated composites are a type of metal matrix composites, consisting of layers metals which they have good of mechanical properties such as high specific strength and stiffness. These kind of materials can be a good choice in transportation industries [1, 2]. There are several methods for the production of metal laminated composites. These can be classified according to the metal state: solid state, liquid state and vapour state processes are thus generally distinguished [2-8]. In the metal laminated composites, the interface of layers affects the properties of them and can be improved by strengthen of reinforcements ceramics. Hence, the whole properties of the composite can be changed by production method, interfaces strength and reinforcement.

In the present work copper-alumina nanoparticles laminated composites were successfully fabricated using alumina suspension method and hot pressed [9-10] and then the effect of amount alumina nanoparticles was studied on tensile and impact strength of the composites

2. Materials and Processing

High purity α-Al_2O_3 nanoparticles (20 nm) were used as starting materials. Slurries with different amount of Al_2O_3 (0-0.1 wt%), 50 cc distilled water and sulphuric acid as a *surfactant* were mixed by a magnetic mixer for 20 min. Rectangular pure copper sheets with $10\times6\times1$ mm^3 in size were cut and cleaned with acetone to remove pollutions. Then the copper sheets were immersed into the slurries to deposit alumina nanoparticles on them. This process was repeated to achieve the required alumina content and distribution. Subsequently, the sheets were gently removed from the slurry and dried

IOP Conf. Series: Materials Science and Engineering **201** (2017) 012026 doi:10.1088/1757-899X/201/1/012026

at100°C for 1 h. Finally, the three layers laminated composites were fabricated at 950 °C and 20 MPa pressure for 5 minutes by pressing method.

3. Experimental

The microstructure of the deposited copper sheets with nanoparticles was studied by a Scanning Electron Microscopy. Also the samples were then cut from the laminated composites using cutter to study microstructure, tensile strength and impact behaviour. After sectioning, the composites were hot-mounted and subsequent polishing of the metallographic specimens. Once the required metallographic finish was obtained, observations were made on unetched samples using a SEM. The tensile tests on the composites were conducted using a Universal testing machine (Instron Model 4206). The room-temperature tensile tests were carried out at a crosshead speed of 1 mm/min. Also, the tensile fracture surfaces of composites samples were investigated by SEM. The interfacial bonding strength of layer was analyzed by impact test.

4. Results and discussions

The results and discussion on this study are presented in this section. First, the microstructure of the copper sheets deposited with alumina nanoparticles before and after pressing is illustrated and discussed. Then, tensile and impact behaviour of the laminated composites were compared. Final, tensile fracture results of the composites are discussed.

4.1 Microstructure

The microstructures of the copper layers surfaces after deposited with alumina nanoparticles are shown in figure 1 and figure 2. The micrograph results showed that distribution and agglomeration of alumina on the copper sheets surfaces depend on wt %t of alumina and surfactant in solution. In higher concentration of alumina in solution causes large size agglomeration. In solution up to 0.05 wt% alumina the nanoparticles were well distributed with small size agglomeration. It should be noted that the initial concentration of particles [11] affects the distribution alumina to the whole laminated composite.

Figure 1. SEM micrograph for 0.1w% alumina agglomerated on copper sheet.

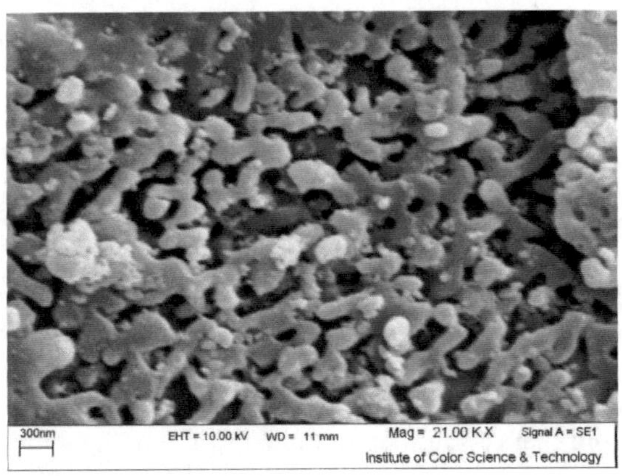

Figure 2. SEM micrograph for 0.1w% alumina copper sheet.

4.2. Tensile and impact behaviour

The results obtained from tensile test are given in figure 3 and figure4. Figure 4 shows representative stress-strain curves of three composites with different wt% alumina in comparison to the unreinforced aluminium. All curves show an increase in strength and the composite with 0.05wt % show higher strength, which is higher than that of matrix at a given strains. It can be concluded that a proper amount of alumina and well distribution in the interfaces leads to significantly higher bonding strength of interfaces which it will be improved fracture strength of the laminated composites.

It can be seen from figure. 5, that the tensile fracture strength was increased with the amount of wt% of alumina from 0 w% to 0.1w%. The delaminating at the interface was no found up to 0.05 wt%. Inhibition of delamination prevents neck formation in the laminated composite. This is mainly due to the well distribution of nanoparticles in the interface which was high enough for Cu sheets to bond significantly. In contrast, the fracture strength of the composite with 0.1wt% of alumina is deceased, due to the agglomeration of alumina during deposition and composite making processes. The fracture surfaces were observed by SEM showed the dimples in figure 5 and figure 6. Hence copper layers in the laminated composites fractured in a rather ductile manner in all composites, but a different manner from that of the monolithic copper and the degree of brittleness was increased. Also the effect of interface strength on the tensile behaviour of the laminated composited were confirmed by others researchers [3, 12-16].

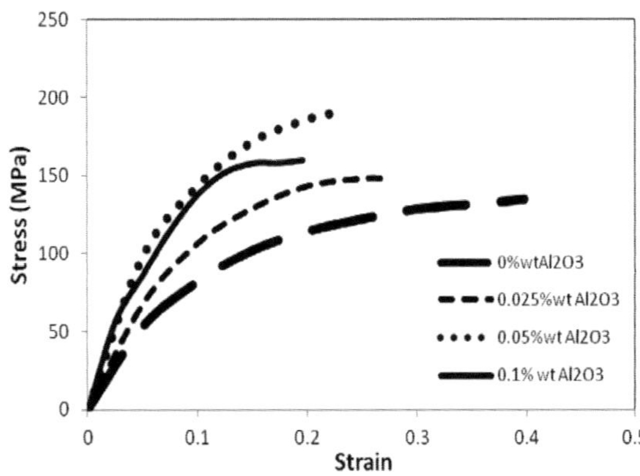

Figure 3. Stress-strain of the composites with different concentration of alumina.

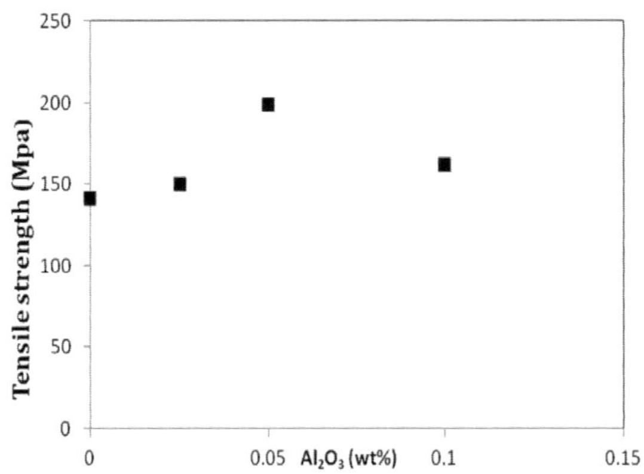

Figure 4. Tensile strength of the composites for different alumina concentration.

The observation in figure 7 is that the total strain to failure results for the composite indicates that the tensile deformation is decreasing with increasing amount of reinforcement. These results have been interpreted in terms of the effect of wt% of alumina on the failure of laminated composites; this trend is attributed to the greater probability for delamination during deformation. Interfacial delamination is increased with increasing agglomeration in higher wt% of alumina. Also the impact test results were showed the same fracture behaviour of tensile test (figure 8).

The fracture surfaces were observed by SEM showed the dimples. Hence copper layers in the laminated composites fractured in a rather ductile manner in all composites, but a different manner from that of the monolithic copper. The delamination at the interface was no found up to 0.05 wt. %. Also the impact test results were showed the same fracture behavior of metal laminated composites [17-18].

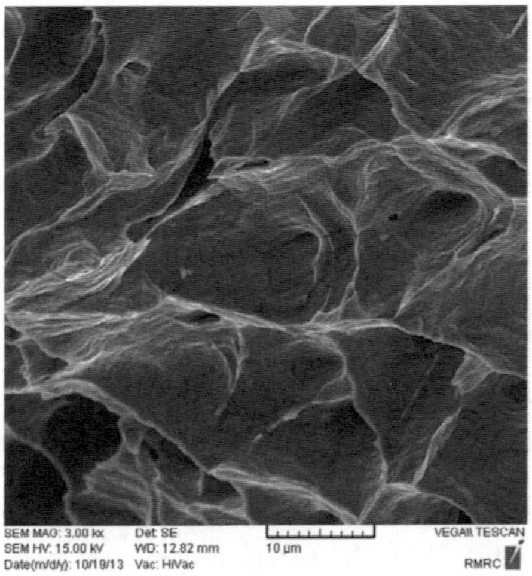

Figure 5. SEM micrograph of tensile fracture surface for the composite with 0.025 wt%alumina.

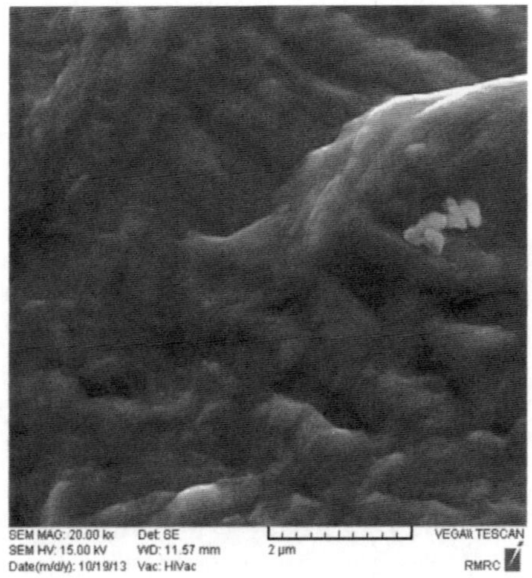

Figure 6. SEM micrograph of tensile fracture surface for the composite 0.05 wt% alumina.

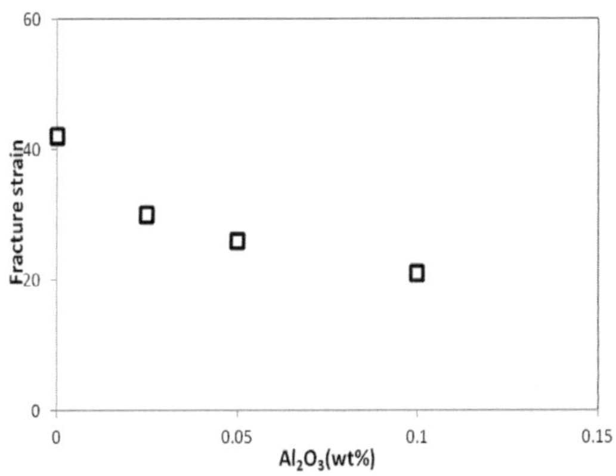

Figure 7. Tensile fracture strain of the composites for different alumina concentration.

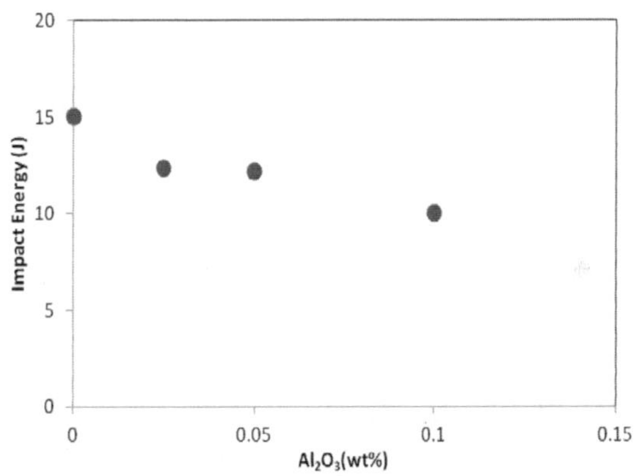

Figure 8. Impact energy versus alumina concentration in the laminated composites.

5. Conclusions

The tensile and impact behavior of the laminated composites was affected by amount of wt% of alumina nanoparticles. The ultimate tensile strength of the composites were increased up to 0.05 wt % reinforcement and in higher concentration, because of agglomeration of the nanoparticleac the strength was decreased .

Acknowledgement

The financial support from the office of research affairs at University of Sistan and Baluchestan under the project 3/9339 for this work is greatly appreciated.

References

[1] A. Mortensen, J. Llorca, Metal Matrix Composites, Annual Review of Materials Research, 40 (2010) 243-270.
[2] K.K.Chawla. N. Chawla, Metal Matrix Composites, Springer, 2006.

[3] C.B. A. Cetin, A. Mortensen, An analysis of the tensile elongation to failure of laminated metal composites in the presence of strain-rate hardening, Acta Mater., 60 (2012) 2265–2276.
[4] A.Ç.a.A.M. Amaël Cohades, Designing laminated metal composites for tensile ductility, Materials and Design, 66 (2015) 412-420.
[5] M. Rosso, Ceramic and metal matrix composites: Routes and properties, Journal of Materials Processing Technology 175 (2006) 364–375.
[6] J.D. Minyu Fan, Kai Jinc, Liang Qin, Shengqiang Cui, Xunzhong Guo,, J.T.a. Naksoo Kimc, Effect of original layer thicknesses on the interface bonding and mechanical properties of Ti\Al laminate composites, Materials and Design, 90 (2016) 535-542.
[7] A.M.a.F.D. Sophie Ryelandt, Al/stainless-invar composites with tailored anisotropy for thermal management in light weight electronic packaging, Materials and Design, 85 (2015) 318-323.
[8] H.A.H.a.J.J. Lewandowski, Laminated nanostructure composites with improved bend ductility and toughness, Scripta Materialia 61 (2009) 1072–1074.
[9] A.E.a.M.T.S. M. Elmasry, Structure and Mechanical Properties of Aluminum Metal Matrix Composite Produced by Hot Pressing Technique, in: 13th International Conference on aerospace sciences & aviation technology. ASAT, Egypt, 2009.
[10] G.F. Hao Wu a, Bo Cheng Jin b, Cui, Meng Huanga, Fabrication and mechanical properties of TiBw/Ti-Ti(Al) laminated composites, Materials and Design, 89 (2016) 697-702.
[11] M.A. Saba Khoramkhorshid , Amir Hossein Taghvaei , Sergio Scudino Microstructure and mechanical properties of Al-based metal matrix composites reinforced with Al84Gd6Ni7Co3 glassy particles produced by accumulative roll bonding, Materials and Design, 90 (2016) 137–144.
[12] H.S.a.N. Pan, Mechanical characterization of the interfaces in laminated composites, Composite Structures 74 (2006) 25–29, 74 (2006) 25–29.
[13] X.C.a.H.C. Youjing Zhang, Fabrication, characterization and tensile property of a novel Ti2Ni/TiNi micro-laminated composite, Materials and Design, 92 (2016) 486-493.
[14] J.D. Minyu Fan, Kai Jinc, Liang Qin, Shengqiang Cui, Xunzhong Guo, Naksoo Kimc, Jie Tao a, Effect of original layer thicknesses on the interface bonding and mechanical properties of Ti\Al laminate composites, Materials and Design, 90 (2016) 535-542.
[15] D.I.e.a.I. Smojver, Explicit multiscale modeling of impact damage on laminated composites – Part I: Validation of the micromechanical model, Composite Structures, 145 (2016) 248–258
[16] Y.H. Hongqiang Duan, Weijie Lu, Liqiang Wang, Jianwei Mao, Di Zhang, Configuration design and fabrication of laminated titanium matrix composites, Materials and Design, 99 (2016) 219–224.
[17] L.C. Liangjun Li, Shangwu Fan, Xiaoju Gao, YuPeng Xie and Litong Zhang, A novel fabrication approach for impact resistance laminated ceramics, Materials and Design, 79 (2015) 26-31.
[18] C.K.S. D. R. Lesuer, O. D. Sherby, J. Wadsworth, J. J. Lewandowski, and W. H. Hunt, Jr, Mechanical behavior of laminated metal composites, International Materials Reviews, 41 (1996) 169-197.

IOP Conf. Series: Materials Science and Engineering **201** (2017) 012027 doi:10.1088/1757-899X/201/1/012027

Influence of stack effect on the heat feedback to the n-heptane pool fires

Xiangyong Yuan, Zihe Gao, Jie Ji*, Lizhong Yang and Jinhua Sun

State Key Laboratory of Fire Science, University of Science and Technology of China, Hefei, Anhui 230026, P. R. China

Email: jijie232@ustc.edu.cn

Abstract: A set of experiments was conducted in a 1/3 scale building model to study the influence of stack effect on the heat feedback to the n-heptane pool fires. The height of opened window and pool size were varied. Results show that the conduction heat feedback fraction slightly decreases and the radiation heat feedback fraction increases with the increase of the strength of stack effect, while the convection heat feedback fraction remains almost invariant.

1. Introduction

In recent years, a lot of high-rise building fires caused great casualties and property loss and attracted more and more attentions. Once a fire occurs, it is easy to form stack effect in high-rise building. Stack effect is the mechanism of air movement caused by the pressure difference generated from the density difference of hot and cold airs inside and outside of the building, respectively [1]. The stack effect can suck more fresh air into the fire room, and may make small fire become larger quickly [2]. The characteristics of smoke movement, the influence of stack effect on fire behaviors in high-rise buildings and the effect of wind on building fire behaviors and heat feedback to pool fires have been studied previously [1-7]. The characteristics of smoke flow and flame behaviors are related to the combustion condition. The heat feedback is a very important parameter reflecting the combustion conditions. But the evolution of heat feedback under the effect of stack effect has not been studied before. To study the heat feedback to n-heptane pool fires in a compartment of high-rise building, a set of experiments was conducted in a 1/3 scale building model.

2. Experimental

Experiments based on Froude modeling were conducted in a facility consisting of a room, an atria and a vertical stairwell, where more details have been described in [2,4], as shown in Figure 1. The 1/3 scale building model with 12 floors is 12.2 m high. The ground floor is 1.2 m high and the other floors are 1.0 m. Each floor has a single window with a size of 0.7 m (width) × 0.9 m (height) in the stairwell. The horizontal cross-sections of stairwell, atria and room are 1.5 m × 1.0 m, 0.8 m × 0.8 m, 0.8 m × 0.8 m, respectively. Each floor has three doors connecting with the stairwell, atria, room and the outdoor, with the size of 0.4 m (width) × 0.6 m (height).

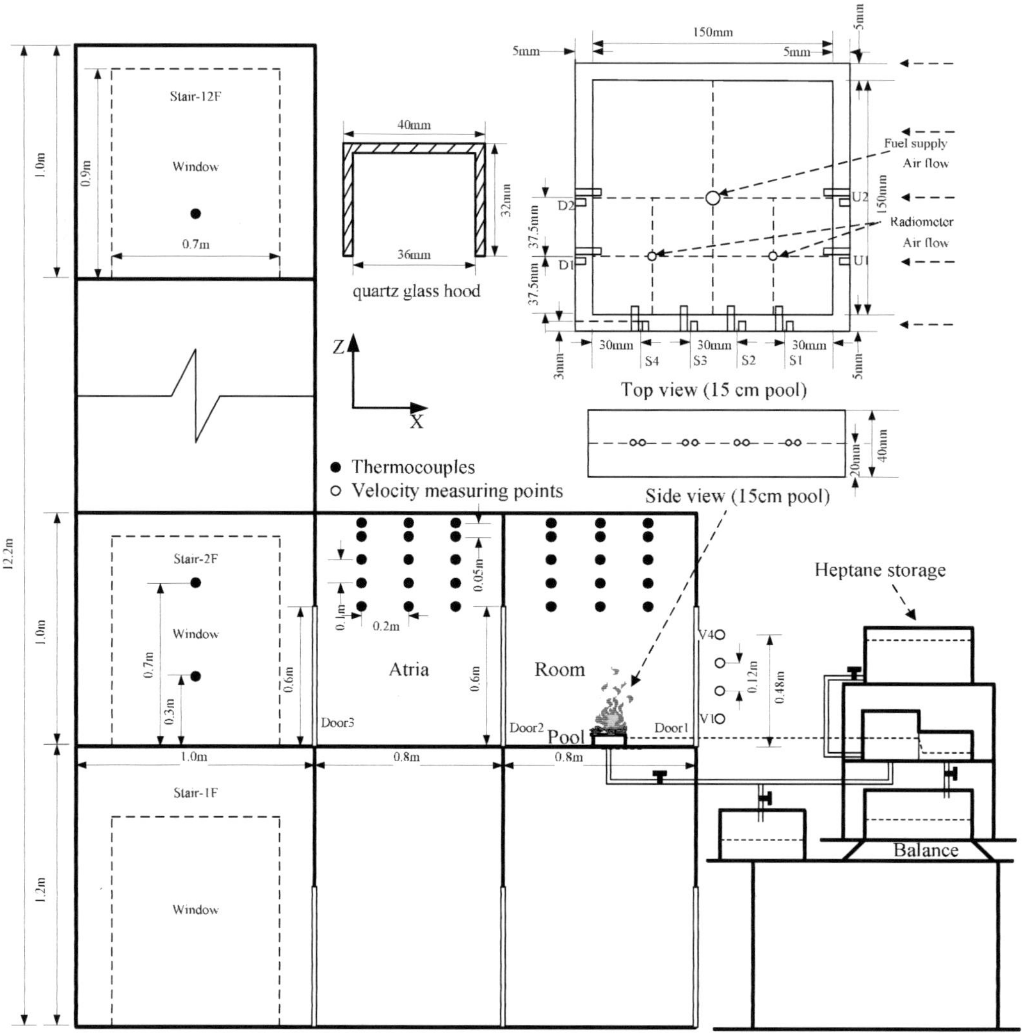

Figure 1. The experimental facility and locations of thermocouples and velocity measuring points

The K-type thermocouples with 1.0 mm in diameter were used in the experiment. Four hot-wire anemometers with a resolution of 0.01 m/s were installed 5 cm away from door 1, as shown in Figure 1. The 10 cm and 15 cm square pools were used with heptane as fuel. The pools are made by 5 mm thick steel plate with inner depth of 4 cm. The fuel surface level is maintained at 3 cm. The fuel was supplied from the bottom of the pool to maintain the height of the fuel surface level. A radiometer was positioned at the bottom of the pool to measure the radiation received by the fuel surface. The radiometer was covered by a quartz glass hood to protect it. The thermocouple with 1 mm diameter for measuring the wall temperature was embedded into a hole and the depth was 3 mm. That for adjacent fuel was 1 cm away from inner wall surface. The positions of the hole supplying fuel, the radiometer and the thermocouples are shown in Figure 1. The experimental facility which maintains the height of the fuel surface level is placed on the first floor, so the fire is placed in the center of the room on the 2nd floor, and the plane connecting the first floor and 2nd floor in the stairwell is blocked by the 8 mm thick fireboard. The fire floor is used as the benchmark and the ground of the fire floor is marked as 0 m in the experiment. The fuel mass was measured using an electronic balance with a resolution of 0.1 g.

During the experiments, door1, door2, door3 were all opened. The ambient temperature was about 293 K. The windows of 4th floor, 6th floor, 8th floor, 10th floor and 12th floor are opened respectively. A digital camera with a frequency of 50 frames per second was used to record the transient flame shapes. Each case was repeated two times to assure its reliability and repeatability.

3. Results and discussion

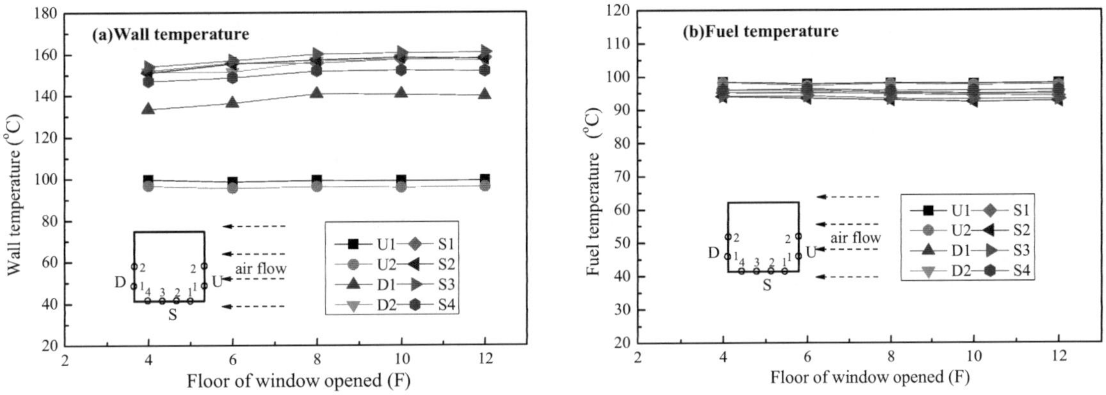

Figure 2. Temperatures of pool wall and adjacent fuel of 15 cm pool

The temperatures of pool wall and fuel of 15 cm pool are shown in Figure 2. It can be found that the upstream side temperature of pool wall remains invariable at about the boiling point of heptane (98.4 °C). The pool wall temperatures of the downstream side and the stream-wise side increase due to its direct heating by the deflected flame [7]. But it does not increase significantly with the increase of the height of opened window. The fuel temperature remains invariable at about the boiling point of heptane (98.4 °C).

The heat flux from the pool side wall to the fuel is calculated by [8]

$$\dot{q}'' = h(T_w - T_l) = \begin{cases} 0.59\dfrac{k_l}{L^{1/4}}\left[\dfrac{g\beta}{\nu\alpha}\right]^{1/4}(T_w - T_l)^{5/4} & 10^4 \le Ra_L \le 10^9 \\ 0.10k_l\left[\dfrac{g\beta}{\nu\alpha}\right]^{1/3}(T_w - T_l)^{4/3} & 10^9 \le Ra_L \le 10^{13} \end{cases} \quad (1)$$

where L is taken here as the constant fuel thickness, 0.03 m. The pool wall conduction heat feedback received by the fuel is calculated by

$$\dot{q}_{cond} = 2\times\dot{q}''_{U_1}Ld/3 + \dot{q}''_{U_2}Ld/3 + 2\times\dot{q}''_{D_1}Ld/3 + \dot{q}''_{D_2}Ld/3 + 2\sum_{i=S_1}^{S_4}\dot{q}''_iLd/4 \quad (2)$$

where d is the pool length, m.

The radiation heat feedback received by the fuel surface is calculated by

$$\dot{q}_{rad} = \sum_{i=1}^{n}\dot{q}''_{rad,i}\cdot s_i \quad (3)$$

The variations of the conduction and radiation heat feedback with the height of opened window are shown in Figure 3. It can be found that both the conduction and radiation heat feedback increase with the height of opened window, but the conduction heat feedback remains almost invariant when the height of opened window is relatively higher.

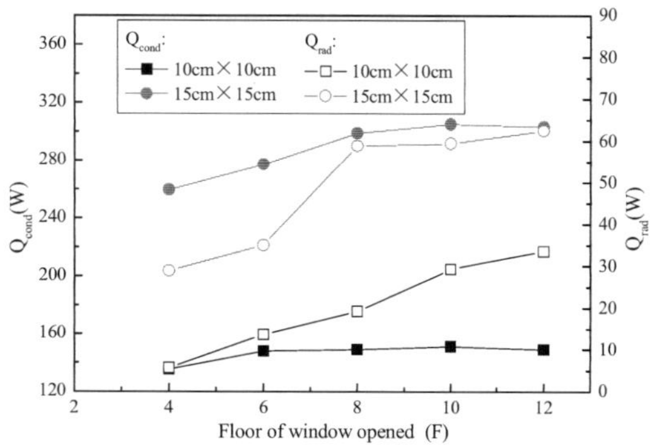

Figure 3. The conduction and radiation heat feedback

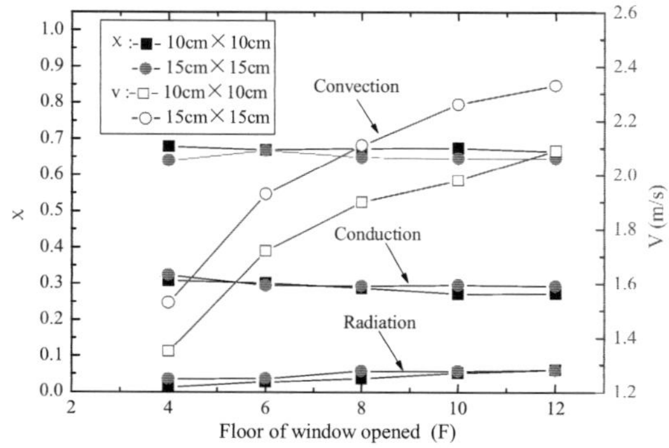

Figure 4. Evolution of heat feedback fractions and corresponding air flow velocity

The heat for the evaporation of liquid heptane can be given by [8]

$$\dot{Q}_{fuel} = \dot{m} \cdot \left[c_p (T_{boil} - T_0) + h_{fg} \right] \tag{4}$$

The wall conduction heat feedback fraction is calculated by $\chi_{cond} = \dot{q}_{cond} / \dot{Q}_{fuel}$. The radiation heat feedback fraction is calculated by $\chi_{rad} = \dot{q}_{rad} / \dot{Q}_{fuel}$. Then we can calculate the convection heat feedback fraction $\chi_{conv} = 1 - (\chi_{cond} + \chi_{rad})$.

The evolution of heat feedback fractions and corresponding air flow velocity with the height of opened window are shown in Figure 4. It can be found that the air flow velocity increases with the height of opened window. The air flow velocity varies from 1.35 m/s to 2.35 m/s. With the height of opened window increasing, the conduction heat feedback fraction decreases from 31.8% to 27.3%, the radiation heat feedback fraction increases from 1.3% to 6.2% and the convection heat feedback fraction remains almost invariant about 67.3% for 10 cm pool size. The conduction heat feedback fraction decreases from 32.4% to 29.4%, the radiation heat feedback fraction increases from 3.6% to 6.1% and the convection heat feedback fraction remains almost invariant about 65.0% for 15 cm pool size. The results of two pool sizes indicate that the convection heat feedback fraction remains almost invariant and it is the dominating one that influences the combustion process most, the conduction heat feedback fraction decreases and the radiation heat feedback fraction increases with the increase of the air flow velocity.

The variation of convection heat feedback fraction with air flow velocity is consisted with the results of Ref. [8] that the convection heat feedback fraction changes little with wind velocity when the Froude number is larger than 1.0. But the variations of conduction heat feedback fraction and the radiation heat feedback fraction with air flow velocity are different from the results of Ref. [8].

The mean flame length and average flame tilt angle are shown in Figure 5. It is found that the mean flame length remains almost invariant for a certain pool size under different opened window heights. That is to say, the projection of the tilting flame shows basically unchanged. But the mass loss rate increases with the height of opened window, as shown in Figure 6. Thus, it can be expected that the width of the flame should increase with the height of opened window and the air flow velocity. And the increase of the flame volume results in the increase of radiation from the flame to the fuel surface and the radiation heat feedback fraction increases. The temperature difference between the pool wall and the fuel has no remarkable increase with different air flow velocities, so the conduction heat feedback fraction decreases.

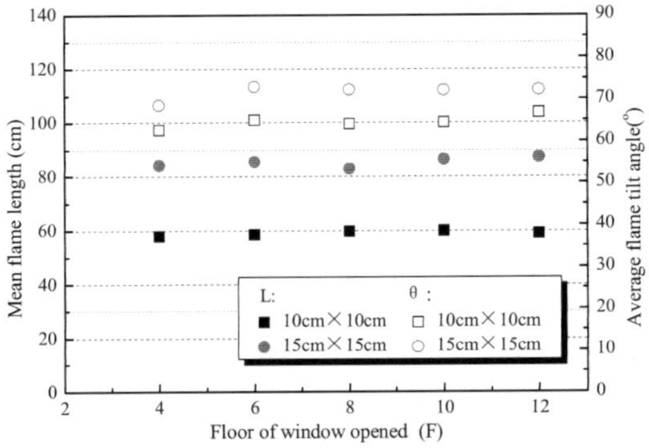

Figure 5. The flame length and flame tilt angle

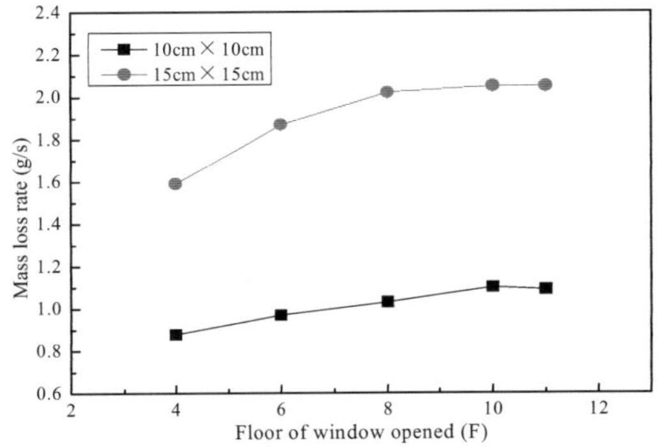

Figure 6. The mass loss rate

4. Conclusions

A set of experiments was conducted in a 1/3 scale building model to study the influence of stack effect on the heat feedback to the n-heptane pool fires. Results show that the convection heat feedback fraction remains almost invariant, the conduction heat feedback fraction slightly decreases and the radiation heat feedback fraction increases with the enhancing of stack effect.

Acknowledgements

This work was supported by the National Key Research and Development Plan under Grant No. 2016YFC0800106 and the Fundamental Research Funds for the Central Universities under Grant No. WK2320000033. Jie Ji was supported by the National Top-Notch Young Talents Program.

References

[1] Barrett R E and Locklin D W 1969 *Fire Technol.* **5** 299-310.
[2] Shi W X, Ji J, Sun J H, Lo S M , Li L J and Yuan X Y 2014 *J. Civ. Eng. Manag.* **20** 121-31.
[3] Zhang J Y, Lu W Z, Huo R and Feng R 2008 *Build. Environ.* **43** (6) 1101-08.
[4] Li L J, Ji J, Fan C G, Sun J H, Yuan X Y and Shi W X 2014 *Energ. Buildings* **68** 108-20.
[5] Ji J, Yuan X Y, Li K Y and Sun J H 2015 *Combust. Flame* **162** 2098-107.
[6] Huang H, Ooka R, Liu N A, Zhang L H, Deng Z H and Kato S 2009 *Fire Saf. J.* **44** 311-21.
[7] Hu L H, Liu S, Xu Y and Li D 2011*Combust. Flame* **158** 586-91.
[8] Hu L H, Hu J J, Liu S, Tang W and Zhang X Z 2015 *Proc. Combust. Inst.* **35** 2511-18.

IOP Conf. Series: Materials Science and Engineering **201** (2017) 012028 doi:10.1088/1757-899X/201/1/012028

Effect of replacing carbon black with surfactant-modified silica clay on mechanical properties of natural rubber composites

N J Pinton and B Pajarito

Polymers Research Laboratory, Department of Chemical Engineering, University of the Philippines-Diliman, Quezon City 1101 Philippines

noeljeffreypinton@gmail.com, bryan.pajarito@gmail.com

Abstract. The effect of replacing carbon black (CB) with surfactant-modified silica clay (SMSC) on hardness and tensile properties of natural rubber (NR) composites was studied using a simplex-lattice design of experiment. SMSC was produced by treating silica clay (SC) with coconut diethanolamine (CDEA) and glycerol monostearate (GMS). Complete replacement of CB with SMSC resulted in the decrease of Shore A hardness and tensile stress at all elongations. At higher CB to SMSC ratio, SMSC modified with CDEA performed better than with GMS. However, at lower CB to SMSC ratio, GMS exhibited larger values in the mechanical properties of NR composites. This can be attributed to the surface morphologies of SMSC and the surface aggregation of surfactant to SC.

1. Introduction

Natural rubber (NR), an elastomer with high molecular weight and viscoelastic properties, has a unique role in rubber technology due to its outstanding high tensile strength and crack growth resistance once vulcanized [1]. Reinforcing fillers are added to NR compounds in order to improve performance of its commercial NR composites. Carbon black (CB) and precipitated silica (PS) are 2 of the most widely used reinforcing fillers for NR composites. However, they have inherent deficiencies [2] and are expensive prompting researchers to search for alternatives. One of the alternatives being studied are mineral fillers, which are being used more in elastomeric compounds [3]. Silica mineral fillers are known to improve important compound properties such as tensile strength and cut growth resistance in NR compounds [4]. High grade silica powder from Siruma Minerals Corporation (Naga City, Philippines), known locally as silica clay (SC), has high silica content (see table 1) making it a potential alternative filler for NR composites.

Table 1. Composition of silica clay

Composition	Content (%)
Silica, SiO_2	98.03
Alumina, Al_2O_3	0.95
Ferric Oxide, Fe_2O_3	0.03
Magnesia, MgO	-
Lime, CaO	0.03
Soda, Na_2O	0.11
Potash, K_2O	0.38

Utilization of SC in NR composites could lead to new marketable products using environment-friendly and low-cost reinforcing materials. However, the surface chemistry and morphology of SC and NR are not compatible. The dipole induced interaction between polar groups on the surface of silica with non-polar groups of hydrocarbon elastomers such as NR are weak compared to dipole-dipole interactions resulting from hydrogen bonding between silanol groups in silica aggregates [5] resulting to poor dispersion of SC in NR matrix.

This study investigates the effect of replacing CB with surfactant-modified silica clay (SMSC) on hardness and tensile properties of NR composites. Coconut diethanolamine (CDEA) and glycerol monostearate (GMS) are used in SMSC synthesis, which are expected to improve dispersion and interface adhesion of SC in NR. Curing characteristics, Shore A hardness and tensile stress at specified elongations (100, 200, and 300%) of NR composites are measured and modeled as function of CB and SMSC loading. The effect of replacing CB with SMSC are discussed from the derived models.

2. Experimental

SC (Siruma Minerals Corporation) was used as received. SMSC was produced by treating SC with CDEA (Fil-Amide 182) and GMS (GMS SE 100), both provided by Chemrez Technologies Inc. (Quezon City, Philippines), at 3 wt.% of SC [6]. Detailed procedure on synthesis of SMSC is found in [7]. SMSC using CDEA and GMS at 3 wt.% of SC (CDEA3-SC and GMS3-SC) were compounded with CB (N330) as mixture filler of NR composites at 56.3 parts per hundred parts (phr) of NR. Table 2 shows the formulations of NR composites containing CB, CDEA3-SC, and GMS3-SC varied at different proportions. Table 2 was derived from a simplex lattice mixture design of experiment (degree of lattice = 3) augmented with central and axial formulations [8]. SC, CDEA3-SC, and GMS3-SC were characterized by attenuated total reflectance Fourier transform infrared spectroscopy (ATR-FTIR, Thermo Scientific Nicolet 6700) and scanning electron microscopy (SEM, Hitachi S-3400N). A modified recipe of camel back was used in compounding of NR composites [9]. The modified recipe is 100 phr NR (Standard Philippine Rubber SPR 5L), 5 phr zinc oxide, 6 phr process oil, 0.65 phr N-tert-butyl-2-benzothiazyl sulfenamide, 2.25 phr sulfur, 1.25 phr antioxidant (Naugard Q), 3 phr stearic acid and 56.3 phr mixture filler. NR composites were prepared by compounding rubber and other ingredients in a laboratory two-roll mill. The NR composites were vulcanized in a compression moulding press at 160 °C for 20 min by Rhodeco Rubber Processing Services (Caloocan City, Philippines).

Table 2a. Formulations of NR composites.

Filler (phr)	Formulations							
	1	2	3	4	5	6	7	8
CB (N330)	56.3	37.5	37.5	18.8	18.76	18.8	0.00	0.00
CDEA3-SC	0.00	18.8	0.00	37.5	18.76	0.00	56.3	37.5
GMS3-SC	0.00	0.00	18.8	0.00	18.76	37.5	0.00	18.8

Table 2b. Formulations of NR composites.

Filler (phr)	Formulations				
	9	10	11	12	13
CB (N330)	0.00	0.00	37.5	9.4	9.4
CDEA3-SC	18.8	0.00	9.4	37.5	9.4
GMS3-SC	37.5	56.3	9.4	9.4	37.5

The Shore A hardness of NR composites was measured using a durometer (Kori Type KA) with 5 replicates for each formulation. Tensile testing was performed on dogbone specimens using a universal testing machine (Shimadzu AGS-5kNX) subjected to a crosshead speed of 500 mm min^{-1} with limited stroke of 200 mm. Tensile stress at 100, 200, and 300% elongations were measured in triplicates for each formulation. Experimental results were first fitted to a full cubic polynomial [8]:

$$y=\sum_{i=1}^{3}\beta_i x_i+\sum\sum_{i<j}^{3}\beta_{ij}x_i x_j+\sum\sum_{i<j}^{3}\delta_{ij}x_i x_j(x_i-x_j)+\sum\sum\sum_{i<j<k}^{3}\beta_{ijk}x_i x_j x_k \quad (1)$$

where y = hardness or tensile stress, β = polynomial coefficients, and x = proportion of filler. β_i, β_{ij}, δ_{ij}, and β_{ijk} are the main, quadratic, full cubic, and special cubic model terms. Subscripts i, j and k indicate CB, CDEA3-SC, and GMS3-SC. Models were improved by removing terms that were statistically insignificant ($p > 0.05$) from analysis of variance (ANOVA). Model terms were removed such that the prediction sum of squares (PRESS) was decreased while values of adjusted and predicted r^2 were increased. Statistical data analysis was performed using Minitab 17 [6].

3. Results and Discussion

Figure 1a shows the ATR-FTIR spectra of SC, CDEA3-SC, and GMS3-SC. The strong peak at 2916 cm^{-1} for C-H stretching, moderate peak at 1170 cm^{-1} for C-O stretching, and another moderate and broad peak at 3400 - 3600 cm^{-1} range for O-H stretching of CDEA3-SC and GMS3-SC indicate the presence of CDEA and GMS. A medium peak at 1620 cm^{-1} represents one R-C(=O)-NR_2 (amide) configuration of CDEA3-SC, while a medium peak at 1728 cm^{-1} constitutes C=O stretching of GMS3-SC. A medium peak at 3690 cm^{-1} representing O-H stretching and the strong sharp peaks at 900 cm^{-1} and 1010 cm^{-1} bands corresponding to Si-O stretching are present in SC [10]. Figure 1b shows the SEM image of SC. After surfactant modification, CDEA3-SC particles (see figure 1c) exhibited a fractal-like structure while no significant change in surface morphology is observed in GMS3-SC particles (see figure 1d).

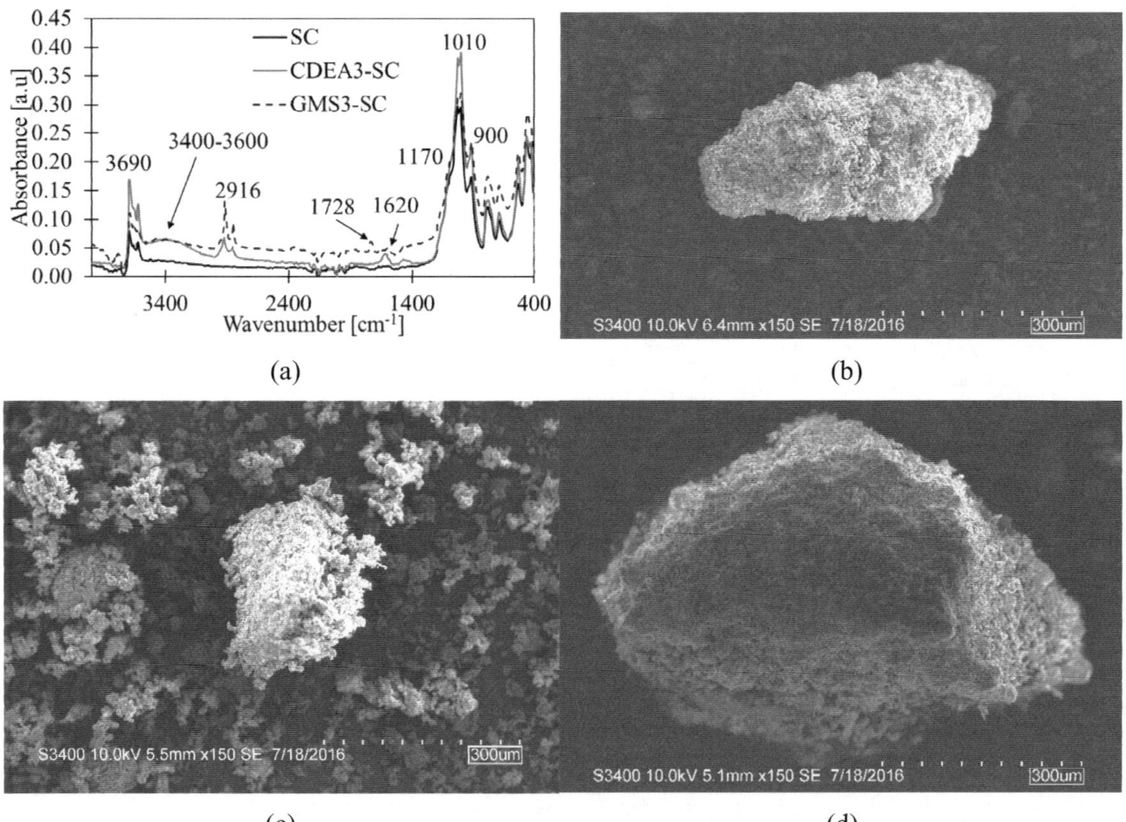

Figure 1. (a) ATR-FTIR spectra of SC, CDEA3- SC, and GMS3- SC; SEM image of (b) SC, (c) CDEA3-SC and (d) GMS3-SC.

The polynomial coefficients of Shore A hardness and tensile stress at 100%, 200% and 300% elongations derived from the results are shown in table 3. Predicted r^2 ranges from 0.9107-0.9546 and adjusted r^2 from 0.8980-0.9461, indicating the models have high correlation with experimental results. The positive values of all β_i indicate that the fillers have linear additive effect. Increasing the CB/CDEA3-SC and CDEA3-SC/GMS3-SC blends reduces the hardness of NR composites ($\beta_{CB,CDEA3\text{-}SC} < 0$, $\beta_{CDEA3\text{-}SC,GMS3\text{-}SC} < 0$). Having all the 3 fillers in the formulation will decrease hardness ($\beta_{CB,CDEA3\text{-}SC,GMS3\text{-}SC} < 0$). Increasing the CB/GMS3-SC blend reduces the tensile stress of NR composites but does not bear significance on hardness ($\beta_{CB,GMS3\text{-}SC} < 0$). Increasing the proportion of CB (CB > CDEA3-SC) in CB/CDEA3-SC blends increase all responses ($\delta_{CB,CDEA3\text{-}SC} > 0$) except at stress at 300% elongation.

Table 3. Polynomial coefficients of hardness and tensile stress.

Coefficients	Responses			
	Shore A Hardness	Stress at 100% elongation (MPa)	Stress at 200% elongation (MPa)	Stress at 300% elongation (MPa)
β_{CB} (response · phr^{-1})	1.128080	0.055978	0.119324	0.191691
$\beta_{CDEA3\text{-}SC}$ (response · phr^{-1})	0.849941	0.017123	0.025177	0.032958
$\beta_{GMS3\text{-}SC}$ (response · phr^{-1})	0.811883	0.018975	0.029664	0.044729
$\beta_{CB,CDEA3\text{-}SC}$ (response · phr^{-2})	-0.002128	-	-	-
$\beta_{CB,GMS3\text{-}SC}$ (response · phr^{-2})	-	-0.000579	-0.001201	-0.001959
$\beta_{CDEA3\text{-}SC,GMS3\text{-}SC}$ (response · phr^{-2})	-0.001913	-	-	-
$\beta_{CB,CDEA3\text{-}SC,GMS3\text{-}SC}$ (response · phr^{-3})	-0.000360	-	-	-
$\delta_{CB,CDEA3\text{-}SC}$ (response · phr^{-3})	0.000154	0.000014	0.000028	-
$\delta_{CB,GMS3\text{-}SC}$ (response · phr^{-3})	0.000092	-	-	-
$\delta_{CDEA3\text{-}SC,GMS3\text{-}SC}$ (response · phr^{-3})	-0.000062	-	-	-

Figure 2 shows the contour plots of Shore A hardness and tensile stress at 100% elongation of NR composites as function of CB, CDEA3-SC and GMS3-SC proportions. Contour plots of tensile stress at 200% and 300% elongations are almost the same with figure 2b. Hardness and tensile stress of NR composites are increased as proportion of CB is increased. CB forms highly branched aggregates network structure in the NR matrix, resulting to increased surface area and good reinforcement filler [11,12]. Meanwhile, SC has irregular shape and size (see figure 1b) resulting to inability to support stress transferred from NR matrix [13,14]. In figure 2, it can be seen that at higher CB to SMSC ratio, the mechanical properties of CDEA3-SC have higher values than GMS3-SC. This can be attributed to the surface morphology of CDEA3-SC having a surface fractal structure (see figure 1d) which usually performs better in reinforcing due to its higher surface area than fillers with smooth surfaces [11]. Thus at lower concentrations of pure SMSC in CB/SMSC blend: CDEA3-SC > GMS3-SC. However, at much lower CB to SMSC ratio (very high SMSC loading) in the NR composite, GMS3-SC performs better than CDEA3-SC. It is the nature of non-ionic surfactants to form surface aggregates at small surface coverage in the material [15]. Coupled with amide linkages making attractive interaction in the adsorbed layer [16,17], CDEA in much higher concentration on the NR composite may have formed aggregates on the surface of SC during modification, resulting to poor dispersion of CDEA3-SC in NR composites compared to GMS3-SC. For comparison, NR composites filled with untreated SC, SMSC with 10 wt.% GMS (GMS10-SC) and SMSC with 15 wt.% CDEA (CDEA15-SC) were also made. The observed hardness of NR composites is as follows: CB (64.0) > CDEA3-SC (47.8) > GMS3-SC (46.0) > SC (45.4) > GMS10-SC (44.8) > CDEA15-SC (38.4). Tukey's test shows that hardness of CDEA15-SC is significantly lower to other fillers while CB is significantly higher than all other fillers. CDEA3-SC GMS3-SC and SC have insignificant difference for hardness of NR composite. For tensile stress at different elongations, the observed trend is (i.e. 100% elongation): CB (3.27 MPa) > SC (1.30 MPa) > CDEA3-SC (1.12 MPa) > GMS3-SC (1.05 MPa) > GMS10-SC (0.97 MPa) > CDEA15-SC (0.73). Tukey's test indicates that all the fillers have insignificant differences except for CDEA15-SC where it is significantly lower to all other fillers while CB is significantly higher than all other fillers.

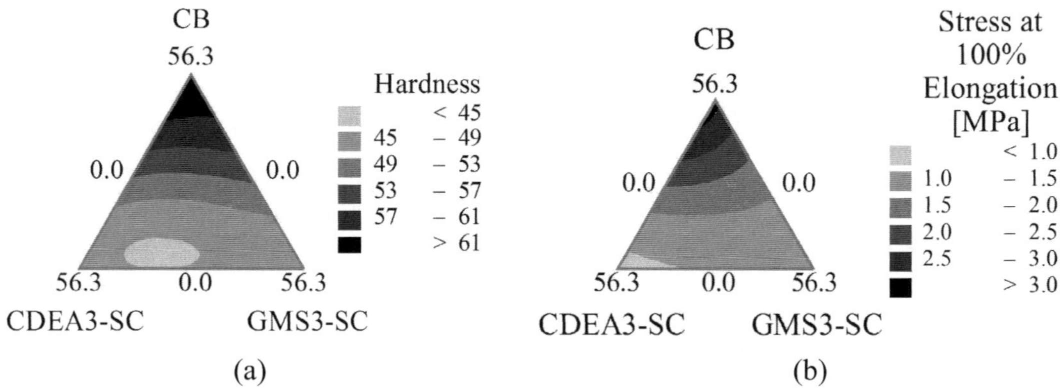

Figure 2. Contour plots of (a) Shore A hardness, (b) tensile stress at 100% elongation.

4. Conclusion

Substituting CB with SMSC generally decreases the mechanical properties of NR composites. Both SMSC fillers have similar performance in hardening the NR composites. At higher CB to SMSC ratio (lower SMSC loading), CDEA3-SC performs better than GMS3-SC which can be attributed to its surface fractal structure. However, at lower CB to SMSC ratio (higher SMSC loading), GMS3-SC is more effective than CDEA3-SC. This can be attributed to the aggregation of CDEA3-SC in the NR composite due to surface aggregation of surfactant in SC resulting to poor dispersion in NR.

Acknowledgement

The authors would like to acknowledge the support of Department of Science and Technology (DOST) through Engineering Research and Development for Technology (ERDT) and Philippine Council for Industry, Energy and Emerging Technology Research and Development (PCIEERD). The authors are also grateful to Rhodeco Rubber Processing Services on this research work.

References

[1] Tohsan, A, and Ikeda, Y 2014 *Chemistry, Manufacture and Applications of Natural Rubber*, pg. 168
[2] Vijay V R, Anitha A M, and Ravindranatha A R 2016 *Polymer Vol.* **89**, 135
[3] Whelan A 1981 *Developments in rubber technology*. (London: Applied Science Publ) pg. 111
[4] Nontasorn P, Chavadej S, Rangsunvigit P, O'Haver J H, Chaisirimahamorakot S and Na-Ranong N 2005 *Chemical Engineering Journal*, **108(3)**, 213-18
[5] Dick J S and Annicelli R A 2001 *Rubber technology: Compounding and testing for perfomance.* (Munich: Hanser) 331
[6] Pinton N J and Pajarito B 2016 *Materials Science Forum* – submitted to ICMSET 2016
[7] Santiago D E, Pajarito B, Mangaccat W F, Tigue M R, and Tipton M 2016 *American Institute of Physics* (2016) **1736**, 020047.
[8] Cornell J A 2002 *Experiments with Mixtures: Designs, Models, and the Analysis of Mixture Data* (John Wiley & Sons, New York). Eq 2.9 pg. 22, 27
[9] Chandrasekaran VC 2007 *Essential Rubber Formulary: Formulas for Practitioners* (William Andrew Publishing, New York), 68
[10] Stuart B 2004 *Infrared Spectroscopy: Fundamental and Applications* (John Wiley & Sons, Ltd.) pg. 28, 109
[11] Kohls D J and Beaucage G 2002 *Current Opinion in Solid State and Materials Science* **6** pg. 183
[12] Witten T, Rubenstein M and Colby R 1993 *J Phys II* France **3**:367
[13] Ismail H, Osman H, and Ariffin A 2005 *Polymer-Plastics Technology and Engineering* **43** pg.

1323–44
[14] Jamil M S, Ahmad I and Abdullah I 2006 *Journal of Polymer Research* **13**: 315-21
[15] Tiberg F 1996 *J. Chem. Soc., Faraday Trans* **92** p. 531
[16] Johansson I 2001 *Kirk-Othmer Encyclopedia of Chemical Technology; Amides, Fatty Acid* (NY, NY: John Wiley & Sons)
[17] Choi S S, Nah C and Jo B W 2003 *Polym Int* **52**:1382–9

IOP Conf. Series: Materials Science and Engineering **201** (2017) 012029 doi:10.1088/1757-899X/201/1/012029

Investigation of the electrical performance of hfo$_2$ dielectrics deposited on passivated germanium substrates

Q Lu[1], Y Mu[1], Y Zhao[1], C Z Zhao[2, a)], S Taylor[1] and P R Chalker[3]

[1] Department of Electrical Engineering and Electronics, University of Liverpool, Liverpool L69 3GJ, UK;
[2] Department of Electrical and Electronic Engineering, Xi'an Jiaotong-Liverpool University, Suzhou 215123, China;
[3] Center for Materials and Structures, School of Engineering, University of Liverpool, Liverpool L69 3GH, UK;
[a)] Corresponding author: cezhou.zhao@xjtlu.edu.cn

Abstract. Propanethiol solution (0.1 M) in 2-propanol, octanethiol solution (0.1 M) in 2-propanol and 20% $(NH_4)_2S$ solution in water were used to passivate the germanium substrates. HfO_2 thin films of 150 ALD cycles were then deposited on the passivated germanium substrates. The morphology of the thin films was investigated by X-ray diffraction and it was found that the morphology of the thin films was not affected by the chemical treatments. A lower leakage current density was observed in the passivated samples compared with the witness one. In addition, the interface quality and long-time stress reliability of the passivated samples were improved when the samples were annealed in forming gas ambient.

1. Introduction

Recently, germanium has emerged as a promising candidate to channel materials for high-speed metal-oxide-semiconductor (MOS) devices, mainly due to its high carrier mobility [1, 2]. However, the hygroscopicity and thermal instability of the native oxide, leading to poor interface quality, would be a potential threat for Ge MOS devices. For the deposition of high quality gate stacks on Ge substrate, a number of methods have been employed to passivate Ge surface, such as nitridation, Al_2O_3 and sulphur passivation [3-5]. In particular, previous research outputs have reported that the introduction of sulfur in the GeO_x can result in superior Ge gate stack [6]. In addition, the forming gas annealing also had an influence on the electrical performance of the MOS devices, such as the suppression of frequency dispersion, reduction of fixed charges and *et al* [7, 8]. However, the direct comparison of the effect of passivation techniques, including the annealing effect, on the electrical performance was seldom reported. Therefore, in this research, high-*k* dielectrics (HfO_2) were deposited on the germanium substrates with three different passivation methods. Then, the MOS capacitors were fabricated and post-metal annealing (PMA) were performed to improve the electrical characteristics and reliability of the devices. The measurement results and the performance of the thin films deposited on Ge substrates were presented and discussed in this paper.

2. Experiment

N-type Ge (100) wafers were cleaned ultrasonically in acetone and ethanol. The native oxide on the surface was removed by diluted 2% HF. Before the deposition of the HfO_2 thin films using ALD, wet chemical treatments were used to passivate the substrates. Two pieces of germanium substrate were immersed in propanethiol solution (0.1 M in 2-propanol) and octanethiol solution (0.1 M in 2-propanol) for 24 hours to form the passivation layer and labeled as 'PT' and 'OT', respectively. Another substrate was dipped in a 20% ammonium sulfide solution in water at 70 ℃ for 15 mins (labelled as 'NS'). The sample without any treatment, labelled as 'WT', was the witness sample. After the chemical treatments, the samples were dried in nitrogen and immediately transferred into the ALD reactor, where 150 ALD cycles were used to deposit the HfO_2 thin films using 120 ℃ $(NMe_2N)_4Hf$ and 50 ℃ DI water. X-ray diffraction (XRD) was carried out to check the morphology of thin films. The aluminium electrode contacts and back contact were deposited by E-beam evaporation. Aglient 4284A precision LCR meter and Agilent B1500A semiconductor analyzer were employed to investigate the electrical properties of the devices with a Faraday Cage surrounding the prober station.

3. Results and Discussions

Figure 1 showed the XRD patterns for the four samples with different passivation methods. For all the samples, no noticeable diffraction peaks were observed, except for the one coming from the substrate centered at around 31.5°. It indicated that all the thin films remained amorphous under these deposition conditions and the morphology of the dielectrics was not influenced by the chemical pretreatments. A relatively smaller leakage current levels were obtained attributed to the amorphous phase of the thin films, which was discussed later [9].

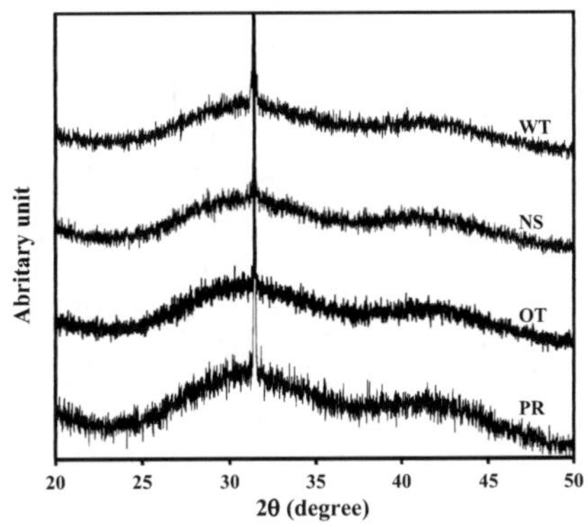

Figure 1. XRD patterns of the samples with different passivation methods.

Figure 2 (a) elucidated the high frequency (1 MHz) capacitance-voltage (CV) characteristics of the four samples with different passivation methods and the corresponding normalized CV curves were presented in **Figure 2** (b). The loop width, or hysteresis, was indicated in the inset of **Figure 2** (b). Although all the samples shared the almost same thickness (about 20 nm), there was a large difference in capacitance. The OT sample had the largest capacitance followed by PR and NS ones. The witness sample had the smallest capacitance. The increase of the capacitance for the passivated samples may partially benefit from a thinner GeO_x layer after chemical treatment. Also, for the samples passivated by organic solution (PR and OT), alkyl chains (or subsequent reaction products) may also enhance the dielectric constant [10]. In addition, The samples with pretreatment had smaller loop width, which was contributed by less interfacial slow traps and/or mobile charges in the oxides [11], especially for the PR and OT ones. In other words, the passivation by organic solution would reduce the number of traps.

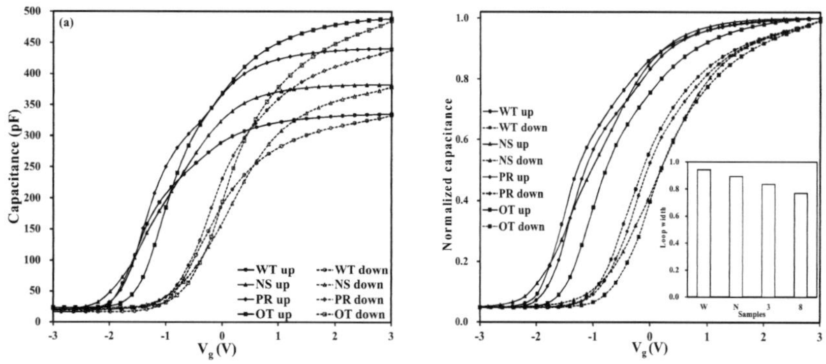

Figure 2. (a) The CV and (b) normalized CV characteristics extracted from the four samples.

Figure 3 showed the leakage current densities of samples treated by different passivation methods. A dramatic decrease of leakage current was observed for the passivated samples. The NS sample had the smallest leakage current density followed by OT and PR ones. In particular, the leakage current of NS samples was one order smaller than that of WT sample. The improvement of leakage current may be due to the suppression of the $HfGeO_x$ formation due to the passivation of the germanium surface [12, 13]. As previous research reported that an increase of leakage current was caused by the formation of $HfGeO_x$ at the interface between HfO_2 and Ge [2]. With regard to a little increase of leakage current density for the PR and OT sample compared with the NS sample, it may be due to the difference in monolayer coverage for the passivation techniques [14, 15].

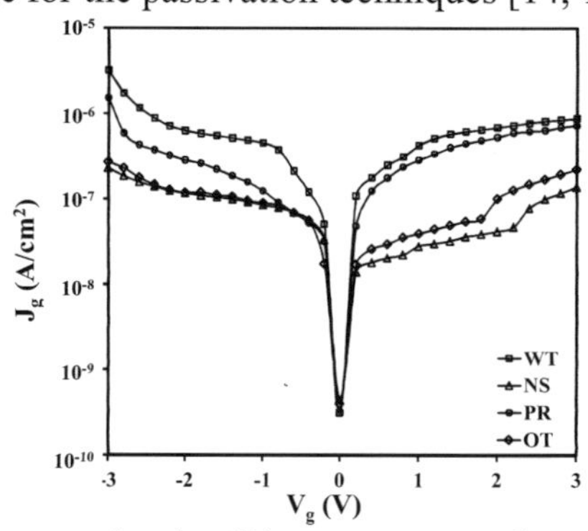

Figure 3. Gate leakage current density (J_g) versus gate voltage (V_g) of the four samples.

Although an improvement in the leakage current was obtained from the passivated samples, a large shift of the CV traces was observed when a long-time bias stress voltage was applied. **Figure 4** (a) illustrated the shift of CV characteristics after a stress time of 21000 seconds (either negative or positive stress) for the NS sample (Other samples had the similarly trend and was not presented here). The shift of the CV curves (ΔV_g) measured from 50% of the maximum capacitance was illustrated in **Figure 4** (b). The change of ΔV_g under positive (negative) gate bias were an indicative of charge trapping in the gate insulator. With the increase of stress time, a CV slope degradation became noticeable, possibly attributed to a secondary instability such as field induced interface state creation [16]. Therefore, the increase of ΔV_g was probably caused by the increase of oxide trap, interface states and border trap density as reported by other research [17]. With regard to a larger change of ΔV_g for the passivation samples compared with the witness sample, it might be related with the stronger Ge–O bond than the Ge–S (or Ge-S(C_nH_{2n+}1)) [18]. Fortunately, after FG annealing, it was observed that there was a great reduction of the ΔV_g shown in **Figure 5** (b), reducing from 1V to 0.25 V, further indicating the high interface quality and fewer traps in the oxides achieved by PMA [19].

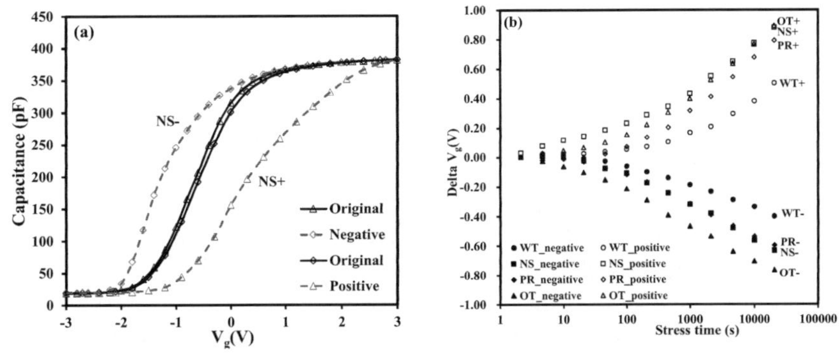

Figure 4 (a) The forward CV curves under positive/negative stress of 21000 seconds for as-deposited NS samples. (b) The gate voltage shift for all the samples with stress time.

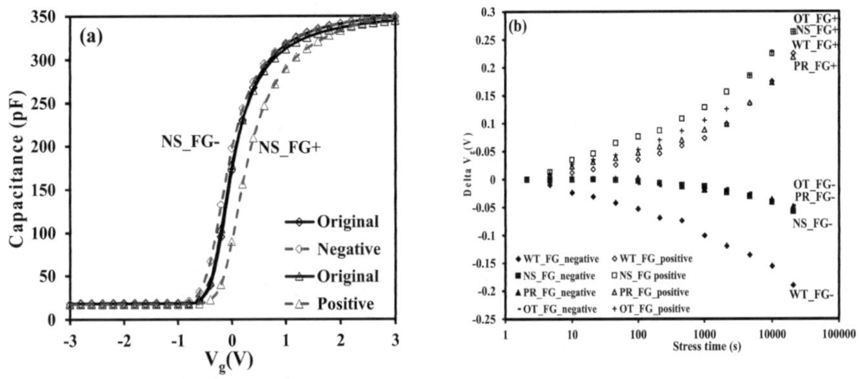

Figure 5 (a) The forward CV curves under positive/negative stress of 21000 seconds for NS after FG annealing (labelled as NS_FG). (b) The gate voltage shift for all the samples with stress time.

Besides the improvement in long-time stress reliability, the loop width was also reduced after FG annealing. **Figure 6** (a) showed the shift of CV traces for the NS samples and changes of the loop width measured from the CV traces were summarized in **Figure 6** (b). It was clear that the loop width of the as-deposited ones was about two time than that of FG samples (0.5 V) for all the samples. This results indicated that a great number of traps can be suppressed by FG treatment. Another point to be mentioned, the positive shift of the CV curves in **Figure 6** (a) after FG annealing was attributed to the decrease of the fixed positive charges as discussed in **Figure 4**. It indicated that the appropriate PMA of films made it possible to recover the negative shift of CV traces without increasing hysteresis as reported by other research output [20].

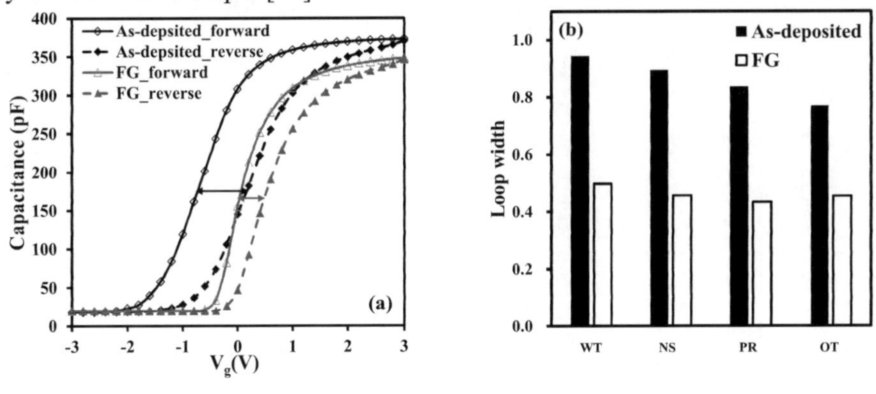

Figure 6. (a) Change of the loop width for the NS samples and changes of the loop width for all the samples were summarized in (b).

4. Conclusion

In this research, the n-type germanium wafers were passivated by propanethiol, octanethiol and $(NH_4)_2S$ solution. After the formation of passivation layer, HfO_2 thin films were deposited on the passivated substrates by atomic layer deposition. The amorphous structure of all the thin films were observed, which implied that the morphology were not affected by the chemical treatments. However, the leakage current density decreased dramatically for the passivated samples. The leakage current of NS samples was one order smaller than that of WT sample. In addition, the interface quality and long-time stress reliability was improved when the samples were annealed in forming gas ambient.

5. Acknowledgments

This research was funded by the National Natural and Science Foundation of China under the grant no. 11375146 and the platform promotion for Suzhou Municipal Key Lab "New Energy and Environmental Protection Techniques".

Reference

[1] Caymax M, Eneman G, Bellenger F, Merckling C, Delabie A, Wang G, Loo R, Simoen E, Mitard J and De Jaeger B 2009 2009 IEEE International Electron Devices Meeting 1-4
[2] Kamata Y 2008 *Materials today* **11** 30-38
[3] Xie Q, Deng S, Schaekers M, Lin D, Caymax M, Delabie A, Qu X-P, Jiang Y-L, Deduytsche D and Detavernier C 2012 *Semicond. Sci. Tech.* **27** 074012
[4] Wu N, Zhang Q C, Balasubramanian N, Chan D S H and Zhu C X 2007 *IEEE Tran. Electron Dev.* **54** 733-741
[5] Chen J J H, Bojarczuk N A, Shang H L, Copel M, Hannon J B, Karasinski J, Preisler E, Banerjee S K and Guha S 2004 *IEEE Tran. Electron Dev.* **51** 1441-1447
[6] Sioncke S, Ceuppens J, Lin D, Nyns L, Delabie A, Struyf H, De Gendt S, Müller M, Beckhoff B and Caymax M 2011 *Microelectron. Eng.* **88** 1553-1556
[7] Xie R and Zhu C 2007 *IEEE Electron Dev. Letters* **28** 976-979
[8] Deng S, Xie Q, Deduytsche D, Schaekers M, Lin D, Caymax M, Delabie A, Van den Berghe S, Qu X and Detavernier C 2011 *Appl. Phys. Lett.* **99** 052906
[9] Lu Q, Zhao C, Mu Y, Zhao C Z, Taylor S and Chalker P R 2015 *Materials* **8** 4829-4842
[10] Rao Y and Wong C 2004 *J. App. Poly. Sci.* **92** 2228-2231
[11] Edon V, Hugon M, Agius B, Miotti L, Radtke C, Tatsch F, Ganem J-J, Trimaille I and Baumvol I 2006 *Appl. Phys. A* **83** 289-293
[12] Moshe H and Mastai Y 2013 *Materials Science-Advanced Topics* chapter 3 pp *64-84*
[13] Hanrath T and Korgel B A 2004 *J. Am. Chem. Soc.* **126** 15466-15472
[14] Bashouti M Y, Stelzner T, Berger A, Christiansen S and Haick H 2008 *J. Phy. Chem. C* **112** 19168-19172
[15] Kosuri M R, Cone R, Li Q, Han S M, Bunker B C and Mayer T M 2004 *Langmuir* **20** 835-840
[16] Libsch F and Kanicki J 1993 *Appl. Phys. Lett.* **62** 1286-1288
[17] Warren W, Fleetwood D, Shaneyfelt M, Schwank J, Winokur P, Devine R and Mathiot D 1994 *Appl. Phys. Lett.* **64** 3452-3454
[18] Li C, Leung C, Lai P and Xu J 2010 *Solid-State Electronics* **54** 675-679
[19] Kawakami D, Yasutake Y, Nishizawa H and Majima Y 2006 *Japan. J. App. Phys.* **45** L1127
[20] Kuriyama A, Ohmi S-i, Tsutsui K and Iwai H 2005 *Japan. J. App. Phys.* **44** 1045

IOP Conf. Series: Materials Science and Engineering **201** (2017) 012030 doi:10.1088/1757-899X/201/1/012030

Formation of $Ni(OH)_2$ hybrid structures on Carbon Cloth

L A Dahonog[1], J D Ocon[2] and M D L Balela[1]

[1]Sustainable Electronic Materials Group, Department of Mining, Metallurgical and Materials Engineering, University of the Philippines Diliman, Quezon City, Philippines 1101
[2]Laboratory of Electrochemical Engineering, Department of Chemical Engineering, University of the Philippines Diliman, Quezon City, Philippines 1101

E-mail: ladahonog@up.edu.ph

Abstract. Nickel hydroxide [$Ni(OH)_2$] structures were successfully grown on carbon cloth via hydrothermal treatment followed by annealing. The $Ni(OH)_2$ structures grown on carbon cloth were characterized using X-Ray Diffraction (XRD) and Scanning Electron Microscope (SEM) analysis. XRD analysis revealed the formation of α and β phases of $Ni(OH)_2$. Microflowers and interconnected grass-like particles were observed on the surface of the carbon cloth. The as-prepared sample could be a promising material for the fabrication of high energy storage devices because of its unique structures.

1. Introduction

Fabrication of low cost, flexible and environment friendly energy storage has been studied for their applications in portable electronic devices, memory devices and electric vehicles. Of all the energy storing devices, supercapacitors have attracted great interest due to their high power density, long cycle life and fast charge/discharge time [1, 2]. Supercapacitors can either store energies via ion adsorption in electrical double layer capacitors (EDLCs) or through surface redox reactions in pseudocapacitors [3]. Transition metal oxides and hydroxides are commonly used electrode materials for pseudocapacitor applications. Among those, ruthenium-based materials are considered the most promising material because of its high specific capacitance [4]. Being rare and toxic, the use of this material for supercapacitor applications is limited. With this, other materials are being studied as alternative electrode for pseudocapacitors.

Nickel hydroxide [$Ni(OH)_2$], having a high theoretical specific capacitance of 2082 F/g, is considered a potential electrode because of its low cost, availability and stability in alkaline electrolytes [5, 6]. Various methods have been employed for the synthesis of $Ni(OH)_2$ such as microwave method [7] and wet chemical method [8] to produce particles of different morphologies. To date, flower-shaped microstructures have been favorable because of high diffusion rate of electrolyte ions during the redox process [9]. However, the electrochemical property of $Ni(OH)_2$ is also limited due to its low conductivity. An effective way to enhance its property is to incorporate the $Ni(OH)_2$ material within carbon-based frameworks to construct a hybrid electrode [5]. Materials grown directly on conductive substrates have also been the resolution to produce a binder-free electrode to avoid the conventional slurry-type method [6]. This method also allows efficient ion transport for electrochemical applications.

IOP Conf. Series: Materials Science and Engineering **201** (2017) 012030 doi:10.1088/1757-899X/201/1/012030

In this work, hybrid structures of $Ni(OH)_2$ were grown on carbon cloth using hydrothermal treatment for 12 h and annealing at 250 °C.

2. Methodology

2.1 Materials

Nickel (II) nitrate ($Ni(NO_3)_2$, RTC Laboratory Services), cetyl trimethylammonium bromide (CTAB, UNILAB) and methanol (CH_3OH, 99% AR, RCL Labscan) were all reagent grade and used as received. Carbon cloths were used as the substrate.

2.2 Hydrothermal Treatment

Carbon cloths (1 cm x 1 cm) were washed with ethanol and water to remove impurities. 2 g of $Ni(NO_3)_2$, 1 g of CTAB and 6 mL of deionized water were mixed in 30 mL methanol. Carbon cloths were immersed in the solution after mixing. The total solution was then transferred in a Teflon-lined steel autoclave reactor and was heated at 180 °C for 12 h. After the hydrothermal treatment, substrates were removed and were washed using ethanol and water followed by drying at 120°C for 8 h. The prepared sample was calcined in air at 250 °C for 4 h.

2.3 Characterizations

The morphology of the prepared samples was observed using scanning electron microscope (SEM, JEOL Ltd., JSM-5310). The phase composition and crystalline structure were determined using X-Ray diffraction (XRD, SHIMADZU MAXima XRD-7000).

3. Results and Discussions

The XRD pattern of $Ni(OH)_2$ – carbon cloth electrode after hydrothermal treatment at 180 °C for 12 h followed by annealing at 250 °C for 4h is given in Figure 1. Two phases of $Ni(OH)_2$ were present which corresponds to the α and β phases. Peak at 14.18° is due to the (003) plane of α-$Ni(OH)_2$ (JCPDS 38-0715) while peak at 17° corresponds to the (001) plane of β-$Ni(OH)_2$ (JCPDS 14-0117). On the other hand, peak at 26° is due to the carbon cloth (JCPDS 75-1621) used in the study. Shown below are the reactions involved in the formation of $Ni(OH)_2$:

$$CH_3OH + H_2O \longrightarrow CH_3OH_2^+ + OH^- \quad (1)$$

$$Ni^{2+} + 2OH^- \xrightarrow{CTAB} \alpha - Ni(OH)_2 \longrightarrow \beta - Ni(OH)_2 \quad (2)$$

Equation 1 shows the decomposition of methanol upon heating releasing hydroxyl ions. Ni^{2+} ions from the nitrate solution react with the OH^- ions to form the $Ni(OH)_2$ phases. Being unstable, some of α phase transformed into the β phases.

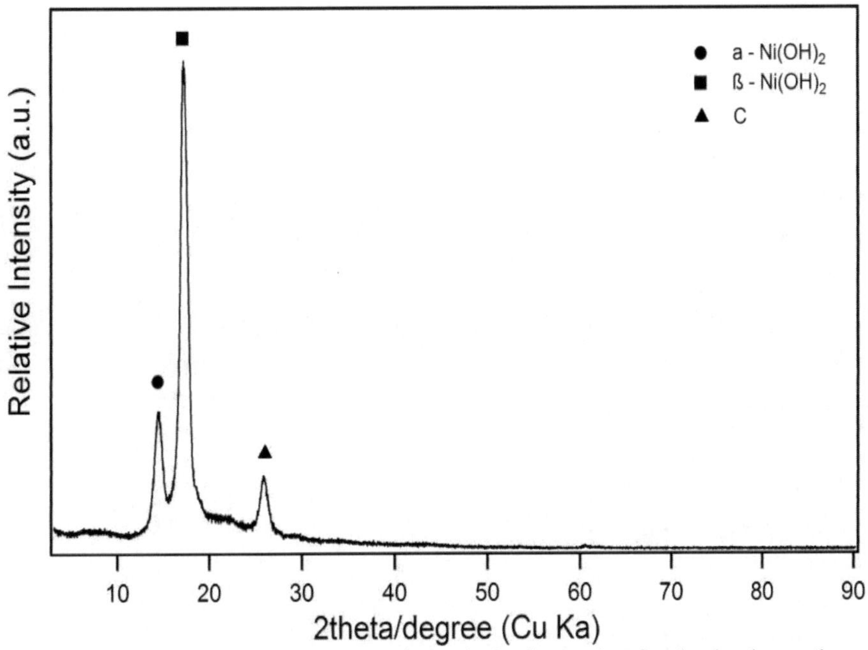

Figure 1. XRD pattern of the Ni(OH)2 – carbon cloth electrode after hydrothermal treatment at 180 °C for 12 h and annealing at 250 °C for 4 h.

Figure 2 shows the corresponding SEM images of the carbon cloth before and after the hydrothermal treatment at 180 °C for 12 h and annealing at 250 °C for 4 h. Fig. 2a shows the pure carbon cloth arranged in an ordered manner. Fig. 2b and 2c describes the smooth surface of each fiber having an average diameter of 10.79 μm. After the hydrothermal treatment and annealing, the ordered structure of the fibers remains and particles were deposited on the fiber based from Fig. 2d. Fig. 2e shows interconnected grass-like particles that were dispersed on the fiber with an average diameter of 2.17 μm. Fig. 2f, on the other hand, illustrates that grass-like structures with thickness of 140 nm were grown on the surface of the fiber. These grass-like structures are formed in an aggregated manner displaying a stable network on the carbon cloth. The presence of CTAB may lead to a slower reaction rate between OH^- and Ni^{2+} and further control the morphology [10]. CTAB molecules tend to form layered structural micelles on the surface of the carbon cloth which then serve as the template for the growth of $Ni(OH)_2$ structures [11].

Figure 2. Corresponding SEM images of (a-c) carbon cloth before treatment at high and low magnifications and (d-f) Ni-carbon cloth sample after hydrothermal treatment at 180 °C for 12 h and annealing at 250 °C for 4 h.

4. Summary and Conclusion

$Ni(OH)_2$ structures were successfully fabricated on carbon cloth via hydrothermal treatment, followed by annealing. The as-prepared sample is composed of microflowers and grass-like structures on the surface of the carbon cloth. α and β phases of $Ni(OH)_2$ were present after annealing, with β as the predominant phase . The presence of hybrid architecture is ideal for efficient ion transportation which is suitable for high-performance supercapacitors.

5. Acknowledgments

This work is supported by the National Research Council of the Philippines under the project "Electrochemical Investigation of the Pseudocapacitive Properties of Ni/NiO Nanowire Nonwoven Electrode Formed Under External Magnetic Field." Project start date: 1 June 2016.

References

[1] Cheng S, Yang L, Liu Y, Lin W, Huang L, Chen D, Wong C.P. and Liu M 2013 *J. Mater. Chem. A* **1** 7709-7716
[2] Wang K, Li L and Zhang H 2013 *Int. J. Electrochem. Sci.* **8** 4785-4791
[3] Wang Y, Guo J, Wang T, Shao J, Wang D and Yang Y-W 2015 *Nanomaterials* **5** 1667-1689
[4] Salunkhe RR, Bastakoti BP, Hsu C-T, Suzuki N, Kim JH, Dou SX, Hu C-C and Yamauchi Y 2014 *Chem. Eur. J.* **20** 3084-3088
[5] Zhang L, Ding Q, Huang Y, Gu H, Miao Y-E and Liu T 2015 *ACS Appl. Mater. Interfaces.* **7** 22669-22677
[6] Xiong X, Ding D, Chen D, Waller G, Bu Y, Wang Z and Liu M 2015 *Nano Energy* **11** 154-161
[7] Mondal AK, Su D, Chen S, Zhang J, Ung A and Wang G 2014 *Chemical Physics Letters* **610-611** 115-120

[8] Meyer M, Bée A, Talbot D, Cabuil V, Boyer JM, Répetti B and Garrigos R 2004 *Journal of Colloid and Interface Science* **277** 309-315
[9] Bastakoti BP, Huang H-S, Chen L-C, Wu KC-W and Yamauchi Y 2012 *Chem. Commun.* **48** 9150-9152
[10] Cheng X, Zhang D, Liu X, Cao D and Wang G 2015 *Ionics* **21** 533-540
[11] Li Q, Liao Y, Liu Y-J, Mu Q-Y and Wang Y-D 2012 *Materials Technology* **27** 104-106

IOP Conf. Series: Materials Science and Engineering **201** (2017) 012031 doi:10.1088/1757-899X/201/1/012031

The use of macroalgae (Gracilaria changii) as bio-adsorbent for Copper (II) removal

Lavania-Baloo[1], Nordin Idayu[1], Ibrahim Umar Salihi[1], Jamari Zainoddin[2],

[1]Department of Civil and Environmental Engineering, Universiti Teknologi PETRONAS (UTP), 32610 Bandar Seri Iskandar Perak, Darul Ridzuan, Malaysia
[2]Fishery Research Institute, Pulau Sayak, Kedah, Malaysia

lavania.baloo@utp.edu.com

Abstract. Biosorption of heavy metals using marine macroalgae biomass can be an effective process and alternative to conventional methods. Activated carbon was developed from macroalgae (Gracilaria changii) and used as adsorbents for the removal of copper (II) from wastewater. Gracilaria changii based activated carbon (GCBAC) was prepared using muffle furnace at a constant temperature of 300 °C for 1 hour. Batch adsorption experiments were conducted to investigate the effets of important parameters such as pH, contact time, initial metal concentration and adsorbent dosage on the removal of Cu (II) from synthetic aqueous solution. Batch adsorption study shows that removal of Cu (II) using GCBAC relied upon pH, contact time, initial metal concentration and GCBAC dosage. The optimum conditions parameters were found to be pH 6.0, time of 60 minutes and GCBAC dosage of 0.3 g, respectively. Adsorption data was described better by Freundlich isotherm model with R^2 value of 0.7936. The maximum Cu (II) adsorption capacity of GCBAC was found to be 0.07 mg/g. The experimental adsorption data obtained fitted well into Pseudo-second-order kinetic model, with R^2 value near unity. Thus, GCBAC can be used as an effective adsorbent for the removal of Cu (II) from aqueous solution.

1. Introduction

Heavy metals are harmful to public health due to their toxicity, non-biodegradable nature and widespread presence in the natural environments. Several process industries such as metal plating, mining operations, tanneries, etc, use a lot of water in their production process, and as such generate large volume of wastewater at the end. The wastewater generated contains different contaminants that are harmful to the environment if discharged without proper treatment. One of the major contaminant of concern is the heavy metals. This is due to their toxicity, and their existence poses environmental-disposal problem [1, 2]. The numerous metals that are significantly toxic to human, plants and the ecological environments includes, copper, chromium, lead, mercury, manganese, cadmium, etc [3]. Copper becomes toxic at elevated level even though it is a required element; as such the level of copper in the natural environment and its availability for biological function is very essential. Since Cu (II) is widely used in production processes, potential or actual sources of Cu (II) pollution are many. Cu (II) can be present as contaminant in food, mainly in mushroom, liver, shellfish, chocolate and nuts [3]. Several conventional methods such as neutralization, chemical precipitation, ion exchange and electrochemical treatment were used to remove heavy metals in wastewater. Most of these methods are more suitable for high concentration of heavy metals. The techniques failed in certain cases, they are not cost-effective, generate excess sludge that requires special disposal system and not environmentally friendly [4]. Further, these techniques are not suitable for the removal of heavy metals especially at lower concentrations in the range of 1-100 mg/L [5]. Adsorption technique is found to be the most effective and efficient method for the removal of heavy metals from aqueous solution [4]. Commercial activated carbon (AC) is the most widely used adsorbent in the adsorption process. However, it is found to be expensive and thus has restricted its widespread usage. In recent time, many researchers have reported the use of industrial by-products as adsorbent to replace the commercial AC. Industrial by-products used for the production of AC include; lignite, agronite shells, rice husk, natural zeolites, kaolinite, lignin, peat, clay, bagasse, etc [6, 7]. Macroalgae or generally known as seaweed, are natural sources that have the potential to be used as bio-adsorbent for heavy metals removal. It is economically feasible to culture seaweed in large scale [8, 9]. It has been used widely in developing industrial product such as chemicals, soil conditioner, food and pharmaceutical products [10, 11]. It has been reported that China alone commercially produces about 4.2 × 106 tons of Laminaria sp.. Macroalgae and considered as the largest producer [12]. Recently, global cultivation and production of macroalgae increases from 95% to 99.6% in year 2012. This is because most of the Asian countries have joined the league of macroalgae production and includes Indonesia (20.6%, 3.9 million tons), Philippines (9.5 %, 1.8 million tons) North Korea (4.7%, 901 700 tons), South Korea (2.3%, 444 300 tons), Japan (2.3%, 432 800 tons) and Malaysia (1.1%, 207 900 tons) [13]. Seaweed consists of three distinct groups; brown, red and green seaweed. This study was carried out to investigate the potential of red seaweed (Gracilaria changii) for the adsorption of Cu (II) from aqueous solution through batch adsorption experiment. Parameters such as pH, initial concentration of Cu (II) ions, contact time, and adsorbent dosage were tested in this study. Langmuir and Freundlich isotherm models were used to fit experimental data. Pseudo-first-order and pseudo-second-order kinetic models were used to evaluate the controlling mechanism of the biosorption process.

2. Material and methods

2.1 Preparations of algal biomass-based activated carbon

The macroalgae sample used was Gracilaria changii and was collected from Fishery Department Langkawi, Malaysia. The material was washed with excess tap water followed by distilled water to get rid of salt and particulate materials. It was then dried in an oven at 70 °C for 24 hours. The dried material was grind and sieved to size 300 μm. The material was then treated chemically using 0.2 M HCl before physical activation at a temperature of 300 °C for 1 hour. The physical activation was carried out using a muffle furnace (Protherm PLF 110/45). The produced Gracilaria changii based-activated carbon (GCBAC) was then cooled and stored in a tight before use.

2.2 Preparation of synthetic solution

A stock solution was prepared by dissolving calculated weigh of Cu (II) chloride salt in 1 L of distilled water to give a concentration of 1000 mg/L. To obtain further required concentrations, the stock solution was diluted appropriately. The chemical employed are of analytical grade, obtained from Merck (Germany).

2.3 Batch adsorption experiments

Batch adsorption experiments were conducted using 0.1 g of GCBAC in 100 mL of Cu (II) solution with predetermined concentrations using 250 mL conical flasks. The flasks were clamped in an orbital shaker (Protech Model 722) and shaken at 150 rpm at room temperature. At the end of predetermined time, the mixture was filtered using Whatman's glass microfiber filter paper. The filtered solution was analysed for residual Cu (II) concentration using Atomic Absorption Spectrometer, AAS (Model AA 6800 Shimadzu). Effect of pH on the adsorption of Cu (II) using GCBAC adsorbent was investigated by varying the initial pH of the solution in the range of pH 1.0 to pH 8.0. The working solutions were adjusted using either 1M HCl or 1M NaOH. 0.1 g of GCBAC was placed into the flasks containing

100 mL of Cu (II) solution and was agitated on a shaker at 150 rpm for 2 hours at room temperature (25±1 °C). The concentration of Cu (II) in the solution was fixed to be 25 mg/L. A pH meter (Model EW 53013) was used to measure the pH of the solution. Batch adsorption experiment was also conducted to investigate the effect of initial metal concentrations and contact time. 100 mL of synthetic Cu (II) solution having initial concentrations of 5, 10, 15 and 20 mg/L and 0.1 g of CGBAC was placed in a series of 250 mL Erlenmeyer flasks and agitated at 150 rpm. Flasks were withdrawn from the shaker at the end of predetermined time interval of 10, 20, 30, 60, 90, 120 minutes, and the solution was analysed for the residual Cu (II) concentration. The effect of CGBAC dosage on the removal of Cu (II) from aqueous solution was investigated by varying the dosages from 0.1, 0.2, 0.3, 0.4, 0.5 and 0.6 g. The pH of the solution was adjusted to the optimum value obtained. The optimum contact time was also adopted. Other parameters such as agitating speed and initial metal concentration were fixed at 150 rpm and 20 mg/L, respectively. Metal removal efficiency and adsorption capacity was calculated were evaluated using Eq. 1 and Eq. 2, respectively.

$$R = \frac{(C_i - C_e)}{C_i} \times 100\% \tag{1}$$

where, R is the Cu (II) removal efficiency (%), Ci is the initial copper concentration and Ce is the residual copper (mg/L) concentrations at equilibrium.

$$Q_c = \frac{(C_i - C_e)V}{W} \tag{2}$$

where, Q_c is the adsorbent adsorption capacity (mg/g), Ci is the initial copper concentration and Ce is the residual concentration copper (mg/L) at equilibrium, V is volume of aqueous copper solution (L) and W mass of CGBAC (g).

3. Results and discussions

3.1 Effect of pH

This study was carried out to investigate the effect of initial pH of the solution on the removal of Cu (II) using CGBAC. The pH of the solution was varied from pH 1.0 to pH 8.0. Fig. 1 shows the plot of Cu (II) removal efficiency against initial pH of the solution. From the graph, it can be observed that Cu (II) removal efficiency is less at low initial pH and then increased continuously as the pH is increased until it reaches an optimum pH of 6.0 and then decline. At low initial pH (pH 2.0 - pH 4.0) there was competition between the Cu (II) ions and hydrogen ions in the solution for an active site on the GCBAC surface. However, as the pH of the solution was increased, the surface of GCBAC becomes less positively charged and the electrostatic attraction between the Cu (II) ions and the GCBAC surface is likely to increases, as such the Cu (II) ions adsorption increased. The decrease in the adsorption of Cu (II) at high initial pH level due to the formation soluble hydroxyl complexes [7]. GCBAC works best at pH value of 6.0 with corresponding Cu (II) removal efficiency of 24.22 %. pH value of 6.0 is selected as the optimum pH and used in the study of the effect of initial metal concentration and contact time (section 3.2).

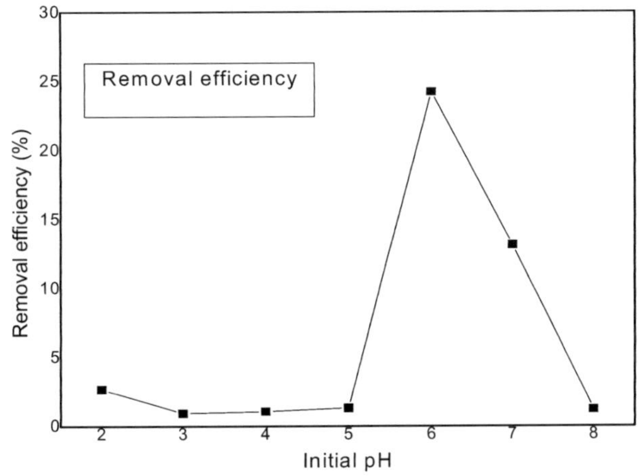

Figure 1. Effect of pH on Cu (II) adsorption (initial Cu (II) concentration 25 mg/L, contact time 2 hours, GCBAC dosage 0.1 g, agitation speed 150 rpm and temperature 25±1 °C).

3.2 Effect of contact time and initial metal concentrations

This study was carried out to investigate the effect of initial metal concentrations and contact time on the removal of Cu (II) using GCBAC. Fig. 2 shows the effect of initial metal concentrations and contact time on adsorption of Cu (II) by GCBAC. It has been observed that there was rapid adsorption of Cu (II) increases during the first 30 minutes of contact time. However, as the contact time increases, the adsorption rate decreases gradually and then become steady until equilibrium removal was achieved. There were more available active sites on the surface of GCBAC at the initial time. However, as the contact time increases the available pores on adsorbents are fully occupied by metals ion and the removal decreases until equilibrium. At this stage, the adsorption is controlled by the rate of transport of the metal ions from the outer to the inner site of the adsorbent material [14]. Equilibrium removal was attained at contact time of 60 minutes. Further significant removal was not observed afterwards. Contact time of 60 minutes was considered to be optimum and used in the study of the effect of GCBAC dosage on Cu (II) removal (section 3.3). The available binding site, thus resulting high removal efficiency whereas, at higher concentration, adsorption is low due to the saturation of binding site of the adsorbent.

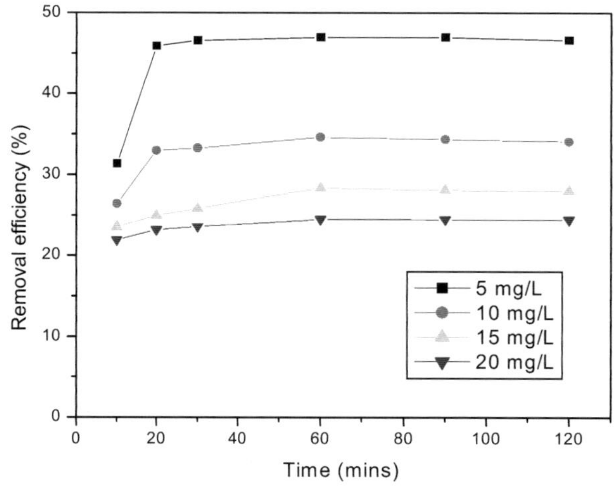

Figure 2. Effect of contact time on Cu (II) adsorption (contact time 2 hours, GCBAC dosage 0.3 g, pH 6.0, agitation speed 150 rpm and temperature 25±1 °C.

It was also observed that Cu (II) removal efficiency decreases with the increase in initial Cu (II) concentration (Fig. 3). This is because, at low initial metal concentration, there were few Cu (II) ions in the solution compared to the available number of active sites on the GCBAC surface; as such the Cu (II) ions can easily be adsorbed. On the other hand, at high initial metal concentrations, there were more Cu (II) ions than the available active sites on the CGBAC surface. Thus, the initial Cu (II) concentration was at 5 mg/L with removal efficiency of 47.0%.

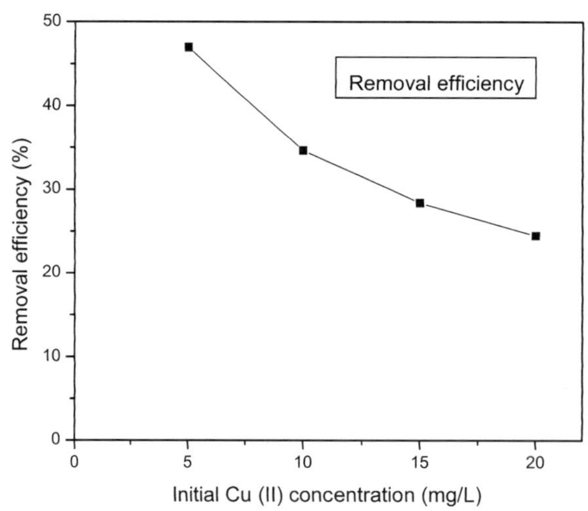

Figure 3. Effect of initial metal concentration on Cu (II) adsorption (contact time 2 hours, GCBAC dosage of 0.3 g, pH 6.0, agitation speed 150 rpm and temperature 25±1 °C).

3.3 Effect of adsorbent dosages

This study was carried out to investigate the effect of adsorbent dosage on the removal of Cu (II) by CGBAC adsorbent. Fig. 4 shows Cu (II) removal efficiency unit weight of GCBAC. From Fig. 4, it is observed that the removal efficiency of Cu (II) increase with the increasing of adsorbent dosage. This is due to more active sites of CGBAC becomes available with every increment of dosage, thus making the Cu (II) ions penetrate easily into the sorption sites. No significant removal was observed when the GCBAC dosage was increased to 0.3 g. At this point, it is expected that the concentration of Cu (II) ions in the solution and the concentration of the adsorbed ions on the CGBAC surface are in equilibrium with each other.

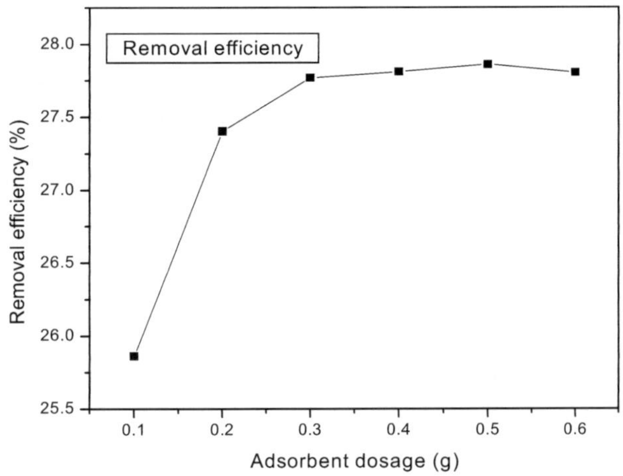

Figure 4. Effect of adsorbent dosage on Cu (II) adsorption (contact time 1 hour, pH 6.0, initial concentration 20 mg/L, agitation speed 150 rpm and temperature 25±1 °C).

3.4 Effect of adsorbent dosages

Langmuir and Freundlich isotherm models were used to interpret the relationship between the GCBAC adsorbent and the Cu (II) ions adsorbed on the adsorbent surface. The models will indicate the metal uptake capacity and the information regarding the differences in metal uptake by various species [18]. According to Wang and Chen [5], Langmuir and Freundlich are represented as in Eq.3 and Eq. 4 below:

$$\frac{C_e}{q_e} = \left(\frac{1}{K_L}\right)\left(\frac{1}{b}\right) + C_e\left(\frac{1}{b}\right) \tag{3}$$

where C_e is equilibrium solute concentration in solution (mg/l), q_e is the amount of metal adsorbed, K_L is the equilibrium Langmuir constant and b is the bonding energy of adsorption required for monolayer formation. Plot of C_e/q_e against C_e will give a straight line with $1/b$ as slope and $1/K_L$ as intercept.

$$\log q_e = \log K_F + \frac{1}{n}\log C_e \tag{4}$$

where log q_e is the amount of adsorbate adsorbed per unit mass of adsorbent (mg/g), K_F is the biosorption equilibrium constant representing the sorption capacity and n is a constant indicative of biosorption intensity. The plot of log q_e against log C_e will yield a straight line with $1/n$ as the slope and log K_F as the intercept. Table 1 shows the coefficient values obtained for Langmuir and Freundlich isotherm models obtained from this study and comparison with other plant-based adsorbents.

Table 1 Freundlich and Langmuir isotherm parameters constant.

Adsorbent	Langmuir				Freundlich	
	Q_{max} (mg/g)	b	R^2	R_L	K_F	R^2
GCBAC	0.07	0.05	0.5691	0.51	3.38×10^{27}	0.8742

From Table 1, it can be seen that Freundlich isotherm model indicates high coefficient of determination R^2 of 0.08742 compared with Langmuir model with low R^2 of 0.5691.This shows that adsorption of Cu (II) using GCBAC follows Freundlich isotherm model. Freundlich isotherm indicates the surface heterogeneity of the adsorbent, thus the surface of Gracilaria changii is made up of small heterogeneous adsorption patches which are similar to each other with respect to adsorption phenomenon [9, 12]. This indicates that the adsorption occurred at the heterogeneous and multilayer surface of the GCBAC adsorbent, with maximum mesopores surface area of 0.8950 m²/g. The Langmuir equilibrium parameter RL for GCBAC was found to be 0.51. This is within the range for favorable adsorption (0 < RL < 1) [14].

3.5 Adsorption kinetics studies

Kinetic study is the function of solute uptake rate has been evaluated for this study. The experimental results obtained were used to evaluate the kinetics of Cu (II) ions adsorption. Pseudo-first-order and pseudo-second-order kinetics models were used to study the rate of kinetics on CGBAC adsorbent. Eq. (5) and Eq. (6) were used to calculate the rate constants for pseudo-first and second-order kinetic models.

Pseudo-first-order: $$\log(q_e - q) = \log(q_e) - \left({k_1}/{2.303} \right) t \quad (5)$$

where q_e is the adsorbed metal at equilibrium in (mg/g), q_t is the metal adsorbed at a given time (t) and k1 is the pseudo-first-order rate constant (1/min). k_1 and qe are determined from the slope and intercept of the plot of log (q_e -q_t) against time (t).

Pseudo-second-order: $$\frac{t}{q_t} = \left(\frac{1}{k_2 q_e^2} \right) + \left(\frac{1}{q_e} \right) t \quad (6)$$

where, q_t and q_e are the amounts of metal ions adsorbed at time t (h) in (mg/g) and at equilibrium, respectively. k_2 is the pseudo-second-order adsorption rate constant. Plot of t/q_t against t will give a straight line with k_2 and q_e are determined from the slope and intercept of the plot. Figures 5 and 6, show the plots of pseudo-first and second-order kinetics. The values of k_1, k_2, $q_{e,exp}$ and $q_{e,\ cal}$ were obtained from the slopes and intercepts of the plots and are presented in Table 2.

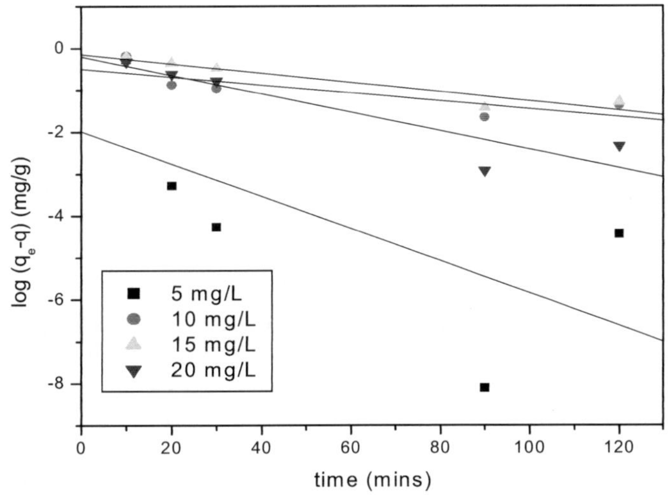

Figure 5. Pseudo-first-order kinetic plot

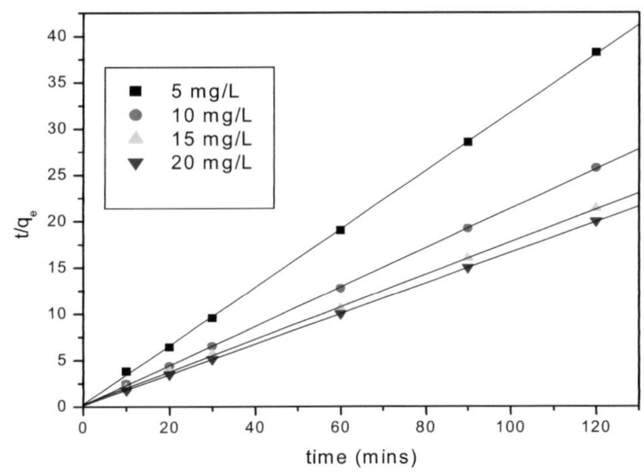

Figure 6. Pseudo-second-order kinetic plot

Table 2 Parameters values of pseudo-first-order and second-order models.

Adsorbent	Initial metals ions concentration (mg/L)		$q_{e, exp}$ (mg/g)	Pseudo first-order model			Pseudo second-order model		
				$q_{e, cal}$ (mg/L)	k_1 (h-1)	R^2	$q_{e, cal}$ (mg/g)	k_2 (g/mg h)	R^2
GCBAC	Cu(II)	5	3.1521	24.7742	0.0836	0.2709	3.1776	0.3147	0.9997
		10	4.7036	2.2065	0.0207	0.3598	4.7169	0.212	0.9998
		15	5.6658	1.0723	0.0246	0.6198	5.7078	0.1752	0.9998
		20	6.0094	1.0271	0.0491	0.6066	6.0643	0.1649	1.0000

As can be seen from Table 2, pseudo-second-order kinetic model presented high values of coefficient of determination (R^2). All the values were close to unity. There was a significant deviation between the experimental $q_{e,\,exp}$ values and the pseudo-first-order calculated $q_{e,\,cal}$ values for GCBAC adsorbent. This indicates a poor fit of the pseudo-first-order model to the experimental data. On the other hand, the calculated $q_{e,\,cal}$ values obtained from pseudo-second-order model agreed effectively with the experimental $q_{e,\,exp}$ values. Therefore, it can be concluded that the adsorption of copper ions onto GCBAC is a pseudo-second-order reaction model. This suggests that the rate limiting step is a chemisorption or chemical adsorption that involves valence forces by means of electrons exchange between the adsorbent (GCBAC) and the adsorbate (copper ions). A similar phenomenon was also reported on the adsorption of copper onto cashew nut shell [15], papaya wood [16] and modified oak sawdust [17].

4. Conclusions

The result obtained from this study indicated the potential use of Gracilaria changii based activated carbon (GCBAC) for the adsorption of Cu (II) ions from synthetically prepared wastewater. The experimental data well correlated with Freundlich isotherm and pseudo second-order kinetic model. pH, initial concentration of Cu(II) ions, contact time, and adsorbent dosage have significant effect on the Cu(II) adsorption by GCBAC. Thus, this study has proven that the use of macroalgae based activated carbon is effective for copper removal. Also, it is abundantly available and therefore would be cheaper and economically feasible activated carbon compared to the commercially available ones.

5. References

[1] Fu F and Wang Q 2011 Removal of heavy metal ions from wastewaters: A review *J. Environ. Mngt.* **92** 407-18.

[2] Ahluwalia S S and Goyal D 2007 Microbial and plant derived biomass for removal of heavy metals from wastewater. *Bioresource Technol.* **98** 2243-57.

[3] Yu B, Zhang Y, Shukla A, Shukla S S and Dorris K L 2001 The removal of heavy metals from aqueous solutions by sawdust adsorption-removal of lead and comparison of its adsorption with copper *J. Hazard. Mater.* **84** 83-94.

[4] Demirbas A 2008 Heavy metal adsorption onto agro-based waste materials: A review. *J. Hazard. Mater.* **157** 220-29.

[5] Wang J and Chen C 2009 Biosorbents for heavy metals removal and their future. *Biotechnol. Adv.* **27** 195-26.

[6] Mohan D, Pittman C U and Steele P H 2006 Pyrolysis of wood/biomass for bio-oil: a critical review. *Energ. Fuel* **20** 848-89.

[7] Salihi I U, Kutty S R M Isa, M H and Aminu N 2016 Zinc removal from aqueous solution using novel adsorbent MISCBA *J. Water Sanit Hyg Dev* **6** 377-88.

[8] Chopin T, Buschmann A H, Halling C, Troell M, Kautsky N, Neori A, Kraemer G P, Zertuche-González J A, Yarish C and Neefus C 2001 Integrating Seaweeds into Marine Aquaculture System: A Key Toward Sustainability *J. Phycol.* **37** 6 975-86.

[9] Neori A, Chopin T, Troell M, Buschmann A H, Kraemer G P, Halling C, Shpigel M, and Yarish C 2014 Integrated aquaculture: rationale, evolution and state of the art emphasizing seaweed biofiltration in modern mariculture *Aquaculture* **23** 361-91.

[10] Troell M, Halling C, Nilsson A, Buschmann A H, Kautsky N, and Kautsky L 1997 Integrated marine cultivation of Gracilaria chilensis (Gracilariales, Rhodophyta) and salmon cages for reduced environmental impact and increased economic output. *Aquaculture* **156** 45-61.

[11] Harrison E 1999 Treatment of metal-contaminated wastes: why select a biological process? *Trends Biotechnol.* **17** 462-65.

[12] Lüning K and Pang S 2003 Mass cultivation of seaweeds: current aspects and approaches *J. Appl. Phycol.* **15** 115-19.

[13] Baloo L 2015 *Semi-closed circulation integrated multi.trophic aquaculture treatment system using aquatic organisms as biofilters to improve shrimp effluent quality* (Malaysia: Universiti Teknologi Malaysia) pp 90-92.

[14] Rafatullah M, Sulaiman O, Hashim R and Ahmad A 2009 Adsorption of copper (II), chromium (III), nickel (II) and lead (II) ions from aqueous solutions by meranti sawdust. *J. of Hazard Mater* **170** 969-77.

[15] SenthilKumar P, Ramalingam S, Sathyaselvabala V, Kirupha S D, and Kirupha S 2011 Removal of copper (II) ions from aqueous solution by adsorption using cashew nut shell. *Desalination* **266** 63-71.

[16] Saeed A, Akhter M W, and Iqbal M 2005 Removal and recovery of heavy metals from aqueous solution using papaya wood as a new biosorbent. *Sep. Purif. Technol.* **45** 25-31.

[17] Argun M E, Dursun S, Ozdemir C and Karatas M 2007 Heavy metal adsorption by modified oak sawdust: Thermodynamics and kinetics. *J. Hazard. Mater.* **141** 77-85.

Acknowledgements

Special gratitude to Dr. Zainoddin Jamari, the Director of Fishery Research Institute (FRI) Pulau Sayak, Kedah, for the technical support. The authors also thankful to Mr. Zaaba and all laboratory technologists from Environmental Laboratory, Civil and Environmental Engineering Department, Universiti Teknologi PETRONAS for the assistance and cooperation given throughout the project. Finally, we are indebtful to Universiti Teknologi PETRONAS for providing the financial assistant (STIRF grant).

Characteristics of Ampel bamboo as a biomass energy source potential in Bali

M. Sucipta[1,2], D. N. K. Putra Negara[1], T. G. Tirta Nindhia[1] and I W. Surata[1]

[1] Department of Mechanical Engineering, University of Udayana, Campus Unud Bukit Jimbaran, Badung, Bali, Indonesia
[2] m.sucipta@me.unud.ac.id

Abstract. Currently, non-renewable fossil energy dominates utilization of the world energy need for many applications. Efforts has been developed to find alternative renewable energy sources, due to fossil energy availability is diminishing. And one of renewable energy source is from biomass. The aim of this research is to determine characteristics of the Ampel bamboo (Bambusa vulgaris) as an energy potential of biomass. The Ampel bamboo's characteristics possessed are evaluated based on its chemical composition; moisture, volatile, ash, and fixed carbon through proximate analysis; and also carbon, hydrogen and nitrogen content through ultimate analysis. From the Thermo-gravimetric analysis (TGA) indicates that Ampel bamboo contains of about 18.10% hemicelluloses, 47.75% cellulose and 18.86% lignin. While from the ultimate analysis results in the content of carbon, hydrogen, and Nitrogen of Ampel bamboo are 39.75%, 5.75% and 0% respectively. With such characteristics, it indicates that Ampel bamboo has an attractive potential as a renewable energy source.

Keywords: Biomass, bamboo, characteristics, potential, renewable energy

1. Introduction

Currently, fossil fuels, which are non-renewable energy sources, are major energy sources of the world for many application sectors. Even more than 80% of the global total energy is obtained by burning fossil fuels, which 58% are consumed by the transport sector [1]. Petroleum is still predicted to dominate the world's primary energy mix by 2050. In Indonesia, fossil fuels contribute for 94% of the national energy mix, with proportion of 47% oil-based, 21% natural gas and 26% coal [2]. However, the availability of fossil-based energy could not compete with the growth of energy consumption by an average of 7% per year [2]. To anticipate shortages or energy crisis in the future, it has been intensively done much effort to look for alternative energy sources such as water, wind and solar energies and energy from biomass. All types of these energies are renewable energy and are easy to find in Indonesia.

In Indonesia, source of biomass energy has a promising prospect due to the amount is abundant with high quality and environmentally friendly. Biomass is biological material produced or material obtained from plants containing lignin and cellulose so called material lignocelluloses. Biomass energy is the only renewable carbon energy with low sulphur, ash, and nitrogen contents as well as CO_2 emissions are close to zero [3]. Utilization of biomass as an energy source can be directly done through combustion to produce heat or be indirectly used after converting to various forms of biofuels [4, 5] either in the form of a solid, liquid or gas [6]. This conversion process can be undertaken through the pyrolysis process, gasification and fermentation of organic material. This last method is commonly used for producing biogas [7]. Biomass has been widely applied in various fields. Their implementation depends on the chemical composition and physical characteristics of the macromolecules of the biomass [8]. Xingxing Gu et al [9] utilized bamboo biomass to manufacture activated carbon that is used as a cathode material of Li-S battery. Huang T et al [10] produced also bamboo activated carbon as a cathode material for super capacitor [10]. Biomass is also used to produce activated carbon and applied for energy storage. Here, activated carbon behaves as an adsorbent for adsorption of methane using adsorbed natural gas (ANG) technology. Coconut shell [11], sorghum and wheat [12] and sugarcane molasses [13] are some examples of biomass converted into activated carbon and used as adsorbent for ANG technology. Longbo Jiang X Y et al [14], investigated pellets made of straw biomass mixed with plant waste. It is concluded that mix of straw and plant waste can produce good quality pellets. Converting of biomass into fuel is summarized by H Zabed et al [15]. In their study is mentioned that ethanol produced by lignocelluloses biomass has some advantages compared to ethanol produced by edible sources (sugars and starch). Lignocelluloses biomasses are rich in carbohydrates and available with low and stable price. In addition, they are mainly waste materials and non-competitive with food chain. Those research indicate that biomass have become an interesting topic investigated in order to provide alternative energy source in the future.

Ampel bamboo (Bambusa vulgaris), as shown in Figure 1, is a biomass that could potentially be used as a source of energy in Bali, Indonesia. This bamboo is very easy to find in almost all districts in Bali, even more this bamboo is also well cultivated in a few region. Potential energy produced by biomass depends on the chemical compositions and properties of the biomass raw material used. In this study, the chemical compositions and properties of Ampel bamboo and its potential as a renewable alternative energy source were evaluated.

Figure 1. Ampel bamboo

2. Method

2.1. Material

The studied bamboo in this research is Ampel bamboo (Bambusa vulgaris). It is collected from the area around Petang district, Badung regency, Bali. Tests conducted include the chemical compositions, proximate and ultimate tests. Bamboo is cut to small pieces specimens (5 x 5 mm), dried under the sunlight for 6 days. Specimens are hence dried in the electric furnace at temperatures of 80°C for 1 hour. The drying process in the electric furnace is repeated until there is no weight change of the specimens. Subsequently, the specimens were formed into powder and prepared each 10 grams for chemical compositions, proximate and ultimate tests.

2.2. Apparatus

The equipment used to determine of carbon, fixed carbon, volatile, ash and moisture contents was Thermo-Gravity Analysis (TGA) 701 machine which is based on ASTM D7582MVA method. Ultimate test was carried out using a CHN628S machine based on ASTM D7582 Biomass. This test is performed to determine the content of carbon, hydrogen and nitrogen. The chemical composition (lignin, hemicellulose, cellulose and silica) is performed using analysis of Van Soest. This method is also known as the USDA (United State Department of Agriculture).

3. Results and discussion

The chemical and properties data of Ampel bamboo are shown in Table 1, Table 2 and Table 3. In order to make a comparison, the properties of coal and common biomass data, which is summarized by Stanislav V V et al [6], is presented in Table 4. From Table 1 it can be seen that Ampel bamboo contains of 47.75% cellulose, 18.10% hemicellulose, 18.86 lignin, and 1.78% silica. According to Li et al [16} and Dence C W [17], in general, the content of alpha cellulose and lignin of bamboos are in the range 40-50% and 20-26% respectively. The cellulose composition of Ampel bamboo is currently on such range, but it is slightly lower for its lignin content. Meanwhile, it has low content of silica. Silica is impurity and undesired in combustion process. Low silica content can also reduce erosion-abrasion and slagging problem during biomass processing. With such chemical compositions, Ampel bamboo has good potential as a source of energy.

Table 1. Chemical compositions of Ampel bamboo.

Sample	Hemi cellulose (%)	Lignin (%)	Cellulose (%)	Silica (%)
Ampel bamboo	18.10	18.86	47.75	1.78

Table 2. Proximate analysis of Ampel bamboo.

Sample	Moisture (%)	Volatile (%)	Ash (%)	Fix Carbon (%)
Ampel bamboo	7.61	74.30	8.69	9.41

Tabel 3. Ultimate analysis of Ampel bamboo.

Sample	C (%)	H (%)	N (%)
Ampel bamboo	39.751	5.752	0

Proximate analysis of Ampel bamboo show that it contains of 7.61% moisture, 74.30% volatile, 8.69% ash and 9.41% fixed carbon. All of these values are included in the range of common biomass values compiled by Stanislav V V et al, as shown in Table 4. Compared with coal, Ampel bamboo has higher volatile but lower fix carbon and carbon. High volatile cause a low ignition temperature; faster oxidation of volatile than char can increase the burning of which produces better and faster burnout material with a lower unburned carbon in the ash [18]. In addition a high volatile content in the biomass provide a more stable flame for mixed fuel [19]. However, combustion of high volatile biomass content is fast, difficult to control and require a larger reactor volume to obtain complete combustion and low emission of pollutants such as CO [20]. Ampel bamboo contains of 8.69% ash which is compared to ash content of common biomass (0.1 - 34.3), as shown in Table 4, it is relatively low. This is an advantage for yielding higher quality of fuel because it can increase the heating value and can be easier in conversions of thermochemical and biochemical [21].

Tabel 4. Proximate and ultimate analysis of coal and common biomass, summarized by Stanislav V V et al [6].

Sample	Proximate (%)				Ultimate (%)		
	Moisture	Volatile	Ash	Fix Carbon	C	H	N
Coal	0.4 - 20.2	12.2 - 44.5	5.0 - 48.9	17.9 - 70.4	62.9 - 86.9	3.5 - 6.3	0.5 - 2.9
Biomass	2.5 - 62.9	30.4 - 79.7	0.1 - 34.3	6.5 - 35.3	42.2 - 60.5	3.2 - 10.2	0.1 - 12.2

The lower carbon content of Ampel bamboo is an advantage in reducing CO_2 emissions during biomass conversion. Carbon dioxide constitutes the main agent of global warming among greenhouse gas emissions, despite that its global warming potential is 25 and 300 times weaker than the CH_4 and N_2O respectively [22]. On the other hand, a low carbon content and higher proportion of H and O of biomass resulted in reduction of energy values of biofuels compared to fossil fuel. This is because C-O and C-H bonds have lower energy contained than C-C bonds. The results of ultimate analysis also showed that the nitrogen content of Ampel bamboo is zero. It is also very advantageous in the combustion process, as it can reduce emissions of NOx (NO, NO_2), N_2O, ammonia (HN_3) and pollution of ozone [18].

4. Conclusion

Biomass is a renewable energy source with a promising prospect. Ampel bamboo is one of biomass with qualified chemical composition and properties as a source of energy. Low ash, nitrogen, silica and moisture contents and high volatile content of Ampel bamboo make it has potential for producing a better quality fuels and friendly environment. In addition, Ampel bamboos are very easy to find due to they thrive and well grow in almost all districts in Bali.

References

[1] Escobar J C, Lora E S, Venturini O J, Yáñez E E, Castillo E F and Almazan O 2009 Biofuels: environment, technology and food security *Renew Sustain Energy Rev.* **13** 1275-87.

[2] Jro W 2012 *Media komunikasi kementerian energi dan sumber daya mineral ESDMMAG* Indra T C, Dian E P, Vagunaldi, Dian L, Arid R A, Laksono N B Jakarta, Kementerian Energi dan Sumber Daya Mineral 9 p 39.

[3] Melike B and Özgür E 2015 An Investigation of the relationship between the biomass energy consumption, economic growth and oil prices *Procedia - Social and Behavioral Sciences* **210** 203-12.

[4] Heinimö J and Junginger M 2009 Production and trading of biomass for energy – an overview of the global status *Biomass Bioenergy* **33** 1310-20.

[5] Naik S N, Goud V V, Rout P K and Dalai A K 2010 Production of first and second generation biofuels: a comprehensive review *Renew Sust Energy Rev* **14** 578-97.

[6] Stanislav V V, Christina G V and Vassil S V 2015 Advantages and disadvantages of composition and properties of biomass in comparison with coal: an overview *Fuel* http://dx.doi.org/10.1016/j.fuel.2015.05.050.

[7] Isabel A A C E, Marta S S L, Pedro M C N and Jos´e P B M 2008 Adsorption of natural gas and biogas components on activated carbon *Separation and Purification Technology* **62** 281–96.

[8] Laurichesse S and Avérous L 2014 Chemical modification of lignins: towards biobased polymers *Prog Polym Sci* **39** 1266-60.

[9] Xingxing G, Yazhou W, Chao L, Jingxia Q, Sheng L, Yang L H, Waedy M, Nazir M and Shanqing Z 2014 *Microporous bamboo biochar for litium - sulfur battery Nano Research* 1-14.

[10] Tianfu H, Zehai Q, Dewu W and Zhibiao H 2015 Bamboo-based activated carbon @ MnO2 nanocomposites for flexible high-performance supercapacitor electrode materials *Int. J. Electrochem. Sci* **10** 6312–23.

[11] Diana C S A, J Cassia S A, Moisés B N, A Eurico B T, Emerson F J and Celio L C 2007 Microporous activated carbon prepared from coconut shells using chemical activation with zinc chloride *Microporous and Mesoporous Materials* **100** 361-64.
[12] Tengyan Z, Walter P W, Fan L T 2010 Grain-based activated carbons for natural gas storage *Bioresource Technology* **101** 1983-91.
[13] Joanna S-N, Weronika K, Beata M, Zvi C K, 2013 Production, characterization and methane storage potential of KOH-activated carbon from sugarcane molasses *Industrial Crops and Products* **47**, 153-59.
[14] Longbo J, Xingzhong Y, Zhihua X, Jie L, Hui L, Liang C, Hou W, Xiaohong C and Guangming Z 2016 A comparative study of biomass pellet and biomass-sludge mixed pellet : energy input and pellet properties *Energy Conversion and Management* **126** 509-15.
[15] Zabed H, Sahu J N, Boyce A N ang Faruq G 2016 Fuel ethanol production from lignocellulosic biomass: an overview on feedstocks and technological approaches *Renewable and Sustainable Energy Reviews* **66** 751–74.
[16] Li X B, Shupe T F, Peter G F, Hse C Y and Eberhardt T L 2007 Chemical changes with maturation of the bamboo species phyllostachys pubescens *Journal of Tropical Forest Science* **19** 6-12.
[17] Dence C W 1992 The determination of lignin, *Methods in Lignin Chemistry* (Berlin Springer-Verlag) pp. 33-61.
[18] Lynch D, Henihan A M, Bowen B, Lynch D, McDonnell K and Kwapinski W 2013 Utilisation of poultry litter as an energy feedstock *Biomass Bioenergy* **49** 197-204.
[19] Wang C, Xu C, Cao Z and Di H 2013 Investigation on minerals migration during co-firing of different straw/coal blending ratios *Energ Convers Manage* **74** 279-85.
[20] Werther J, Saenger M, Hartge E-U, Ogada T and Siagi Z 2000 Combustion of agricultural residues *Prog Energ Combust* **26** 1-27.
[21] Huber G W, Iborra S and Corma A 2006 Synthesis of transportation fuels from biomass: chemistry, catalysts, and engineering *Chem Rev.* **106** 4044–98.
[22] Oke D and Olatiilu A 2011 Carbon storage in agro ecosystems: a case study of the cocoa based agroforestry in Ogbese forest reserve, Ekiti state, Nigeria *J Environ Protect* **2** 1069–75.

Acknowledgments
We are deeply gratitude to Department of Research and Community Service (LPPM), University of Udayana for funding this research through a Letter of Assignment Agreements in the Framework of the Implementation of Decentralization and National Competitive Research, Ministry of Research, Technology, and Higher Education, Republic of Indonesia, fiscal year 2016, Number: 486.138/UN14.2/PNL.01.03.00/2016.

IOP Conf. Series: Materials Science and Engineering **201** (2017) 012033 doi:10.1088/1757-899X/201/1/012033

Chemical, strength and microstructure characterization of Balinese bamboos as activated carbon source for adsorbed natural gas application

D N K Putra Negara[1,2,3], T G Tirta Nindhia[2,3], I W Surata[2] and M Sucipta[2]

[1] Doctoral Student of Engineering Science of Udayana University, Denpasar, Bali, Indonesia
[2] Mechanical Engineering Department of Udayana University, Street of Bukit Jimbaran, Badung, Bali, Indonesia
[3] devputranegara@gmail.com, nindhia@yahoo.com

Abstract. Bamboo is one of promising biomass raw materials that has been widely researched, developed, manufactured as activated carbon (AC) and applied in many fields of life. However, there were only a few references associated to the use of bamboo AC as an adsorbent for Adsorbed Natural Gas (ANG) application. The purpose of this study was to characterize chemical, strength and microstructure of two types of local Balinese bamboo that were Tabah bamboo (Gigantochloc Nigrociliata) and Tamblang bamboo (Schizostachyum Brachycladum) as AC source. Characterization was carried out by undertaking proximate, ultimate, tensile tests, Van Soest analysis and microstructure observation. The results have showed that Tabah bamboo has 22.9151 % lignin, 44.9456 % cellulose, 84.56 % volatile and 44.47 % carbon, which were those characteristics higher than Tamblang bamboo. Furthermore, Tabah bamboo also has higher bond density vascular bundles, higher tensile strength (240.85 ± 17.53 Nmm-2) and fracture strength (182.39 ± 17.46 Nmm-2), lower ash (2.92 %), silica (1.84 %) and nitrogen (0.95 %) compared to Tamblang bamboo. Due to such characteristics, Tabah bamboo has greater potential as an AC precursor than Tamblang bamboo.

1. Introduction

Activated carbon (AC) is carbon material having a high porosity [1], high adsorption capacity [2], high mechanical strength [3] and a very large surface area [4]. Activated Carbons (ACs) are widely used in industry such as to separate the various types of pollutants such as dyes and metal ions from industrial waste [5], as gas storage [6-10] and others. Commercial ACs are generally derived from coal which is a non-renewable carbon source with relatively expensive price [11]. As an alternative, it has been extensively researched and developed ACs from biomass, such as coconut shell [12-13], palm [14], and bamboo [5, 14-17].

Bamboo is a biomass material widely studied as a source of activated carbon. Its growth is very fast; maximum altitude can be reached in just a few months [18] and matures in three years. Bamboo constitutes a tropical plant and grows naturally in all continents except Europe [19]. In Asia, bamboo grows in China, Thailand, Vietnam [20] and Indonesia. It can be converted into activated carbon through carbonization and activation processes. Carbonization process is intended to enrich material carbon content by eliminating non-carbon elements using thermal decomposition [21] and to reduce the content of volatile and converting into char with higher fix carbon content. While the purpose of activation (physical or chemical activation) is to develop advanced porosity and to produce regular structure that will eventually form the high porosity of solid activated carbon [7].

Some researchers have developed bamboo ACs for various applications such as for separation of heavy metals [22], for purification of waste water [14], as a cathode material of super capacitor [17] and others. Although bamboo ACs have been studied and applied in many fields, almost no references were found relating to their application as an adsorbent for ANG. In ANG technology, adsorbent serves to stored methane gas. Methane gas molecules caught inside the pores of the adsorbent due to their strong surface tension force (Van der Walls force). Distance molecules in the pores of the adsorbent is much closer than in the gas phase at the same pressure and temperature that causes the density of the methane molecules in the adsorbed phase become like liquid phase and has a greater density than the gas phase at equilibrium condition. The result is an increase in the fix volume of the system adsorption due to larger amount of adsorbed gas than the volume gas which was replaced by adsorbent [23]. As an ANG adsorbent, ACs must have high micro pore and large surface area. As a material source of ACs, bamboo has great properties to fulfil those requirements. It has high carbon content (48.64 %) and low contents of nitrogen (0.14 %), sulphur (0.11 %) and hydrogen (6.75 %) [24]. The chemical compositions of bamboo are similar to wood with containing cellulose, lignin and hemicellulose more than 90% of its total mass. In general, the content of alpha cellulose of bamboo is 40-50%, while the content of alpha cellulose of softwood and hardwood are 40-52 % and 38-56 %, respectively [25]. The lignin content of bamboo ranges from 20% to 26%. With such chemical compositions, bamboo has a good criterion as a raw material of ACs. Bamboo-based activated carbon has four times bigger micro pore and 10 times greater surface area when it is compared to carbon-based wood [26]. In addition, bamboo is very easy to find and has a short cutting-age cyclic (3-5 years) [25]. One of the factors affecting the characteristics and qualities of ACs is the chemical compositions of raw material used. In this study, the preliminary characterization and evaluation of two types of Balinese bamboo were carried out.

2. Method

The biomass raw materials used were two local Balinese bamboos: Tabah bamboo (*Gigantochloc Nigrociliata*) and Tamblang bamboo (*Schizostachyum Brachycladum*). Bamboo was cut into small pieces, dried during 10 days under the sunshine and heated in the electric furnace for 1 hour at a temperature of 110^{0}C until no changes in weight. Specimens were formed into powder each 10 grams for chemical composition, proximate and ultimate tests. Proximate test was carried out by using TGA 701, 0.02 % RSD Precision to find out of moisture, volatile, fix carbon and ash of both bamboos. The ultimate test was based on ASTM D7582 Biomass and using CHN628S machine to investigate the contents of carbon, hydrogen and nitrogen of the raw materials. Van Soest Analysis, also known as USDA (United State Department of Agriculture), was used to determine the content of lignin, hemicellulose, cellulose and silica. Tensile test was based on ASTM D 143-94 using hydraulic tensile machine with a capacity of 3,000 kg. In order to observe microstructure, bamboo was cut to size 2 x 1 cm, polished, etched and observed by using Microscope Optic (10-1,200 optical zoom).

3. Results and discussion

3.1. Chemical compositions

Chemical compositions of both bamboos are presented in table 1. It can be seen that the main content of bamboo was composed by cellulose, hemicellulose and lignin. Tabah bamboo contained 44.94 % cellulose that higher than the Tamblang bamboo, which was 42.52 %. Tabah bamboo also contained higher hemicellulose and lignin than Tamblang bamboo but it implied lower silica. Some researchers had carried out the chemical composition test of different types of bamboo. Scurlock et al reported the cellulose content of a Japanese bamboo around 43.30 % [27]. This result was close to the cellulose value of Tabah and Tamblang bamboos. Li et al examined the cellulose content of bamboo taken from Kisatchie National Forest, Louisiana, USA and obtained value in the range of 46.08 - 47.91% [25].

Luz et al [28] got the cellulose content of 4 types *Guadua* bamboo (*Macana, Cebolla, Rayada and Castila*) varies from 37% to 44%. These values were also similar to cellulose content of Tabah and Tamblang bamboos.

Table 1. Chemical compositions of tabah and tamblang bamboos

Materials	Hemicellulose (%)	Lignin (%)	Cellulose (%)	Silica (%)
Tabah bamboo	16.99	22.91	44.94	1.84
Tamblang bamboo	16.73	21.23	42.52	1.96

3.2. Proximate and ultimate analysis

Table 2 shows the results of proximate and ultimate analysis of both bamboos. Tabah and Tamblang bamboos have a fix carbon 3.22 % and 4.61 %, respectively. Fix carbon is a carbon that still bound by the binding elements such as H, N and O. The ultimate analysis showed that Tabah bamboo contained higher carbon (42.47 %) than Tamblang bamboo (41.26 %). The increase of these carbons due to elements such as H, N and O separated from fix carbon during ultimate analysis. Furthermore, after the carbonization and activation processes, the fix carbon and carbon content are expected to increase. The volatile content of Tabah bamboo (84.56 %) was higher than Tamblang bamboo. Meanwhile, Tabah bamboo contained lower ash (2.92 %) and nitrogen (0.95 %) than Tamblang bamboo and relatively had almost the same hydrogen content. Low ash content will result in minimal effect of impurities on the formation of pores during the activation process. According to Mahamin et al, a good activated carbon source should be low ash and high carbon and volatile [15]. Ash consists of mostly minerals such as silica, alumina, iron, magnesium and calcium that are undesirable due to all of them are impurities. In general, a material with the lowest ash content will generate the most active carbon [15].

Table 2. Proximate and ultimate analysis of tabah and tamblang bamboos

Materials	Moisture (%)	Volatile (%)	Ash (%)	Fix Carbon (%)	C (%)	H (%)	N (%)
Tabah bamboo	9.27	84.56	2.92	3.22	42.47	6.10	0.95
Tamblang bamboo	7.67	83.48	4.26	4.61	41.26	6.09	1.03

Table 3. Ultimate and fracture strength of tabah and tamblang bamboos

Materials	Tensile strength (Nmm^{-2})	Fracture strength (Nmm^{-2})
Tabah bamboo	240.85 ± 17.53	182.39 ± 17.46
Tamblang bamboo	185.24 ± 20.58	112.98 ± 19.88

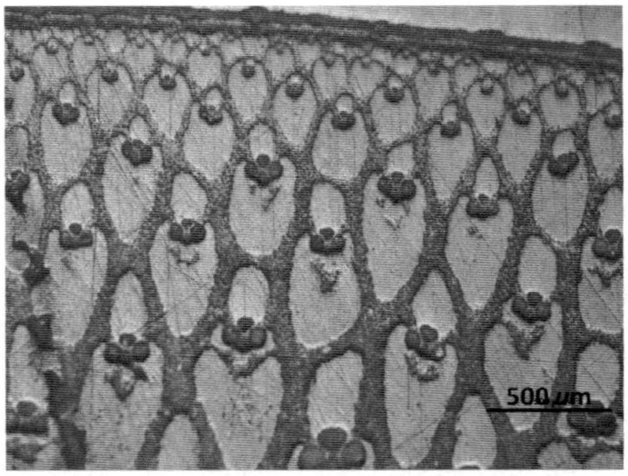

Figure 1. Vascular bundles of Tabah bamboo

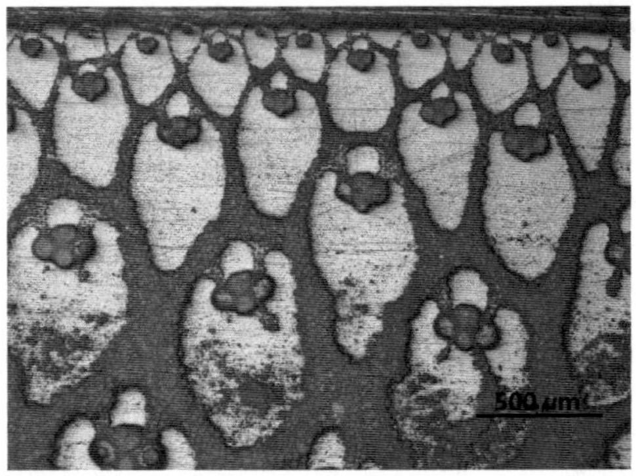

Figure 2. Vascular bundles of Tamblang bamboo

3.3. Vascular bundles and ultimate strength of tabah and tamblang bamboos

Table 3 shows the tensile and fracture strength of both bamboos. Tabah bamboo has higher ultimate and fracture strength than Tamblang bamboo. These higher values may related to their vascular bundles as shown in figures 1 and 2. Figure 1 and figure 2 illustrate the microstructure showing the vascular bundles in the cross section area of Tabah and Tamblang bamboos, respectively. There were differences in the vascular bundles between adjacent parts of the epidermis and part away from the epidermis. The vascular bundles adjacent parts of the epidermis have smaller size compared to the away from the epidermis and Tabah bamboo has higher bond density vascular bundles than Tamblang bamboo. This produced a higher tensile and fracture strength of Tabah bamboo compared to Tamblang bamboo, as shown in table 3. According to Sahu et al [3], raw material with higher ultimate strength has better potency to produce better quality of activated carbon.

4. Conclusion

Both Tabah and Tamblang bamboos were qualified as a raw material for activated carbon. However, Tabah bamboo will potentially produce better quality of activated carbon than Tamblang bamboo. It is due to its higher cellulose, carbon, and volatile contents, higher bond density vascular bundles and lower ash, silica and nitrogen contents. The next process is the selection of the appropriate carbonization and activation process parameters that can produce AC with high micro pores and surface area. So that it can be applied for ANG application. In addition, in terms of the availability, Tabah bamboo is obtained more easily compared to Tamblang Bamboo. Tabah bamboo is original species of Balinese bamboo and cultivated in almost all districts in Bali [29] while Tamblang bamboo mainly grow sporadically in Tamblang, Buleleng regency and also in Bangli regency, Bali, Indonesia. So that Tamblang bamboo is difficult to obtain in a large scale.

References

[1] Jin X J, Yu Z M and Wu Y 2010 Preparation of activated carbon from lignin obtained by straw pulping by KOH and K_2CO_3 chemical activation *Cellul. Chem. Technol.* **46** 79-85.

[2] Hu Z and Srinivasan M P 2001 Mesoporous high-surface-area activated carbon *Micro. Meso. Mater.* **43** 267-75.

[3] Sahu J N, Acharya J and Meikap B C 2010, Optimization of production conditions for activated carbons from tamarind wood by zinc chloride using response surface methodology *Bioresource Technology* **101** 1974-82.

[4] Idris S, Iyaka Y A, Dauda B E N, Ndamitso M M and Umar M T 2012 Kinetic study of utilizing ground nut shell as an adsorbent in removing chromium and nickel from dye effluent *Am. Chem. Sci. J.* **2** 12–24.

[5] Cheung W H, Lau S S Y, Leung S Y, Ip A W M and McKay G 2012 Characteristics of chemical modified activated carbons from bamboo scaffolding *Chinese Journal of Chemical Engineering* **20** 515-35.

[6] Isabel A A C E, Marta S S L, Pedro M C N and Jos´e P B M 2008 Adsorption of natural gas and biogas components on activated carbon *Separation and Purification Technology* **62** 281–96.

[7] Inomata K, Kanazawa K, Urabe Y, Hosono H and Araki T 2002 Natural gas storage in activated carbon pellets without a binder *Carbon* **40** 87-93.

[8] Diana C S A, J Cassia S A, Moisés B N, A Eurico B T, Emerson F J and Celio L C 2007 Microporous activated carbon prepared from coconut shells using chemical activation with zinc chloride *Microporous and Mesoporous Materials* **100** 361-64.

[9] Zhang T, Walawender W P and and Fan L T 2010 Grain-based activated carbons for natural gas storage *Bioresource Technology* **101** 1983-91.

[10] Joanna S N, Weronika K, Beata M and Zvi C K 2013 Production, characterization and methane storage potential of KOH-activated carbon from sugarcane molasses *Industrial Crops and Products* **47** 153-59.

[11] Patil B S and Kulkarni K S 2012 Development of high surface area activated carbon from waste material *International Journal of Advanced Engineering and Studies (IJAERS)* **1** 109-13

[12] Dipa D, Debi P S and and Meikap B C 2015 Preparation of activated carbon from green coconut shell and its characterization *J. Chem. Eng. Process Technol.* **6** 248 doi:10.4172/2157-7048.1000248

[13] Ahmadpour A, Okhovat A, and M J Darabi M 2013 Pore size distribution analysis of activated carbons prepared from coconut shell using methane adsorption data *Journal of Physics and Chemistry of Solids* **74** 886-91.

[14] Koo W K, Gani NA, Shamsuddin M S, Subki N S, and Sulaiman M A 2015 Comparison of wastewater treatment using activated carbon from bamboo and oil palm: an overview *Journal of Tropical and Resource Sustainable Science* **3** 54-60.

[15] Mahanim S M A, Asma I W, Rafidah J, Puad E and Shaharuddin H 2011 Production of activated carbon from industrial bamboo waste *Journal of Tropical Forest Science* **23** 417-24.

[16] Ademiluyi F T and Braide O 2012 Effectiveness of Nigerian bamboo activated with different activating agents on the adsorption of *BTX J. Appl. Sci. Environ. Manage.* **16** 267-73.

[17] Tianfu H, Zehai Q, Dewu W and Zhibiao H 2015 Bamboo-based activated carbon @ MnO2 nanocomposites for flexible high-performance supercapacitor electrode materials *Int. J. Electrochem. Sci.* **10** 6312 - 23.

[18] Itoh T and Shimaji K 1981 Lignification of bamboo culm (phyllostachys pubescens) during its growth and maturation *Bamboo Production and Utilization Proc. XVII IUFRO Congress Group 5.3. Ed. T. Higuchi Kyoto Japan* **10** 101-4

[19] Lus A S C, Giovanna A, Diana R and Mario E R G 2014 Correlation between chemical compounds and mechanical response in culms of two different ages of guadua angustifolia kunth, *Madera Bosques* **20** 87-94.

[20] Keith K H C, John P B and Gordon M 2005 Production of activated carbon from bamboo scaffolding waste-process design, evaluation and sensitivity analysis *Chemical Engineering Journal* **109** 147–65.

[21] Norhusna M N, Lau L C, Lee K T and Abdul R M 2013 Synthesis of activated carbon from lignocellulosic biomass and its applications in air pollution control: a review *Journal of Environmental Chemical Engineering* **1** 658–66.

[22] Sheng F L, Song Y W, Ming J T and Lang D L 2012 Adsorption capacity and removal efficiency of heavy metal ions by moso and ma bamboo activated carbons *Chemical Engineering Research and Design* **90** 1397–406.

[23] Lozano-Castello D, Alcan˜iz-Monge J, M. A. de la Casa-Lillo, Cazorla-Amoro´s D and

Linares-Solano A 2002 Advances in the study of methane storage in porous carbonaceous materials *Fuel* **81** 1777-803.

[24] Edward L K M, Cheung W H, Vinci K C L and Gordon M 2008 Kinetic study on bamboo pyrolysis *International Engineering Chemical Resources* **47** 5710-22.

[25] Li X B, Shupe F T, Peter G F, Hse C Y and Eberhardt T L 2007 Chemical changes with maturation of the bamboo species phyllostachys pubescens *Journal of Tropical Forest Science* **19** 6-12.

[26] Zhao R S, Yuan J P, Jiang T, Shi J B, and Cheng C C 2008 Application of bamboo charcoal as solid-phase extraction adsorbent for the determination of atrazine and simazine in enviromental water samples by high-performance liquid chromatography-ultraviolet detector *Talanta* **76** 956-59.

[27] Scurlock J M O, Dayton D C and Hames B 2000 Bamboo: an overlooked biomass resource *Biomass & Bioenergy* **19** 229-44.

[28] Sánchez-Echeverri L A, Aita G, Robert D and Garcia M E R 2014 Correlation between chemical compounds and mechanical response in culms of two different ages of guadua angustifolia kunth *Verano* **20** 87-94.

[29] Putra Negara D N K, Nindhia T G T, Surata I W and Sucipta M 2016 Development and application of bamboo activated carbons and their potency as adsorbent material for adsorbed natural gas (ANG); an overview *Key Engineering Materials* **705** 126-30.

Adsorption of Lead ions onto Activated Carbon derived from Sugarcane bagasse

I. U Salihi[1*], S.R.M Kutty[1], M. H Isa[1]

[1]Department of Civil and Environmental Engineering, Universiti Teknologi PETRONAS (UTP), 32610 Bandar Seri Iskandar Perak, Darul Ridzuan, Malaysia
[*]iusalihi@gmail.com

Abstract. In this study, activated carbon was developed from sugarcane bagasse and its effectiveness in adsorbing lead (Pb^{2+}) ions from synthetic aqueous solution was examined. Sugarcane bagasse activated carbon (SCBA) was developed in a tube furnace at a temperature of 900 °C, a heating rate of 10 °C/min, residence time of 3 hours, and at a nitrogen flow rate of 100 mL/min. Batch adsorption experiments were carried out to investigate the effects of pH and SCBA dosages on the adsorption process. The batch adsorption test showed that extent of Pb^{2+} adsorption by SCBA was dependent upon pH and SCBA dosage. The optimum pH for Pb^{2+} adsorption was found to be at pH 5.0. Maximum Pb^{2+} removal efficiency obtained from the batch studies was 87.3 % at SCBA dosage of 10 g/L. Equilibrium adsorption data was described by Langmuir model with a coefficient of determination (R^2) of 0.9508. Maximum adsorption capacity according to Langmuir model was evaluated to be 23.4 mg/g. The adsorption capacity of the SCBA was compared with that of other plant-based adsorbents. SCBA is an effective adsorbent for the removal of Pb^{2+} from aqueous solution.

1. Introduction

Wastewater containing heavy metals is increasingly release into the environment directly or indirectly, particularly in developing countries [1]. Unlike other contaminants found in wastewater, heavy metals are known to be toxic, non-biodegradable and have long half-life [2]. The existence of heavy metals in contaminated water has been a matter of interest to the public because of their toxicity effect in the discharging environment [3]. Elimination of these heavy metals from wastewater has become a point of concern, especially in places where recycling of the water is mainly practiced, the recovery of these heavy metals from wastewater is a significant environmental debate in recent time [4]. Treatment of wastewater containing heavy metals can be accomplished through conventional treatment techniques such as ion exchange, floatation, chemical precipitation, membrane filtration, electrochemical process, adsorption and biological systems [5-10]. These techniques were found to have some drawbacks, which includes incomplete metal removal, production of excess toxic sludge which requires special disposal method, high initial and operation cost, and demand for high energy [11]. Lately, several channels have been explored for the formation of inexpensive and more efficient technologies; this is to reduce the quantity of wastewater generation and also to enhance treated effluent quality. Adsorption method was found to be one of the most suitable substitutes for wastewater treatment compared to the existing treatments. Commercial activated carbon (AC) is the most effectively and widely used adsorbent in the adsorption process. Application of AC in the adsorption process was found to have some setbacks. AC is found to be very expensive; its cost is related to its quality, the finer it is, the more costly it becomes. Recently, the search for inexpensive adsorbents with the ability for metal binding has been strengthened [12]. The adsorbents can be of agricultural waste, industrial by products, biomass, biological and organic origin, polymeric materials and zeolites [13]. In this study, sugarcane bagasse activated carbon (SCBA) was produced as an adsorbent through thermal incineration process. Adsorption experiments were conducted to investigate the possibility of using SCBA as an alternative adsorbent for removal of lead ions from aqueous solution. Parameters such as initial pH of the solution and SCBA dosages were investigated to find their effects on the adsorption of Pb^{2+}.

2. Material and methods

2.1 Preparations of adsorbent material (SCBA)

Sugarcane bagasse was collected from a local outlet (Pasar Malam) night market at Seri Iskandar, Tronoh, Malaysia. The bagasse was manually cut to the average size of 3 cm and thoroughly washed with tap water to remove ligneous, cellulose and trapped impurities. Bagasse was further washed with deionized water to achieve further purity and for decontamination purpose. The purified bagasse was then dried in an oven at 105 ºC for a period of 24 hours until constant weight was achieved. 10 g of the dried bagasse was weighed and placed in a tube furnace; the furnace was then heated to a temperature of 900 °C, at a heating rate of 10 °C/min. The furnace temperature was maintained at 900 °C for 3 hours. Nitrogen gas at a flow rate of 100 mL/min was used as inert gas to deplete oxygen within the system. The resulting material (SCBA) was then grinded and made into powder form with an average sieved size of 150 μm. SCBA was then pulverized and immersed in a weak acid (0.5% H_2SO_4, pH 4.25) at room temperature for 24 hours; this was to ensure the adsorbent is kept within an acidic state. The material was filtered from the solution, washed and dried in an oven at 70 °C. SCBA was stored in an airtight container before use.

2.2 Preparation of adsorbate solution

A stock solution of lead chloride (Pb^{2+}) was prepared by dissolving its calculated weight of salt in 1 L of distilled water to obtain a concentration of 1000 mg/L. Further working concentrations were obtained by diluting the stock solution to the required strength. The chemical employed are of analytical grade, obtained from Merck (Germany).

2.3 Experimentations

Series of 250 mL Erlenmeyer flasks containing 100 mL of synthetic Pb^{2+} solution at predetermined concentrations and 0.5 g SCBA dosage were used in the adsorption experiments. An orbital shaker (Protech Model 722) was used to agitate the flasks at a constant speed of 150 rpm at room temperature (26 ± 1 °C), for a scheduled period of time. At the end of the scheduled agitation time, the flasks were removed from the orbital shaker and the mixture was filtered through Whatman's glass microfiber filters paper (GF/C). Atomic Absorption Spectrophometer, AAS (Model AA 6800 Shimadzu) was used to analyze the filtrate for residual Pb^{2+} concentration in mg/L.

The effect of pH on the adsorption of Pb^{2+} was studied by varying the initial pH of the solution in the range of pH 1 to pH 8. The pH of the solution was adjusted using either 2.0N HCl or 1.5N NaOH. Other parameters such as initial Pb^{2+} concentration, contact time, adsorbent dosage and agitation speed were kept constant at 10 mg/L, 2 hrs, 0.5 g and 150 rpm, respectively. The pH of the solution was measured using a pH meter (Model EW 53013, Hach Sension 1).

Batch experiments were also carried out to investigate the effect of SCBA dosages on the adsorption of Pb^{2+} ions. The SCBA dosage was varied from 2 g/L to 16 g/L until equilibrium Pb^{2+} removal was achieved. The pH of the solution was initially adjusted to the optimum value obtained. Other parameters such as contact time, initial metal concentrations and agitation speed were fixed at 3.5 hrs, 100 mg/L and 150 rpm, respectively. At the end of the scheduled time, the mixture was filtered and analyzed for residual Pb^{2+} concentration in mg/L. Eq. 1 and Eq. 2 were used to compute metal adsorption efficiency and adsorption capacity of SCBA, respectively.

$$R = \frac{(C_i - C_e)}{C_i} \times 100\% \qquad (1)$$

where, R is the Pb^{2+} removal efficiency (%), Ci is the initial Pb^{2+} concentration and Ce is the residual Pb^{2+} concentrations (mg/L) at equilibrium.

$$Q_c = \frac{(C_i - C_e)V}{W} \qquad (2)$$

where, Q_c is the adsorbent adsorption capacity (mg/g), C_i is the initial Pb^{2+} concentration and C_e is the residual Pb^{2+} concentration (mg/L) at equilibrium, V is the volume of aqueous Pb^{2+} solution (L) and W is the mass of SCBA (g).

2.4 Adsorption isotherms studies

Freundlich isotherm model is a commonly used model for describing adsorption equilibrium based on the empirical equation. Freundlich isotherm model has the ability to unfold adsorption related to both organic and inorganic composites on a wider range of adsorbents inclusive of biosorbent [14]. Freundlich isotherm model can be written in the following linearized form of Eq. 3 [15].

$$\log A_q = \log K_F + \frac{1}{n} \log C_e \qquad (3)$$

where $\log A_q$ is the quantity of the adsorbed pollutant per gram of the adsorbent (mg/g), C_e is the equilibrium concentration (mg/L), K_F and n are Freundlich constants. The plot of $\log A_q$ against $\log C_e$ has $1/n$ as slope and $\log K_F$ as its intercept.

Langmuir model is another frequently used model which describes heavy metals sorption unto biosorbent. Langmuir model operates based on three distinct assumptions viz; that all surface sites are the same and can occupy one adsorbed atom, it is limited to monolayer coverage adsorption and capability of a molecule adsorbed at a given site is not dependent on the occupancy of its adjoining sites [14]. Eq. 4 below describes Langmuir isotherm model in its linear form [16].

$$\frac{C_e}{A_q} = \left(\frac{1}{K_L}\right)\left(\frac{1}{b}\right) + (C_e)\left(\frac{1}{b}\right) \qquad (4)$$

where A_q is the quantity of pollutant adsorbed per gram of adsorbent (mg/g), K_L is the Langmuir constant for equilibrium, b is the adsorbate quantity that is required for a monolayer formation and C_e is the concentration of residual metal at equilibrium (mg/L). Thus, the plot of C_e/A_q against C_e will be a straight line with a slope of $1/b$ and intercept of $1/K_L$.

3. Results and discussions

3.1 Effect of initial pH

This experiment was conducted to investigate the effect of initial pH on the removal of Pb^{2+} using SCBA. Figure 1 shows the plot of Pb^{2+} removal efficiency against initial pH of the solution.

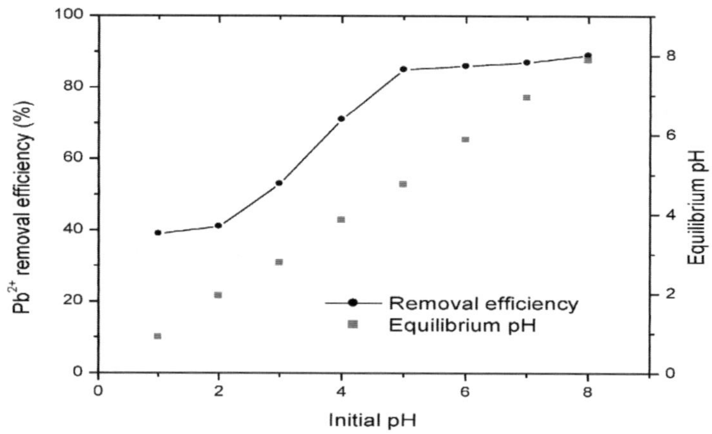

Figure 1. Pb^{2+} removal efficiency against initial pH (Pb^{2+} concentration 10 mg/L, contact time 2 hours, speed 150 rpm, adsorbent dosage 5 g/L and temperature 26 ± 1°C).

From Fig. 1, it can be observed that Pb^{2+} removal efficiency increased with increasing pH of the solution. At lower pH level, the removal efficiency is low compared with higher pH level. This was because of the presence of high concentration of hydrogen ions (H^+) at lower pH compared to their less concentration at higher pH level. As the pH of the solution was increased, the SCBA surface becomes more negatively charged, as such, the electrostatic attraction between the SCBA surface and the Pb^{2+} ions increased. This led to the adsorption of more Pb^{2+} ions and the release of H^+ ions from the SCBA surface back into the solution. At the end of the experiments and for all the samples, the final pH of the solutions was found to be less than the initial pH values. The changed in pH was due to the ion exchange process that took place between the hydrogen ions and the lead ions as the pH was increased. Qaiser et al. [17] reported that lead usually present as Pb^{2+} at a pH range of 2 to 6, and at pH above 6.0, it will hydrolyze to $Pb(OH)_2$ and $PbOH^+$ [17]. For this reason, the pH value of 5.0 was considered to be optimum and was used in the study of the effect of adsorbents dosages on the adsorption of Pb^{2+} ions by SCBA.

3.2 Effect of adsorbent dosages

This experiment was conducted to investigate the effect of SCBA dosages on the removal of Pb^{2+} from aqueous solution. The results obtained were plotted in Fig. 2 below.

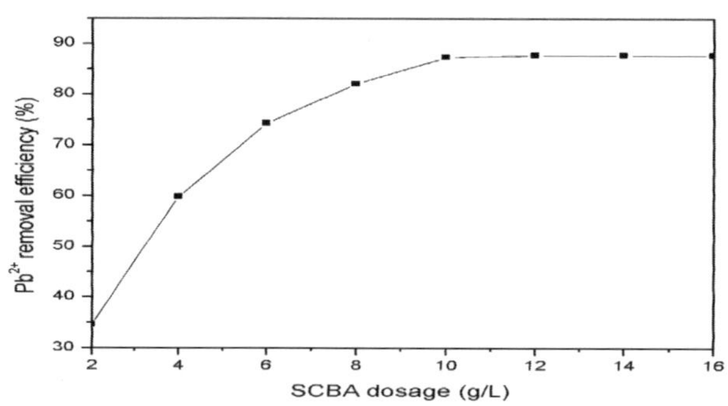

Figure 2. Pb^{2+} removal efficiency against SCBA dosages (Pb^{2+} concentration 100 mg/L, contact time 3.5 hrs, agitation speed 150 rpm, pH 5.0 and temperature 27 ± 1°C).

From Fig. 2, it can be observed that the removal of Pb^{2+} ions increases with increase in the SCBA dosages from 2 to 10 g/L . This was due to the presence of more active sites and large surface area of SCBA that are readily available and easily be accessible for Pb^{2+} ions adsorption. The incremental removal of Pb^{2+} ions was not significant when the SCBA dosage was increased beyond 10 g/L. At this stage, it was expected that the concentrations of Pb^{2+} ions at the surface of SCBA and the concentrations of the Pb^{2+} ions remaining in the solution are in equilibrium with each other. SCBA dosage of 10 g/L was considered to be optimum for this study. The corresponding removal efficiency was found to be 87.3%. This was in accordance with the result reported by Ribeiro et al on the removal of Pb^{2+} using modified egg shell waste [18].

3.3 Adsorption isotherms studies (performance of SCBA)

This study was carried out to establish the relationship between the concentration of Pb^{2+} ions and its degree of adsorption onto the surface of the SCBA adsorbent at a fixed temperature. The adsorption isotherm constants obtained from the plots of the linearized Langmuir and Freundlich models are presented in Table 1.

Table 1 Langmuir and Freundlich isotherm for Pb^{2+} and its comparison with other studies

Adsorbent	Langmuir				Freundlich			Reference
	Q_o	b	R^2	R_L	K_F	R^2	$1/n$	
SCBA	**23.4**	**0.29**	**0.9508**	**0.26**	**1.73**	**0.8629**	**0.52**	**This study**
Cedar leaf ash	7.23	0.17	0.998	N/E	1.56	0.992	0.43	[19]
Peat	82.31	0.45	0.992	N/E	18.72	0.525	0.54	[20]

From Table 1, it can be seen that coefficients of determination (R^2) for SCBA from the Langmuir isotherm model was found to be 0.9508. This indicates that Langmuir isotherm model satisfactorily described the adsorption of Pb^{2+} onto SCBA. This indicates the monolayer coverage of Pb^{2+} ions on the homogenous surface of SCBA. The Langmuir equilibrium parameter R_L value for SCBA was found to be 0.26. This is within the range for favorable adsorption ($0 < R_L < 1$) [21]. The maximum uptake capacity for SCBA was found to be 23.4 mg/g. This is better than some of the adsorption capacities for agricultural by-products reported in the literature. It can be seen that the Freundlich R^2 value from SCBA was found to be 0.8629, (Table 1). This indicates that the adsorption data was not sadequately described by Freundlich isotherm model. The adsorption intensity *1/n* value for Pb^{2+} ions adsorption was found to be 0.52. This indicates a favorable adsorption process with increasing adsorption capacity and the appearance of new adsorption sites [22]. The Freundlich and Langmuir expressions for the removal of Pb^{2+} ions by SCBA are shown in Eqn. 5 and Eqn. 6, respectively.

Freundlich equation,

$$q_e = 1.73C_e^{0.52} \tag{5}$$

Langmuir equation,

$$q_e = \frac{6.77C_e}{1 + 0.29C_e} \tag{6}$$

4. Conclusions

The present study has shown that activated carbon produced from the sugarcane bagasse (SCBA) using tube furnace can be engaged as an environment friendly and low cost adsorbent for the removal of Pb^{2+} ions from synthetic aqueous solution. The study on the effect of pH revealed that the competition between H^+ and metal ions at low pH values is the main leading factors that affect the adsorption characteristics of SCBA. Optimum removal efficiency was achieved at pH 5.0. The number of adsorption sites increased due to an increase in SCBA dosage and optimum removal efficiency of 87.3% was achieved at SCBA dosage of 10 g/L.

Acknowledgements

The authors wish to acknowledge Fundamental Research Grant Scheme (No. 0153AB-I60) for their financial support. We wish to also extend our profound gratitude to Miss Yusyawati Bnt Yahaya who has worked hard in operating the atomic absorption spectrophotometer. Nevertheless, our immense appreciation goes to Universiti Teknologi PETRONAS (UTP) through whom the grant was tenable and for providing a conducive and enabling working environment for this research work.

References

[1] F. Fu and Q. Wang, "Removal of heavy metal ions from wastewaters: a review," *Journal of Environmental Management,* vol. 92, pp. 407-418, 2011.

[2] I. U. Salihi, S. R. M. Kutty, M. H. Isa, and N. Aminu, "Zinc removal from aqueous solution using novel adsorbent MISCBA," *Journal of Water Sanitation and Hygiene for Development,* vol. 6, pp. 377-388, 2016.

[3] M. E. Hodson, "Effects of heavy metals and metalloids on soil organisms," in *Heavy metals in soils*, ed: Springer, 2013, pp. 141-160.

[4] S. Y. Chen and Q. Y. Huang, "Heavy Metals Recovery From Wastewater Sludge Of Printed Circuit Board Industry By Thermophilic Bioleaching Process," *Journal of Chemical Technology and Biotechnology,* 2013.

[5] A. Y. Dursun, "A comparative study on determination of the equilibrium, kinetic and thermodynamic parameters of biosorption of copper (II) and lead (II) ions onto pretreated< i> Aspergillus niger</i>," *Biochemical Engineering Journal,* vol. 28, pp. 187-195, 2006.

[6] L. Deng, X. Zhu, X. Wang, Y. Su, and H. Su, "Biosorption of copper (II) from aqueous solutions by green alga Cladophora fascicularis," *Biodegradation,* vol. 18, pp. 393-402, 2007.

[7] M. A. Hanif, R. Nadeem, H. N. Bhatti, N. R. Ahmad, and T. M. Ansari, "Ni (II) biosorption by< i> Cassia fistula</i>(Golden Shower) biomass," *Journal of Hazardous Materials,* vol. 139, pp. 345-355, 2007.

[8] B. Preetha and T. Viruthagiri, "Batch and continuous biosorption of chromium (VI) by< i> Rhizopus arrhizus</i>," *Separation and Purification Technology,* vol. 57, pp. 126-133, 2007.

[9] D. Satapathy and G. Natarajan, "Potassium bromate modification of the granular activated carbon and its effect on nickel adsorption," *Adsorption,* vol. 12, pp. 147-154, 2006.

[10] K. Vijayaraghavan, K. Palanivelu, and M. Velan, "Biosorption of copper (II) and cobalt (II) from aqueous solutions by crab shell particles," *Bioresource technology,* vol. 97, pp. 1411-1419, 2006.

[11] H. Eccles, "Treatment of metal-contaminated wastes: why select a biological process?," *Trends in biotechnology,* vol. 17, pp. 462-465, 1999.

[12] W. Leung, M. Wong, H. Chua, W. Lo, P. Yu, and C. Leung, "Removal and recovery of heavy metals by bacteria isolated from activatedsludge treating industrial effluents and municipal wastewater," *Water Science & Technology,* vol. 41, pp. 233-240, 2000.

[13] T. A. Kurniawan, G. Chan, W.-h. Lo, and S. Babel, "Comparisons of low-cost adsorbents for treating wastewaters laden with heavy metals," *Science of the Total Environment,* vol. 366, pp. 409-426, 2006.

[14] J. Febrianto, A. N. Kosasih, J. Sunarso, Y.-H. Ju, N. Indraswati, and S. Ismadji, "Equilibrium

and kinetic studies in adsorption of heavy metals using biosorbent: A summary of recent studies," *Journal of Hazardous Materials,* vol. 162, pp. 616-645, 2009.

[15] I. U. Salihi, S. R. M. Kutty, M. H. Isa, A. Malakahmad, and U. A. Umar, "SORPTION OF ZINC USING MICROWAVE INCINERATED SUGARCANE BAGASSE ASH (MISCBA) AND RAW BAGASSE," *Jurnal Teknologi,* vol. 78, 2016.

[16] I. U. Salihi, M. Kutty, S. Rahman, M. Hasnain Isa, U. A. Umar, and E. Olisa, "Sorption of Copper and Zinc from Aqueous Solutions by Microwave Incinerated Sugarcane Bagasse Ash (MISCBA)," in *Applied Mechanics and Materials*, 2016, pp. 378-385.

[17] S. Qaiser, A. R. Saleemi, and M. Umar, "Biosorption of lead from aqueous solution by Ficus religiosa leaves: Batch and column study," *Journal of Hazardous Materials,* vol. 166, pp. 998-1005, 7/30/ 2009.

[18] A. Ribeiro, J. Carvalho, and C. Vilarinho, "Lead (II) adsorption by modified eggshell waste," in *WASTES 2015–Solutions, Treatments and Opportunities: Selected papers from the 3rd Edition of the International Conference on Wastes: Solutions, Treatments and Opportunities, Viana Do Castelo, Portugal, 14-16 September 2015*, 2015, p. 253.

[19] L. D. Hafshejani, S. B. Nasab, R. M. Gholami, M. Moradzadeh, Z. Izadpanah, S. B. Hafshejani, *et al.*, "Removal of zinc and lead from aqueous solution by nanostructured cedar leaf ash as biosorbent," *Journal of Molecular Liquids,* vol. 211, pp. 448-456, 11// 2015.

[20] P. Bartczak, M. Norman, Ł. Klapiszewski, N. Karwańska, M. Kawalec, M. Baczyńska, *et al.*, "Removal of nickel(II) and lead(II) ions from aqueous solution using peat as a low-cost adsorbent: A kinetic and equilibrium study," *Arabian Journal of Chemistry*.

[21] M. Rafatullah, O. Sulaiman, R. Hashim, and A. Ahmad, "Adsorption of copper (II), chromium (III), nickel (II) and lead (II) ions from aqueous solutions by meranti sawdust," *Journal of Hazardous Materials,* vol. 170, pp. 969-977, 10/30/ 2009.

[22] M. Hasnain Isa, L. Siew Lang, F. A. H. Asaari, H. A. Aziz, N. Azam Ramli, and J. P. A. Dhas, "Low cost removal of disperse dyes from aqueous solution using palm ash," *Dyes and Pigments,* vol. 74, pp. 446-453, // 2007.

IOP Conf. Series: Materials Science and Engineering **201** (2017) 012035 doi:10.1088/1757-899X/201/1/012035

Effect of volume fraction of alpha and transformed beta on the high cycle fatigue properties of bimodal Ti6Al4V alloy

Shital Jadhav[1], Amit Powar[2], Sandip Patil [3], Ashish Supare[4], Bhagwan Farane[5], Dr. Rajkumar Singh[6]

[1-6] Kalyani Centre for Technology and Innovation; Bharat Forge Ltd. Pune, 411036, India

[a]shital.jadhav@bhratforge.com, [b]amit.powar@bharatforge.com

Abstract. The present study was performed to investigate the effect of volume fraction of alpha and transformed beta phase on the high-cycle fatigue (HCF) properties of the bimodal titanium Ti6Al4V alloy. The effect of such morphology on mechanical properties was studied using tensile and rotating bending fatigue test as per ASTM standards. Microstructures and fractography of the specimens were studied using optical and scanning electron microscopy (SEM) respectively.Ti6Al4V alloy samples were heat treated to have three distinctive volume fractions of alpha and transformed beta phase. With an increase in quench delay from 30,50 and 70 sec during quenching after solutionizing temperature of 967°C, the volume fraction of alpha was found to be increased from 20% to 67%. Tests on tensile and rotating bending fatigue showed that the specimen with 20% volume fraction of alpha phase exhibited the highest tensile and fatigue strength, however the properties gets deteriorate with increase in volume fraction of alpha.

1. Introduction

Due to the excellent properties (high specific strength, high fatigue strength, good corrosion resistance, etc.), titanium components (particularly Ti–6Al–4V) are often used for manufacturing critical systems such as airfoils, undercarriage components, and airframes instead of heavy steel components. During these applications titanium structures are often exposed to fatigue loading [1]. Fatigue fracture is an important failure mode for these structures. Ti-alloys are categorized into three groups; alpha (α)- alloy, β (beta)-alloy, and α-β (alpha-beta) alloy systems. The composition of α and β can be adjusted in such a way that the optimum combination of creep, fatigue, and yield strength can be achieved based upon the component and its application. The microstructure and, therefore, the mechanical properties of α + β titanium(Ti) alloys can be tailored for specific applications by carefully choosing a sequence of thermomechanical processes and heat treatments to obtain certain morphologies and spatial distributions of the α phase[2-5].

Till date, cold-end components of a gas turbine engine are dominated by Ti-alloys. Ti-6Al-4V is widely used α – β alloy in the aerospace industry (50 per cent market share for various aero applications). Gas turbine disk is subjected to high centrifugal loading cycle to increase the compressibility of fan and compressor. Hence Ti-6Al-4V can be tailor-made to impart optimum combination of yield strength and fatigue property [3].As per the stringent airworthiness certification requirements, any rotating component needs to be forged to improve the better properties and structural integrity of the components at critical operating conditions. Normally Ti-6Al-4V forging is carried out in α - β condition below β transus temperature. The final microstructure of the forging is dictated by the volume fraction of α and β, which decides the mechanical property of Ti-6Al-4V [2-7].

Much of the research centered on improving the microstructure and attaining appropriate mechanical properties [1-10]. Earlier work has shown that the initial microstructure such as the fraction of α and β phases, the morphology and thickness of α-lathes as well as the size of α colonies (i.e. the geometrical arrangement of α and β-phases) have significant effects on the mechanical properties of the Ti64alloy[11-12].Although there are many investigations reporting the relationship between the HCF and strength of titanium alloy, however the relationship between the HCF, strength and volume fraction in bimodal microstructure has rarely been investigated. Therefore, in the present study, the specimens with three different volume fractions of alpha and transformed beta were prepared and their HCF properties were investigated

2. Experimental work

The material used in this study was Ti-6Al-4V alloy rod of size Ø85 mm x 210 mm. Material was supplied by the TIMET Metal Corporation. The Material received (VAR - Vacuum Arc Remelted) was in rolled condition. The β transition temperature for the material was 995°C as given by supplier considering top, bottom and middle portion of the ingot. The chemical composition of the input material is given in table 1. As received material was deformed at a temperature of 935°C (1 Hr. soaked) to a size of 202x189x31mm. To achieve different volume fraction of alpha and transformed beta phase, the deformed specimens were undergone to Solutionizse at 967°C and then quenched with quench delay of 30, 50 and 70 sec, in water followed by aging at 730°C. In this study laboratory-scale muffle furnace was used. The uniformity achieved inside the furnace was +/- 3 °C. It was a temperature controlled furnace heated by electric coils and could achieve a maximum of 1200 °C and having facility of varying the heating rate up to max.12 °C /min.

Table 1. Chemical composition

Element	Al	V	Fe	Sn	O_2	Cu	Mo	Ti
Weight[%]	6.43	4.11	0.14	<0.04	0.2	<0.04	<0.04	Balance

Vickers hardness tests were performed under a 500 g load. Tensile specimen (diameter 6.25 and gage length of 25 mm) and fatigue (gage length of 15 mm and width of 6 mm) specimens were machined parallel to the longitudinal direction as per ASTM standards. Tensile tests were performed at constant crosshead speed with an initial strain rate of 0.005 min-1. Fatigue behaviour was examined in a rotating bending fatigue test according to the DIN50113 standard on the INSTRON make RBF test machine with a frequency of 50 Hz and R = -1 with a sinusoidal wave form; three samples were tested on each stress condition. Then stress-life (S–N) curves were drawn, and fatigue limits were measured when the specimens did not fail at 10^7 cycles.

3. Results and discussion

3.1 Microstructure and tensile properties

Fig. 1a shows the as-received microstructure of the deformed condition prior to heat-treatment. It is seen that bimodal structure composed of equiaxed α embedded in transformed beta matrix. After solution treatment at 967°C for 1hr, water quenching and ageing at 730 for 2hr of the as received deformed samples, it was noted that the volume fraction of alpha and transformed beta varied with the quench day (Fig.1b-d) and size of alpha grains gets reduced. With the increase of quench delay from 30, 50 and 70 sec, the volume fraction of alpha increased from 20% to 67%.This is due to increasing quench delay after solution treating which allows the alpha phase to precipitate and grow, which further reduced the tensile properties

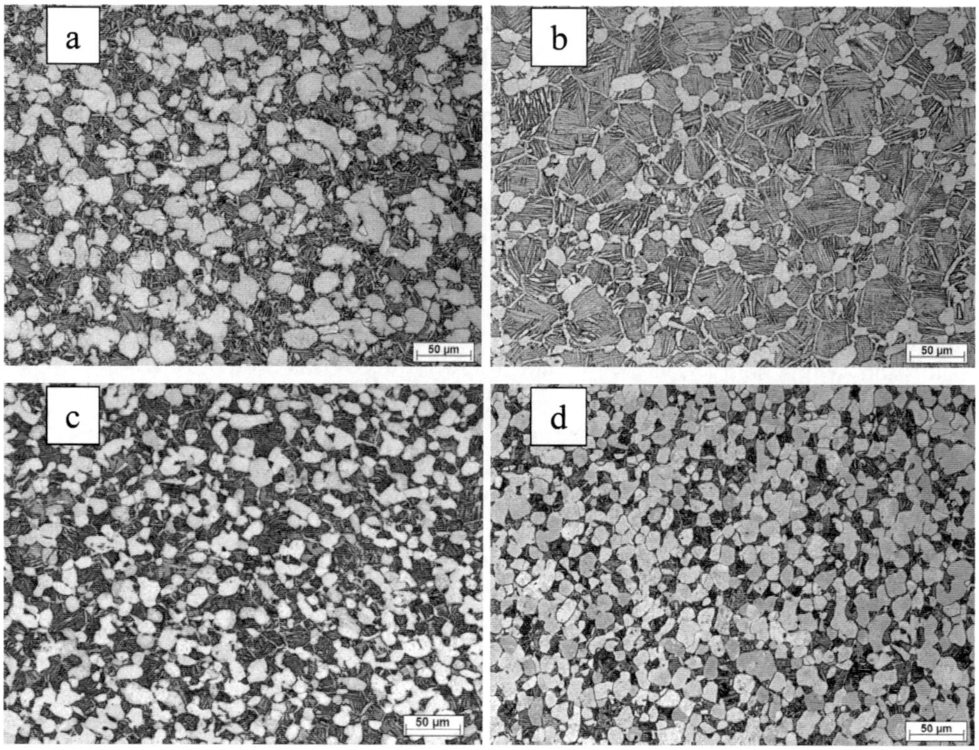

Figure.1 Optical micrograph of the specimens: (a) as-received deformed specimen (b) 30 sec quench delay (20% alpha phase) (c) 50 sec quench delay (43% alpha phase) and (d) 70 sec quench delay (67% alpha phase).

Table.2 Mechanical properties of specimen used in this study

Specimen (Quench Delay, Sec)	Alpha Phase (%)	Transformed Beta phase (%)	Ultimate tensile Strength (MPa)	Yield Strength (MPa)	Hardness (HRC)	Fatigue Limit (MPa)
30	20	80	1073	1034	37.2	630
50	43	57	1006	998	35.6	610
70	67	33	993	950	32.5	585

Table 2 shows the mechanical properties and the volume fraction of specimen used in this study. Tensile strength and hardness values of heat-treated specimens were higher in lower alpha phase, and decrease with the increase of the volume fraction of alpha.

3.2 Fatigue properties

Fig. 2 shows the S–N data of the specimens containing various volume fraction of alpha and transformed beta.Corresponding fatigue strength values measured at 107 cycles were presented in Table 3. The specimen having the highest strength (containing 20% volume fraction of alpha) represented higher fatigue limit. For bi-modal microstructures is the alloy element partitioning effect increasing with increase in volume fraction of alpha and leading to a lower basic strength within the lamellar part of the bi-modal microstructure as compared to a fully lamellar structure [13]. Thus HCF strength is lowered with increasing volume fraction of alpha phase. The fatigue cracks are nucleated in lamellar grains of bi-modal structure. These lamellar grains are softer than alpha grains as a consequence of alloy partitioning effect. The continuous decline in HCF strength with increasing volume fraction of alpha shows that, at lower stress amplitudes with limited slip activity over short distances, the decline in basic strength of lamellar regions (alloy element partitioning effect) is stronger than the positive contribution due to the reduced alpha colony size[13]. Despite the similar slip lengths in the equiaxed grains and α colonies, the cracks initiated consistently within primary α grains, but short crack propagation by faceted growth occurred through both α grains and α colonies. Elemental partitioning during α + β processing[13] results in a higher concentration of Al in the equiaxed α as compared to lamellar α. This causes increased slip planarity and preferential crack nucleation in the primary α grains.

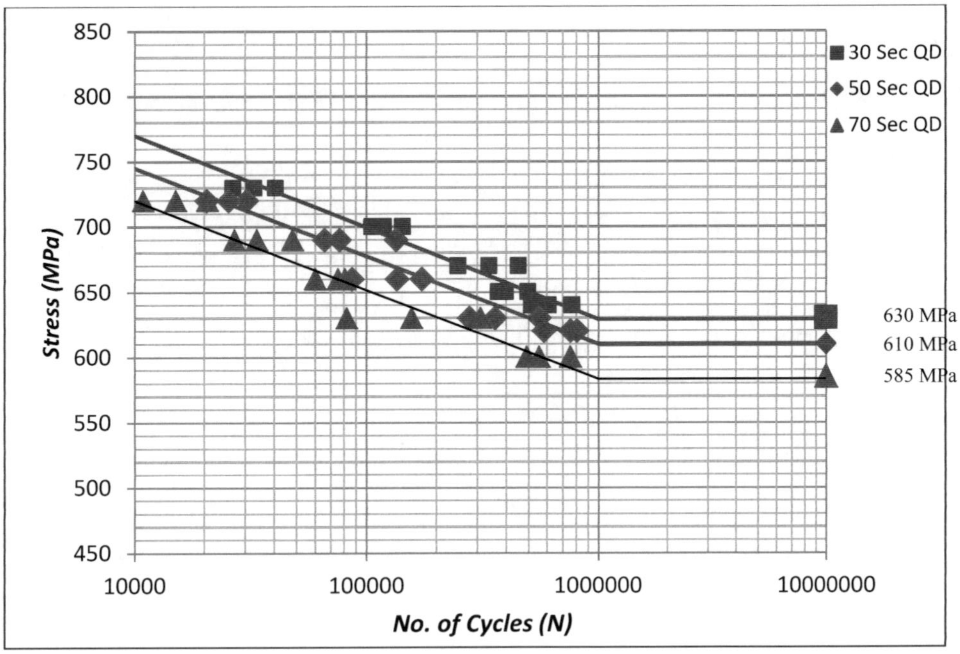

Figure.2 S–N curve for the specimens containing various volume fraction of alpha and transformed beta

3.3 Fractography

In the fatigue damage stage, cyclic stress attacks both surface and internal defects, and it is considered that the site which is selected for main crack initiation is always competitive and depends on the size of the potential sites and the stress level. SEM analysis shows similar behavior for all samples that the transition of fatigue crack initiation originates from surface to the interior. The entire fatigue fracture surface can be separated into three typical zones (see Fig. 3(a)) as crack initiation, crack propagation and final fracture. Fig3 (b) showed the image of crack initiation zone. Several facets (as marked by the arrow) usually formed in globular primary α can be commonly observed. F. Bridier et al. [12] observed such facets always formed near the basal plane and it propagated more rapidly than prismatic cracks, indicating that basal slip played an important role in crack initiation during high cycle fatigue. Fig. 3(c) showed the morphology of the steady crack propagation zone, where slow crack growth resulted in the formation of very fine striation-like features and the space expanded gradually with the increase of the distance from crack initiation area [13]. High magnification observation of this region (Fig. 3c) revealed a large amount of very fine microscopic cracks distributed parallel to the major stress axis. The fast crack growth zone was associated with dimples (Fig. 3d), which indicated the locally ductile failure mechanism. Neither inclusions nor pores were detected at the initiation site of the investigated fracture surface.

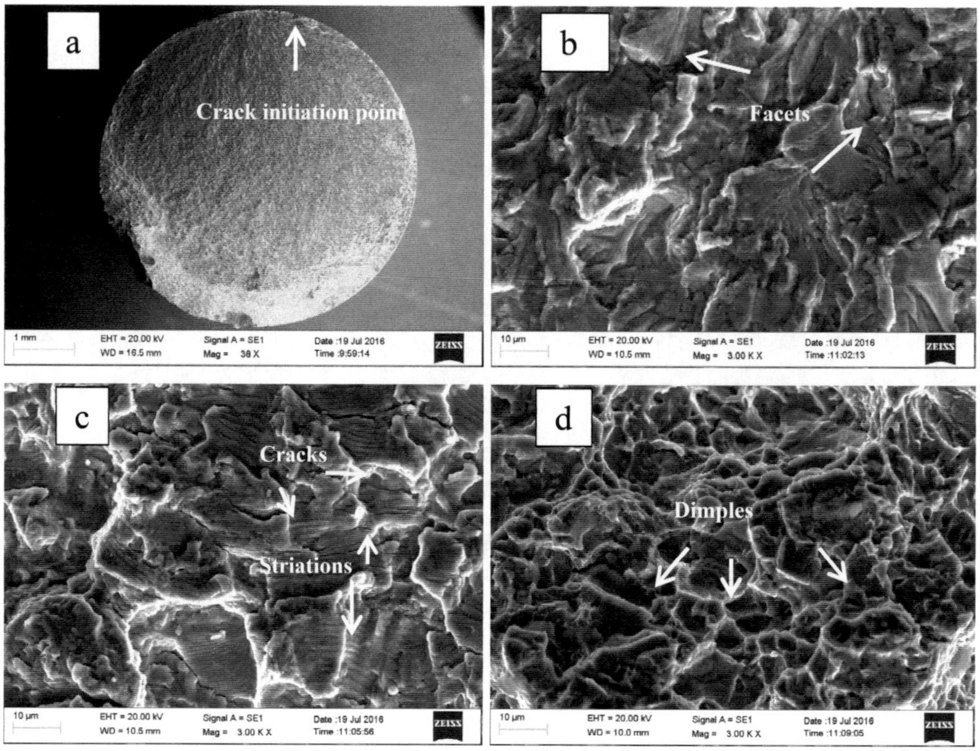

Figure.3 SEM micrographs of a) Fatigue fracture surface b) Crack initiation zone c) Crack Propagation zone d) Final fracture zone

4. Summary

The high cycle fatigue properties of bimodal Ti6Al4V titanium alloy were investigated in this study in relation to the volume fraction of Alpha and transformed beta. An increased in quench delay from 30sec, 50sec to 70sec during Solution Treatment process resulted into variable volume fractions of alpha and transformed beta, where alpha percentage was found to be increase to 20%, 43% and 67%, respectively. Reduction in tensile strength and hardness was recorded with an increase in quench delay and the subsequent alpha percentage. Excess increase in volume fraction of alpha in comparison with transformed beta contributed to reduced fatigue strength. The entire fatigue fracture surface can be separated into three typical zones as crack initiation, crack propagation and final fracture. There was no major difference notified on the effect of change in volume percentage of alpha on the fatigue behavior phenomenon. The transition of fatigue crack initiation was observed from surface to interior for all samples.

5. Acknowledgment

The authors gracefully acknowledge Kalyani Centre for Technology and Innovation (KCTI) – R & D institute of Bharat Forge Ltd for providing the facilities for heat treatment, metallography and microstructural analysis.

6. References

[1] G.Q. Wua, C.L. Shi, W. Sha, A.X. Sha, H.R. Jiang, Effect of microstructure on the fatigue properties of Ti–6Al–4V titanium alloys, *Materials and Design* **46** (2013) 668–674.

[2] Anuradha Nayak Majila, D. Chandru Fernando, S.N. Narendra Babu, B.V.A. Patnaik, and N.E. Prasad, Evaluation of Tensile Properties and their Correlation with Microstructural Characteristics of a Closed Die Forging of Iso-symmetrical Aerospace Grade Ti-6Al-4V Alloy,

Defence Science Journal, Vol. **65**, No. 2, March 2015, pp. 171-178.

[3] Babu, J.; Dutta, Abhijit; Kumar, Amit & Raghu, T. Flow behaviour of Ti-6Al-4V subjected to step temperature isothermal forging. *Int. J. Emerging Tech. Adv. Eng.*,2012, **2**(2), 321-325.

[4] Mithun, Kuruvilla; Srivatsan,T.S.; Petraroli, M. & Lisa, Park. An investigation of microstructure, hardness, tensile behaviour of a titanium alloy: Role of orientation. Sadhana, 2008, 33(3), 235-250.

[5] Nalla, R.K.; Boyce, B.L.; Campbell, J.P.; Peters, J.O. & Ritche, R.O. Influence of microstructure on high-cycle fatigue of Ti-6AI-4V: bimodal vs lamellar structure. *Metallurgical Material Trans*. A., 2002, **33**(3), pp. 899.

[6] Whittaker, M.T.; Evans, W.J.; Lancaster, R.; Harrison, W. & Webster, P.S. The effect of microstructure and texture on material properties of Ti6-4. *Int. J. Fatigue*, 2009, **31**(11-12), 2022-30.

[7] Filip, R.; Kubiak, K.; Ziaja, W. & Sieniawski, J. The effect of microstructure on the mechanical properties of two-phase titanium alloy. *J. Mater. Process Techno,*2003, **133**(1-2), 84-89.

[8] Ding, R.; Guo, Z.X. & Wilson, A. Microstructural evolution of a Ti-6Al-4V alloy during thermomechanical processing. *Mater. Sci. Eng.,* 2002**, 327**(2), 233-245.

[9] M. Jane£eka, F.Novýb, P. Harcubaa, J. Stráskýa, L. Trkoc, M. Mhaeded and L. Wagnerd, The Very High Cycle Fatigue Behaviour of Ti6Al4V Alloy, *Proceedings of the International Symposium on Physics of Materials (ISPMA13),*2015,497-502

[10] K. Kubiak , J. Sieniawski, Development of the microstructure and fatigue strength of two phase titanium alloys in the processes of forging and heat treatment, *Journal of Materials Processing Technology* **78** (1998) 117–121.

[11] H. Zuoa, Z.G. Wang , E.H. Han ,Effect of microstructure on ultra-high cycle fatigue behavior of Ti–6Al–4V, *Materials Science and Engineering A* **473** (2008) 147–152.

[12] F. Bridier, P. Villechaise, J. Mendez, Slip and fatigue crack formation processes in an α/βtitanium alloy in relation to crystallographic texture on different scales. *Acta Mater*, **56** (2008),3951-3962.

Reference to a book:

[13] G.Lutjering, J.C.Willaims, *Engineering materials and processes, Titanium, second edition,* springer Berlin Heidelberg New York.

Correlation analysis of heat flux and fire behaviour and hazards of polycrystalline silicon photovoltaic panels

Xiaoyu Ju[1], Xiaodong Zhou[1], Fei Peng[1], Zhibo Wu[1], Dimeng Lai[1], Yue Hu[1] and Lizhong Yang[1, 2,*]

[1] State Key Laboratory of Fire Science, University of Science and Technology of China, Hefei, People's Republic of China
[2] Collaborative Innovation Center for Urban Public Safety, Hefei, People's Republic of China

yanglz@ustc.edu.cn

Abstract. This work aims to gain a better understanding of fire behaviour and hazards of PV panels under different radiation heat fluxes. The cone calorimeter tests were applied to simulate the situations when the front and back surfaces are exposed to heat flux in a fire, respectively. Through comparison of ignition time, mass loss rate and heat release rate, it is found that the back-up condition is more hazardous than face-up condition. Meanwhile, three key parameters: flashover propensity, total heat release and FED, were introduced to quantitatively illustrate fire hazards of a PV panel.

1. Introduction

Currently, many countries like the United States, Germany and China proposed "solar roof" projects, where Building integrated photovoltaic (BIPV) is an important representative of them. As an emerging technology installed on residential and commercial buildings, the security of BIPV is a prime concern, that is, it should be ensured that photovoltaic (PV) panels will not cause any damage to the buildings nor harm to the residents. However, fires in such constructions are relatively common. In the past few years, severe building fires involving PV arrays have been reported, such as the fires in Munich in 2012 and in Apple in 2015, both of which the big fires began with a small fire from the rooftop PV system. PV panels attached to buildings are closely related to construction safety. Consequently, there have been lots of efforts to establish rigorous safety standards to mitigate the potential risks [1-5]. As to researches on fire performance of PV panels, there are seldom investigations involved in it. The objective of the present study is to contribute to the understanding of the correlation between external heat flux and flammability and fire hazards of PV panels.

2. Experimental setup

As shown in figure 1, this study selected small typical polycrystalline silicon PV panel, 100mm$\times$100mm$\times$3.5mm in size. The initial mass of a panel is equal to 135±2g. Every PV panel has a five-layer structure. The experiment is divided into two conditions, face up (figure 1(a)) and back up (figure 1(b)), to simulate the situations when the front and back surfaces are exposed to heat flux in a fire, respectively.

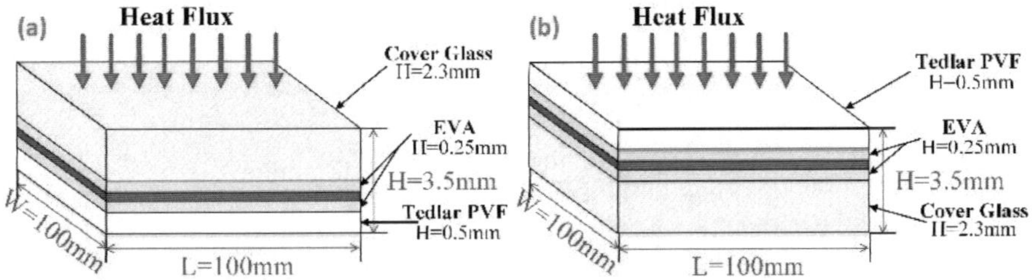

Figure 1. PV panels in the test: (a) face up and (b) back up.

The Cone calorimeter tests (CCTs) were performed according to ISO 5660 standard procedures. According to preliminary experiments, 25, 30, 35, 40 and 45kW/m^2 were chosen in the experiments.

3. Results and discussion

Figure 2 presents the evolution of ignition times (t_{ig}) of PV panels as a function of external heat flux ($\dot{q}''_{ext}$). It can be seen that the increase of radiation heat flux induces a rapid decrease in ignition time.

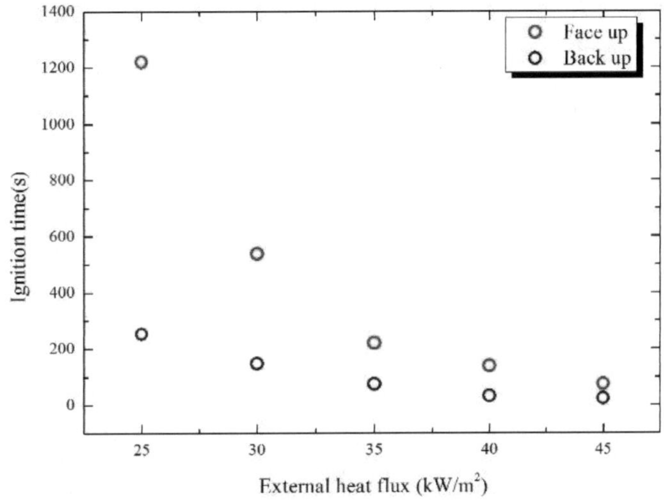

Figure 2. Evolution of ignition time versus heat flux.

Based on previous theory, for the thermally-thick materials, $t_{ig}^{-0.5}$ is proportional to $\dot{q}''_{ext}$ [6]:

$$t_{ig}^{-0.5} \propto \dot{q}''_{ext} \tag{1}$$

While for the thermally-thin materials, t_{ig}^{-1} is proportional to $\dot{q}''_{ext}$ [7]:

$$t_{ig}^{-1} \propto \dot{q}''_{ext} \tag{2}$$

Figure 3(a) and (b) displays t_{ig}^{-1} and $t_{ig}^{-0.5}$ versus external heat flux, respectively. It can be seen the plot of t_{ig}^{-1} as a function of the external heat flux is not linear, while $t_{ig}^{-0.5}$ is linear. The fitted lines in figure 3(b) are written as:

$$t_{ig}^{-0.5} = 0.0046\dot{q}''_{ext} - 0.09477 \tag{3}$$

$$t_{ig}^{-0.5} = 0.0077\dot{q}''_{ext} - 0.14421 \tag{4}$$

Deduced from equations (3) and (4), the critical heat fluxes for PV panels are 20.8kW/m^2 (face up) and 18.8kW/m^2 (back up), respectively. The observations allow us to conclude that the PV panel is thermally thick. Meanwhile, the back-up condition is more hazardous.

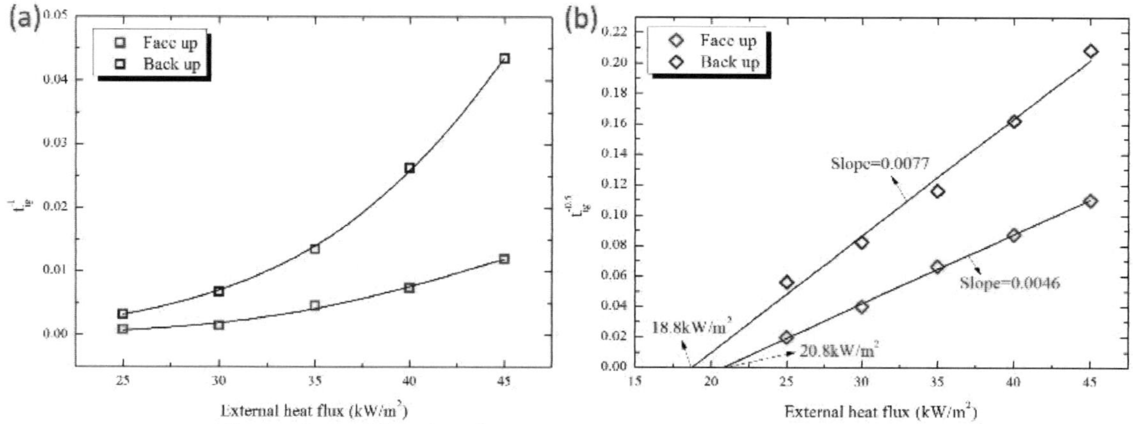

Figure 3. Ignition data for PV panels exposed to different heat fluxes: (a) inverse of the ignition time; (b) square root of the inverse of the ignition time.

Generally, larger mass loss rate means rapider release of pyrolysis gas, faster flame propagation and greater risk to be ignited for the material. Figure 4 presents the transient evolutions of mass loss rates (MLR), where MLR peak amplitude increases and MLR peak width decreases with an increase of the irradiance level. Comparing face-up and back-up curves, it is found that the latter burns more violently, which shows shorter experimental times and higher MLR peaks. This is because the cover glass hinders the burning of solar panels in the face-up condition.

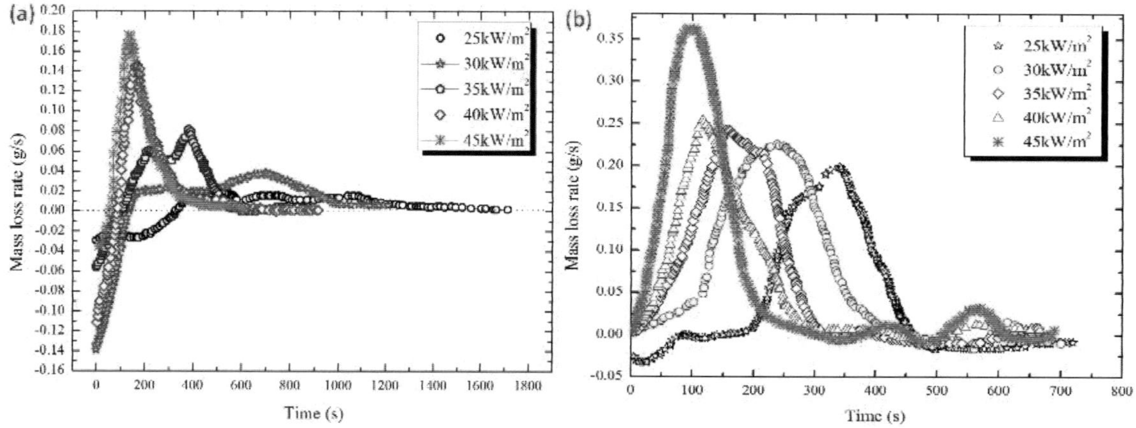

Figure 4. Mass loss rates for PV panels: (a) face up and (b) back up.

In addition, the averaged specific mass loss rate (SMLR) also increases with heat flux shown in figure 5. These observations show that the increase of heat flux accelerates the decomposition of the composite and decreases its thermal resistance. Meanwhile, back-up condition is more hazardous than face-up condition.

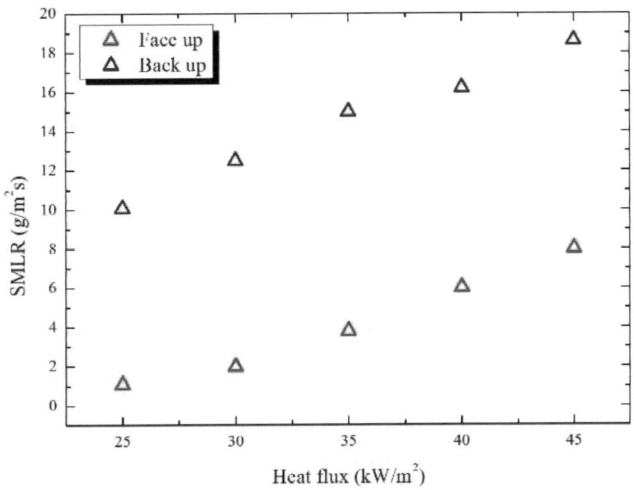

Figure 5. Specific mass loss rates for PV panels.

Figure 6 shows the heat release rates of PV panels at five irradiance levels. It is observed that peak heat release rate (PHRR) increases with external heat flux, which implies that an increase of radiation level enhances the risk factor of the panels. Meanwhile, the PHRRs of back-up conditions are higher than that of face-up conditions under the same heat fluxes, which means back-up condition is more hazardous.

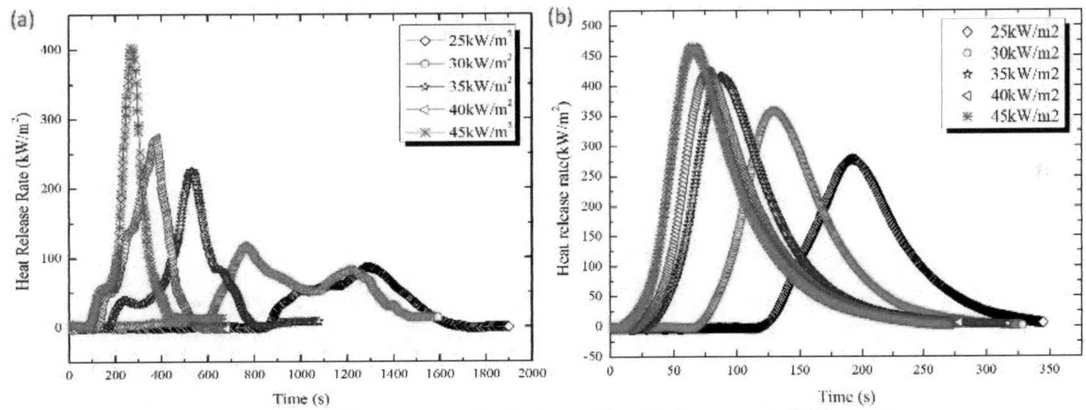

Figure 6. Heat release rates for PV panels: (a) face up and (b) back up.

In order to quantitatively illustrate fire hazards of a PV panel under different radiation level, Petrella mentioned that a combination of flashover propensity and total heat release (THR) would give reasonable indications of fire thermal hazard of the material [8]. Flashover propensity (kW/m^2s) is given as the ratio of PHRR to ignition time. The rating system presented by Petrella is shown in [8]. Figure 7 shows the flashover propensity and total heat release of PV panels. For face-up condition, flashover propensities of PV panels are low risk when heat flux is less than or equal to $35kW/m^2$ but intermediate risk at $40kW/m^2$ or more and the heat contributions are intermediate risk. For back-up condition, flashover propensities of PV panels are intermediate risk when heat flux is less than or equal to $40kW/m^2$ but high risk at $45kW/m^2$ or more and the heat contributions are intermediate risk. In general, the back-up condition have higher values of both parameters, which indicates a more dangerous situation.

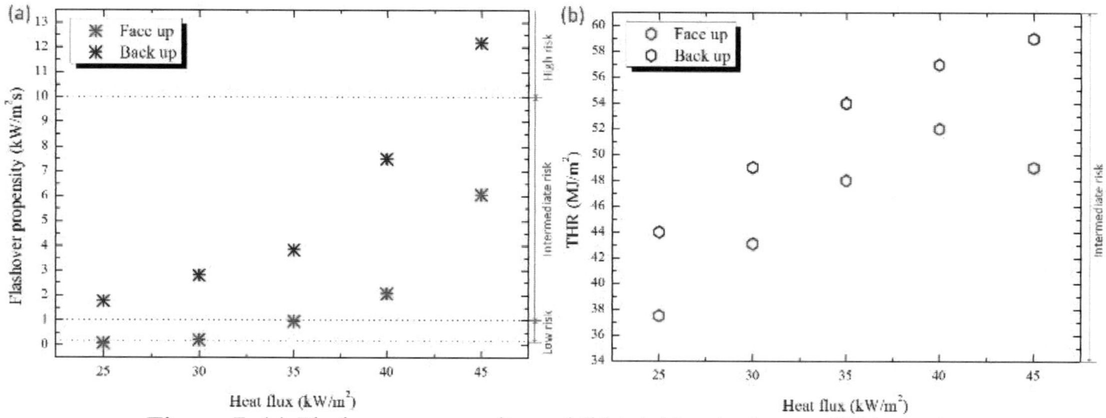

Figure 7. (a) Flashover propensity and (b) total heat release of PV panels.

In addition, smoke toxicity is equally important in hazards assessment of a material. The peak fractional effective dose (FED) can be estimated by the peak concentration of CO and its toxic potency LC_{CO} [9, 10]:

$$FED = \frac{peak[\mathrm{CO}]}{5000} \tag{5}$$

The measured transient carbon monoxide concentration, denoted by [CO] (in ppm), is shown as a function of time and external heat flux in figure 8 and CO concentration is higher in face-up condition. This is because the cover glass hinder the spread of fire and thus a less complete combustion occurred in this condition.

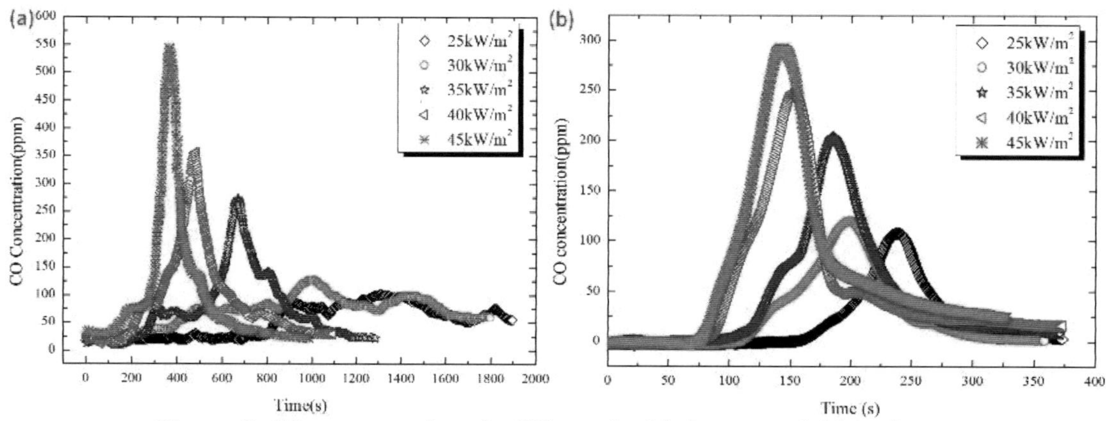

Figure 8. CO concentrations for PV panels: (a) face up and (b) back up.

Deduced from equation (5) and figure 8, the calculated *FED* value is plotted in figure 9. It is observed that the face-up condition has a higher FED value, which indicates that it is more hazardous for this condition. Nevertheless, both FED values are within the range of 0.020-0.110. Since toxic gases can cause death when FED value is about 1 [11], it means that CO emitted by PV panels is negligible.

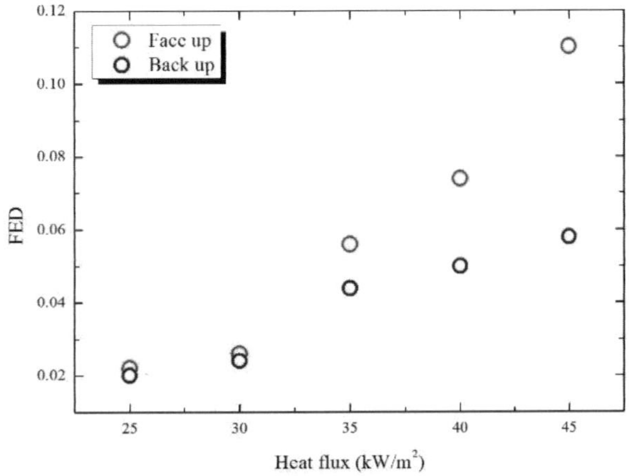

Figure 9. FED of PV panels.

4. Conclusions

This work aims to gain a better understanding of fire behaviour and hazards of PV panel under different radiation heat fluxes. The cone calorimeter tests were applied to simulate the situations when the front and back surfaces are exposed to heat flux in a fire, respectively. Through the analysis of ignition time, mass loss rate and heat release rate, it is found that the back-up condition is more hazardous than face-up condition. Meanwhile, three key parameters: flashover propensity, total heat release and FED, were introduced to quantitatively illustrate fire hazards of a PV panel. In addition, both FED values of the two conditions are within the range of 0.020-0.110, which means that CO emitted by PV panels is negligible.

Further, from the analysis of the studied parameters, it can be seen that PV panels are highly sensitive to the change of external heat flux. Thus, it is of great significance for scientists and manufacturers to improve the flame retardant properties of PV panels to make them safer in practical use.

Acknowledgments

This work was supported by National Natural Science Foundation of China [No.51476158], and the Fundamental Research Funds for the Central Universities [No.WK2320000033]. The authors deeply appreciate the supports.

References

[1] International Electrotechnical Commission 2004 IEC 61730-2 *Photovoltaic (PV) Module Safety Qualification-Part 2: Requirements for Testing*

[2] Underwriters Laboratories Inc. 2004 UL 1703 *Flat-Plate Photovoltaic Modules and Panels*

[3] Underwriters Laboratories Inc. 2002 UL 1256 *Fire Test of Roof Deck Constructions*

[4] Cancelliere P 2014 PV electrical plants fire risk assessment and mitigation according to the Italian national fire services guidelines *Fire Mater.* **26** 2029-2035

[5] National Fire Protection Association 2014 NFPA70 *National Electrical Code*

[6] Hopkins Jr D and Quintiere J G 1996 Material fire properties and predictions for thermoplastics *Fire Saf. J.* **26** 241-268

[7] Patel P, Hull T R, Stec A A and Lyon R E 2011 Influence of physical properties on polymer flammability in the cone calorimeter *Polym. Adv. Technol.* **22** 1100-1107

[8] Petrella R V 1994 The assessment of full-scale fire hazards from cone calorimeter data *J. Fire Sci.* **12** 14-43

[9] Babrauskas V 1997 Sandwich panel performance in full-scale and bench-scale fire tests *Fire Mater.* **21** 53-65

[10] Han S S and Chow W K 2005 Calculating FED and LC_{50} for testing toxicity of materials in

bench-scale tests with a cone calorimeter *Polym. Test.* **24** 920-924
[11] Institution of British Standard *Determination of the lethal toxic potency of fire effluents*

IOP Conf. Series: Materials Science and Engineering **201** (2017) 012037 doi:10.1088/1757-899X/201/1/012037

Life prediction of l6 steel using strain-life curve and cyclic stress-strain curve by means of low cycle fatigue testing

Sanket Inamdar[1*], Manoj Ukhande[2], Prashant Date[3], Dattaprasad Lomate[1], Shyam Takale[2], RKP Singh[1]

[1]Kalyani Centre for Technology & Innovation, Bharat Forge Ltd., Pune, India -411036
[2]R&D CDFD Engineering, Bharat Forge Ltd., Pune, India-411036
[3]Indian Institute of Technology Powai, Mumbai. India-400076

Email: sanket.inamdar@bharatforge.com

Abstract. L6 Steel is used as die material in closed die hot forging process. This material is having some unique properties. These properties are due to its composition. Strain softening is the noticeable property of this material. Due to this in spite of cracking at high stress this material gets plastically deformed and encounters loss in time as well as money. Studies of these properties are necessary to nurture this material at fullest extent. In this paper, numerous experiments have been carried on L6 material to evaluate cyclic Stress - strain behavior as swell as strain-life behavior of the material. Low cycle fatigue test is carried out on MTS fatigue test machine at fully reverse loading condition R=-1. Also strain softening effect on forging metal forming process is explained in detail. The failed samples during low cycle fatigue test further investigated metallurgically on scanning electron microscopy. Based on this study, life estimation of hot forging die is carried out and it's correlation with actual shop floor data is found out. This work also concludes about effect of pre-treatments like nitro-carburizing and surface coating on L6 steel material, to enhance its fatigue life to certain extent.

1. Introduction

Forging is a metal forming process used to manufacture critical geometry components which are difficult to manufacture by other modes of production. Metal Forming, it is a process that has been constantly revolutionizing and evolving since the concept of fire. In today's era, the markets of automotive, aerospace, defense and industrial products are controlled by the countries which are producing the accurate metal parts using most cost efficient process control [1]. As long as there is no material stronger cheaper, and abundant then metallic material, the world will forever revolve around those, who can control and form it the best. Even though forging is not a new science. It has been constantly getting evolved with the help of super computers and finite element modeling. Recently it has become more popular and inexpensive way of crafting net shaped and near to net shaped metal products with less design time and cost. As technology is progressing, the spectrum of physical size of product being generated is becoming larger with smaller tolerances. In that forging is standing well ahead of all.

Wear of forging dies is a predominant issue and that is uncontrollable. Another unwanted phenomenon is fatigue failure and plastic deformations, occurring frequently in hot forging process. Prediction of die failure due to plastic deformation will be an enormous step for the forging industry [2].

2. Experimentation

2.1 Pre-Processing on Material
The die material selected for present investigations is L6 steel. Cylindrical shaped samples were used for performing thermal fatigue experiments. Sample preparation involved following stages;

1. Heat treatment of raw material (annealed) with different austenizing temperatures.
2. Machining to dimension as per ASTM E606, with surface roughness of 0.3 to 0.5 Ra.
3. Nitriding using ammonia gas nitriding to obtain desired case depth

Heat treatment was carried out using two different austenizing temperatures followed by triple tempering to obtain final hardness values of 42 HRC; Later these samples were machined to finished dimension as mention above and subject to surface nitriding treatment using ammonia gas nitriding technique.

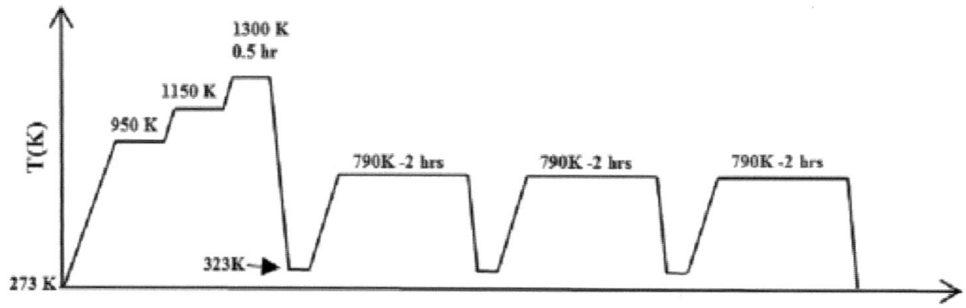

Figure. 1: Heat Treatments and Triple Tempering for 42 HRC.

(A) Heat Treatment, (B) Triple Tempering.

2.2 The governing equations used for heating and cooling are as below
I. Heat transfer equation,

$\nabla(-k\nabla T)=Q-(\rho Cpu.\nabla T)$

Where, k – Thermal Conductivity,

Q – Heat source,
ρ – Density,
C_p – Specific heat capacity at constant pressure &
T – Temperature

II. Induction heating with high frequency

$Qv=\sqrt{3}\ V\ I \cos\emptyset$

Where, Q_v – Power per unit volume of heat source,
V – Voltage,
I – Current,
φ – Power factor,

2.3 Boundary Conditions
1. Heating

$I = 220$ A for $(0+it) < t < (5+it)$ where, t = Time
$V = 320$ V for $0 < t < 5$ $t = 0 – 4200$ seconds
$\varphi = 85 – 90$ % $i = 0:1:300$
$f = 12$ kHz where, f = Frequency [5]

2.4 Low Cycle Fatigue Test on MTS:
Low cycle fatigue testing is carried out in Kalyani Centre for Technology and Innovation, to generate strain life curves for L6 Steel Material at various conditions. These conditions are as follows.

IOP Conf. Series: Materials Science and Engineering **201** (2017) 012037 doi:10.1088/1757-899X/201/1/012037

1. Plain sample without any surface treatment at room temperature (23 C)
2. Plain sample without any surface treatment at elevated temperature (350 C)
3. Nitrided sample at room temperature (23 C)
4. Nitrided sample at elevated temperature (350 C)

Elevated temperature is taken 350 C as this temperature is measured immediately after forging of a component. Nitriding parameters are same as of sample S2C1. Parameters specified for tests are as follows. The specimen, used for this test is dumbbell shaped threaded sample, as shown in figure 2

Table 1. Parameters used for Low cycle fatigue testing

Sr. No	Parameter	Room Temperature	Elevated Temperature
1	Frequency	3Hz	3Hz
2	No. of cycles	20000	20000
3	Strain	0.001-0.1	0.001-0.1

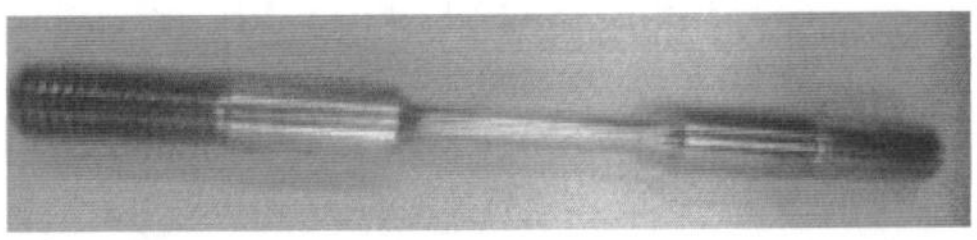

Figure 2.: Specimen for Low Cycle Fatigue Test

This specimen is loaded on MTS machine and strain value is provided to software of this machine. With extensometer attached to gauge length of the sample we can control and measure strain in specimen. If sample completes 20000 axial cycles then sample is removed and next sample is mounted on machine with increase in strain rate. Loaded sample on machine is shown in figure 5.16.

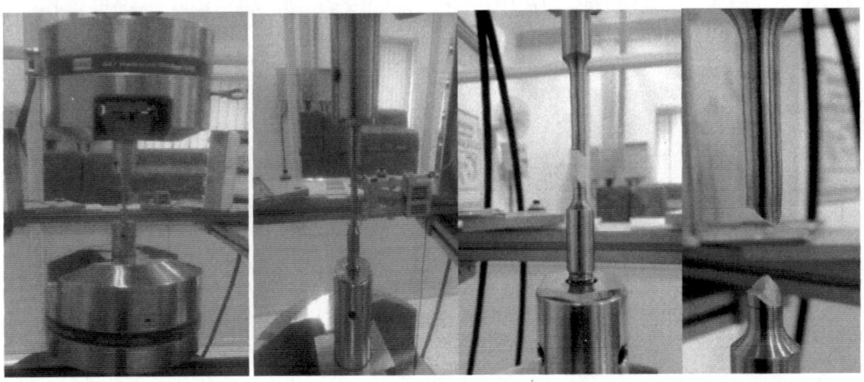

Figure 3. Specimen for Low Cycle Fatigue Test mounted on MTS

3. Results and Discussion

3.1 Tensile Test results and analysis

Tensile test has been carried out to find out material properties and also to analyze deformation of the material. Following figure 3.4 shows stress strain curve for die steel material. By this curve we can say that yield point of this material is 1000 MPa and UTS is around 1250 MPa. This curve helps to differentiate between plastic phase and elastic phase.

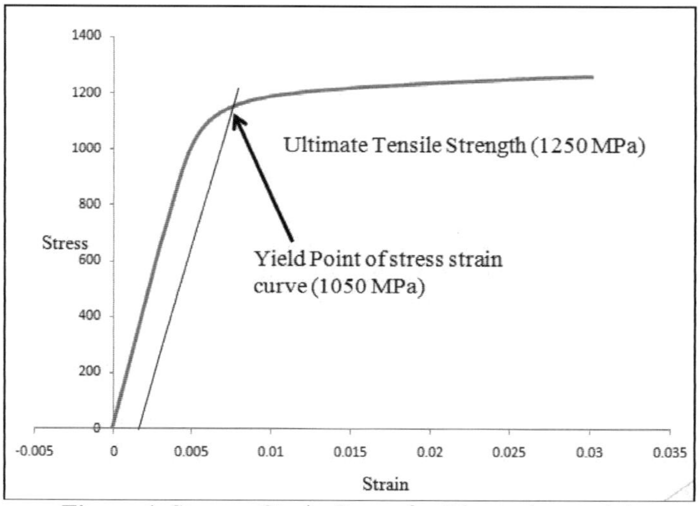

Figure 4: Stress – Strain Curve for Die steel material

From Stress – Strain Curve it is concluded that, if stresses in die are above yield point (1050 MPa), plastic deformation is prone to happen in die. So maintaining die stresses below 1050 MPa leads to only elastic deformation of the die. This elastic deformation cannot be controlled as this is prone to happen in the die. Limiting this die deflection to a certain limit is also important. Maximum deflection of the die in above case is 0.89 mm, maximum strain in die is 0.007 and maximum stresses are around 1145 MPa. Stresses up to yield point i.e. 1000 MPa is consider as elastic stresses so equivalent strain to 1000 MPa is 0.005. Considering these values are limiting values for elastic deformation. And above that can be considered as plastic deformation. Now to analyze the case for front axle beam die up to 1000 MPa it is elastic deformation and for next 145 MPa stress leads to Plastic deformation. Till yield point there is very less plastic strain present in the material. Similarly strain values till 0.005 are considered as elastic strain while next 0.002 strain will cause plastic deformation, hence 0.002 is considered as plastic strain. This plastic strain leads to plastic deformation. In front axle beam die. Elastic and plastic combined deflection is 0.89 mm. By stress strain curve 0.83 mm is elastic deformation while 0.06mm deformation is plastic deformation.

Table 2. Actual Stresses and Strains Present in Specimen by Low cycle fatigue testing

Strain	Elastic strain	Plastic Strain
0.002	0.00198	0.00002
0.003	0.00295	0.00005
0.004	0.00385	0.00015
0.0051	0.0041	0.001
0.0062	0.0042	0.002
0.0075	0.0041	0.0034

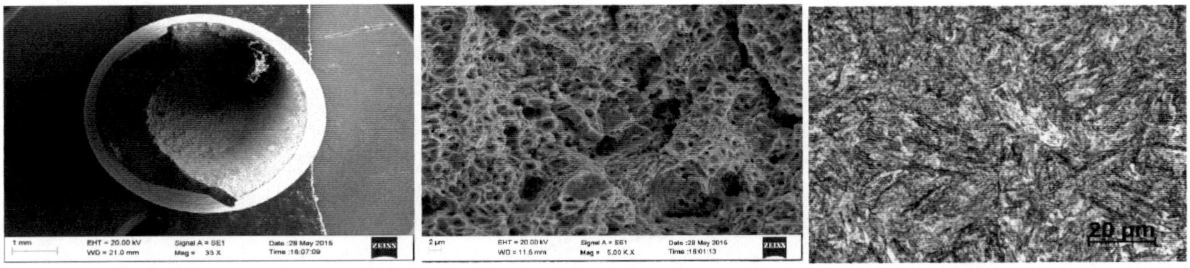

Figure 5: Failed Specimen of die material, SEM images and Microstructure

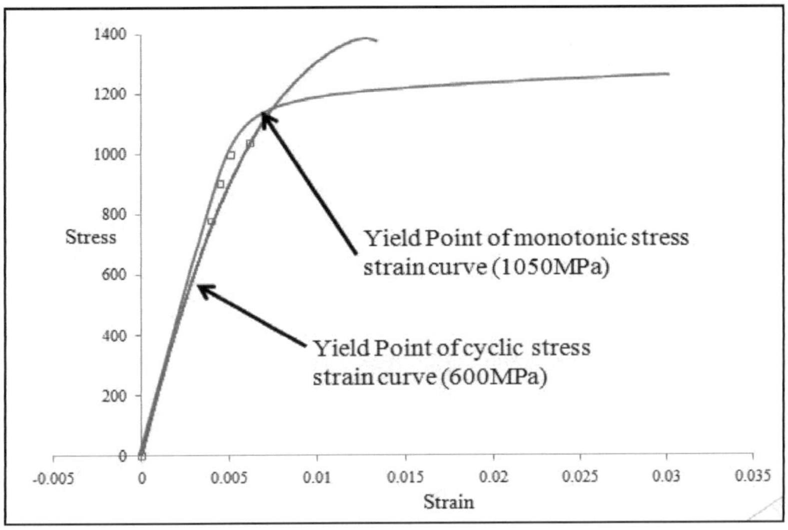

Figure 6. Cyclic Stress – Strain Curve for Die steel material

3.2 Cyclic Testing results and Analysis

To analyze actual deformation in hot forging dies it is important to observe this process from start. Because, any forming process is a cyclic process. And to analyze this process we need to generate cyclic stress-strain curve and give it as an input to analysis software so that results from that analysis will be more accurate.

For doing the same low cycle fatigue testing of die material was conducted and stress strain curves from the same is generated. By observing this stress strain curve, we can say that die material is softening at high stress zone. So to conclude this test we can consider it is a property of special steel like die material to get strain softened. Also its yield point reduces and it is almost 60-62 % of original yield point.

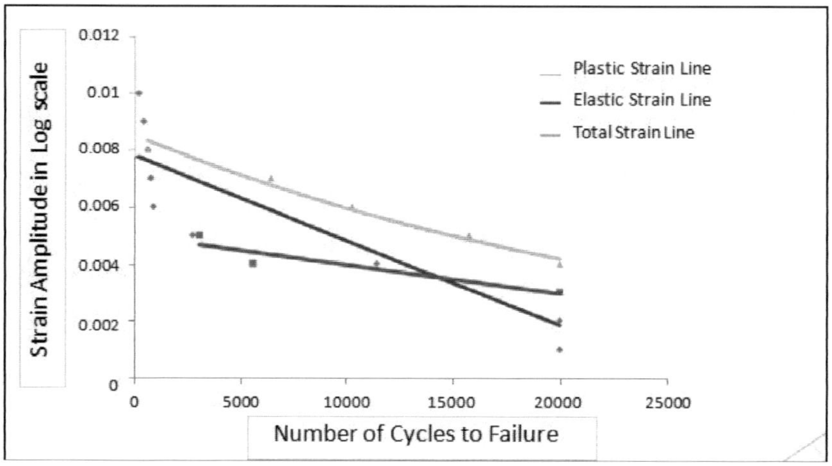

Figure 7. Strain life curve of Material

So SEM and Micro structural analysis of this material is carried out. This shows ductile failure of the material. This is mainly due to low resistance from dislocations inside the material at high stresses. Following are the results for analysis of the die with cyclic stress strain curve.

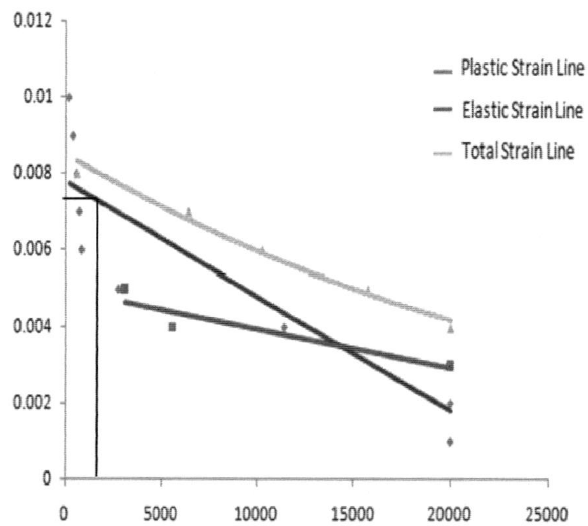

Sr. No.	Actual number of Parts produced	Predicted number of parts	Correlation %
1	480	590	77%
2	870	990	76%
3	1500	7650	20%

Figure 8: Life prediction with strain life curve

4. Conclusions

Metal forming is a cyclic process, so analysis of this process by monotonic stress-Strain curve fails to give accurate values of deformation. So to do accurate analysis instead of using monotonic stress strain curve, cyclic stress-strain curve is used. These values we get from are very nearer to actual shop floor values. Brittle material like die steel fails ductile while loaded in cyclic manner. It concludes that, Strain Softening is prone to happen in the material. Strain softening is occurring due to carbon rejection in BCT phase tempered martensite transforms in to Soft martensite. This reduces yield point of the material from 1050 MPa to 600MPa. So to analyze the deformation, cyclic stress strain curve is useful rather than using monotonic stress strain curve.

Acknowledgment

The authors gratefully acknowledge the extended support provided in providing the facility and infrastructure to this work by KCTI (Kalyani Centre for Technology and Innovation). The authors also acknowledge the support provided by Bharat Forge Ltd, Pune and DSIR, Govt. of India. The authors would also like to express special thanks and gratitude to review committee and top management of Bharat Forge Ltd. for granting the permission to publish/present the research work.

Referances:-

[1] Bannantine, Julie A., Comer, Jess J, Handrock, James L. "Fundamentals of Metal Fatigue Analysis". Prentice Hall. Englewood Cliffs N.J. 1990.
[2] Bay, N., P. Skov-Hansen, J. Gronbaek, P. Brondsted. "Fatigue in Cold Forging Dies: Tool Life Analysis." Journal of Materials Processing Technology. Vol. 95. 1999: 40-48.
[3] Bayer, Raymond G., "Wear Analysis for Engineers.", New York: HNB Publishing, 2002.
[4] Berger, Paul D. & Maurer, Robert E., "Experimental Design: With Applications in Management, Engineering, and the Sciences.", Duxbury Thomson Learning, 2002.
[5] Caddell, Robert M., & William F. Hosford. Metal "Forming: Mechanics and Metallurgy.", PTR Prentice Hall Inc. Englewood Cliffs, N.J. 1993.
[6] Castro, J.T.P., M.A. Meggiolaro. "Statistical Evaluation of Strain-Life Fatigue Crack Initiation Predictions.", International Journal of Fatigue. Vol. 26. (2004). 463-476.
[7] Collins, Jack A., "Failure of Materials in Mechanical Design: Analysis, Prediction, Prevention", Wiley-Inter science Publication 1993.
[8] Dean, T.A., D.J. Jeong, D.J. Kim, J.H. Kim, B.M. Kim. "Effects of Surface Treatments and Lubricants for Warm Forging Die Life.",Journal of Materials Processing Technology. Vol. 113. 2001: 544-550.
[9] Dobychin, M.N., I.V. Kragelsky, V.S. Kombalov., "Friction & Wear: Calculation Methods." Pergamon Press. USSR. 1982.
[10] Engel, U., Falk, B., & Geiger, M., "Estimation of Tool Life in Bulk Metal Forming Based on Different Failure Concepts.", Journal of Materials Processing Technology. Vol. 80-81. 1998: 602-607.
[11] Farahmand, Bahram.,"Fatigue and Fracture Mechanics of High Risk Parts.", Chapman & Hall International Thomson Publishing. 1997.
[12] Fatemi, Ali, Fuchs, Henry O., Stephens, Ralph I., & Stephens, Robert R., "Metal Fatigue in Engineering.", 2nd edition, John Wiley & Sons. 2001.

IOP Conf. Series: Materials Science and Engineering **201** (2017) 012038 doi:10.1088/1757-899X/201/1/012038

Embedding of copper sulfate and copper oxide on multipurpose paper

D L V Almanza[1], J L A de Luna[2], and M U Herrera[1]

[1] Institute of Mathematical Sciences and Physics, University of the Philippines Los Baños, College Laguna, 4031 Philippines
[2] San Pablo City Science High School, San Pablo City, 4000 Philippines

dlvalmanza@up.edu.ph

Abstract. Copper sulfate salts were embedded on multipurpose paper using simple soaking technique while copper oxide particles were embedded using *in-situ* technique. In simple soaking technique, the papers were simply soaked in copper sulfate solution in order for the copper salts to be incorporated in the paper. In the *in-situ* technique, the copper sulfate-embedded papers were soaked in sodium hydroxide solution for reactions to occur that will lead to the formation of copper oxide. Copper sulfate-embedded papers have blue green color while copper oxide-embedded papers have brown color. The copper sulfate-embedded paper shows excellent antimicrobial property against *Staphylococcus aureus*. Meanwhile, the copper oxide-embedded paper shows small zone of inhibition against *Escherichia coli* and *Staphylococcus aureus*.

1. Introduction

Copper-based materials [1-12] such as copper metal, copper sulfate, copper oxide, have antimicrobial properties. Copper sulfate is a salt, which has high solubility in water. This salt has many applications; they are used as antifungal and antiseptic agent, they are also used for the prevention and cure of foot rot in animals, eliminate algae in pools and many more. Meanwhile, copper oxide is a p-type semiconductor with a band gap of 1.7 eV and has a brown color. It has promising application in the creation of solar cells, gas sensor, photodetectors, supercapacitor, photocatalysis, and antimicrobial products. Copper oxide can be synthesized by reacting copper salts (e.g., copper sulfate) with precipitating agent (e.g., sodium hydroxide) to form metastable copper hydroxide, which in turn could oxidize to copper oxide.

Paper is a flexible substrate, which can be easily tailor-fitted to many different forms to suit different applications. The use of multipurpose paper (copier/bond paper) would be very convenient and practical since it is very cheap and widely available almost anywhere. The incorporation of material on paper [13-18] would create product, which would hybridize the flexibility of the paper and functionality of the immobilized material. Copper oxide embedded on paper has many possible applications such as creation of antimicrobial sheets that can be used as patches, covering, lining and the likes. The use of copper-based materials is much cheaper than the use of silver-based particles [19-22], which are commonly used as inorganic antimicrobial agents.

The objective of this research is to embed copper sulfate on 90-gsm multipurpose paper using soaking technique and copper oxide using *in-situ* technique. In soaking technique, the paper is simply immersed in the copper sulfate solution in order for copper ions to penetrate in the pores. Meanwhile, in *in-situ* technique, the copper sulfate-embedded paper is immersed in sodium hydroxide solution for reactions to happen to form the embedded copper oxide particles. The use of copper sulfate and sodium hydroxide as reacting materials are commercially advantageous since they are cheap and widely available.

2. Methodology

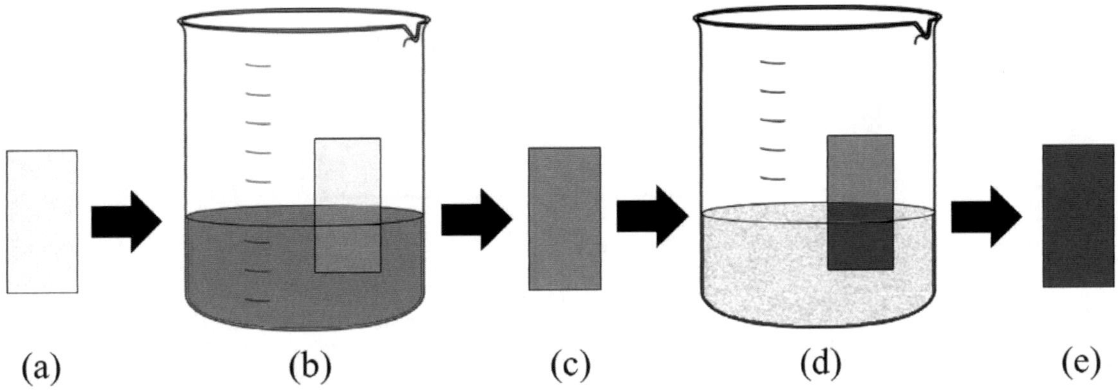

Figure 1. Embedding of copper sulfate salts (a-c) and copper oxide particles (c-e) on multipurpose paper.

Copper sulfate salts were embedded on the paper using soaking technique (Figure 1a-c) while copper oxide particles were embedded using in-situ technique (Figure 1c-e).

In the soaking technique, the paper was immersed in copper sulfate solution for 180 minutes. The copper sulfate solution was done by mixing copper sulfate pentahydrate ($CuSO_4 \cdot 5H_2O$) in distilled water. After drying, the soaking would result to a blue-colored paper indicating the presence of copper sulfate salt.

In the in-situ technique, the copper sulfate-embedded paper was immersed in sodium hydroxide solution for 180 mins. After drying, the treated paper changed color to brown as indication of the formation of copper oxide particles.

The images of copper sulfate-embedded and copper oxide-embedded papers were taken using HP Desket 2510 Series. The color of the images was analyzed using ImageJ software, in such, the color of the samples was resolved into three components: red, green, and blue (RGB). The RGB values ranges from 0 to 255 wherein the colors become lighter when it approaches 255 while it became darker when it approaches 0.

Furthermore, agar diffusion test was used to determine the antimicrobial property of the 6-mm diameter samples. In this test, the zone of inhibition (ZOI) of the samples was taken.

IOP Conf. Series: Materials Science and Engineering **201** (2017) 012038 doi:10.1088/1757-899X/201/1/012038

3. Results

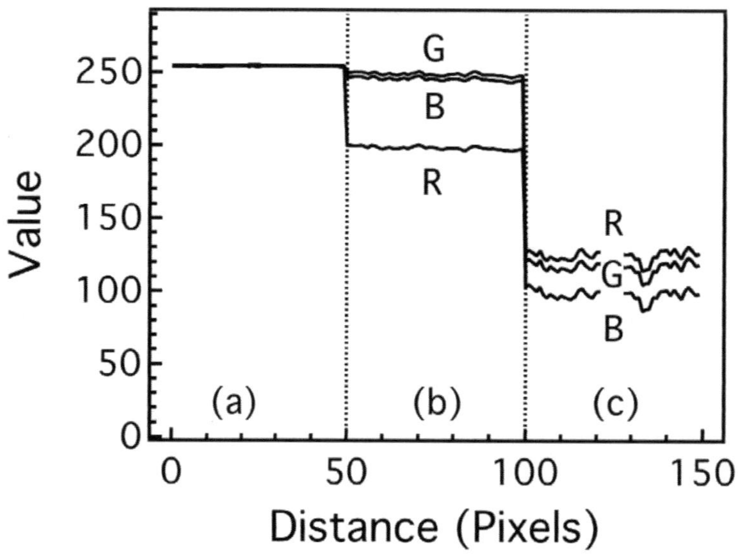

Figure 2. Red, Blue and Green (RGB) values of (a) multipurpose paper (0-49 pixel), (b) copper sulfate-embedded paper (50-99 pixel), and (c) copper oxide-embedded paper (100-149 pixel). The values are taken from different 50-pixel segments.

Table 1. Red, Blue and Green (RGB) values of the different samples and the unevenness of their distribution. The unevenness values are taken to be the standard deviation of RGB values on the 50-pixel segment present in Figure 1.

Sample	RGB values			Unevenness (standard deviation)		
	R	G	B	R	G	B
a. Multipurpose Paper	254	255	255	0.2	0.2	0.2
b. Copper sulfate-embedded in paper	198	249	246	1.0	1.0	1.1
c. Copper oxide-embedded in paper	125	117	98	4.0	4.0	4.0

Table 2. Antimicrobial property of 6-mm diameter copper sulfate-embedded paper against *Escherichia coli* and *Staphylococcus aureus*.

Samples	Zone of inhibition (mm)	
	E. coli (gram negative)	*S. aureus* (gram positive)
Copper sulfate-embedded in paper	0.0	20.3
Sterile distilled water (- control)	6.0	6.0
Streptomycin (+ control)	19.7	15.3

Table 3. Antimicrobial property of 6-mm diameter copper oxide-embedded paper against *Escherichia coli* and *Staphylococcus aureus*.

Samples	Zone of inhibition (mm)	
	E. coli (gram negative)	*S. aureus* (gram positive)
Copper oxide-embedded in paper	8.0	13.3
Sterile distilled water (- control)	6.0	6.0
Streptomycin (+ control)	16.0	16.3

4. Discussion

We have embedded copper sulfate salts and copper oxide particles on multipurpose paper. Figure 2 shows the Red, Blue and Green (RGB) values of (a) multipurpose paper (0-49 pixel), (b) copper sulfate-embedded paper (50-99 pixel), and (c) copper oxide-embedded paper (100-149 pixel). For presentation purpose, the RGB values of the three types of paper are condensed in one illustration, each occupying 50-pixel segment. Table 1 summarizes the average RGB values of the 50-pixel segment. Meanwhile, the standard deviations of RGB values at the 50-pixel segment are taken to be a measure of unevenness or non-uniformity.

4.1. Embedding of copper sulfate salts on multipurpose paper

Figure 2b shows the RGB values of the multipurpose paper, after it was soaked in copper sulfate solution. The RGB values changed from 254, 255 and 255 (Table 1a) to 198, 249, and 246 (Table 1b), this is from white color to blue color. The bluish color indicates the presence of embedded copper sulfate salt on the paper. In aqueous solution, copper sulfate salt dissolves into ions enabling them to penetrate to the pores and small spaces on the soaked paper. The ions recrystallized into salt when the paper was removed from the solution and then dried. The unevenness of the RGB color distribution are 1.0, 1.0 and 1.1 (Table 1b), respectively. These values are small, which indicates uniformity in color distribution.

Table 2 shows the antimicrobial properties of the copper sulfate-embedded paper against *Escherichia coli* and *Staphylococcus aureus*. The zone of inhibition of the 6-mm diameter paper sample against *S. aureus* (gram-positive) is 20.3 mm; this value is greater than that of the positive control (Streptomycin) which is 15.3 mm. However, no zone of inhibition was formed against *E. coli* (gram-negative). This was attributed to the stronger cell wall of this gram-negative bacteria.

4.2. Embedding of copper oxide on multipurpose paper

Figure 2c shows the RGB values of the copper sulfate-embedded paper, after it was soaked in sodium hydroxide solution. The RGB values changed from 198, 249, and 246 (Table 1b) to 125, 117, and 98 (Table 1c), this is from blue to brown color. This brownish color indicates the presence of embedded copper oxide particles on the paper. Sodium hydroxide reacts with the embedded copper sulfate salt to form metastable copper hydroxide, which oxidized to form copper oxide particles. However, the unevenness of the RGB color widens to 4.0, 4.0, 4.0 (Table 1c) indicating higher non-uniformity of color distribution.

Table 3 shows the antimicrobial properties of the copper oxide-embedded paper against *Escherichia coli* and *Staphylococcus aureus*. The zone of inhibition of the 6-mm diameter paper sample against *S. aureus* (gram positive) is 13.3 mm; this is lower than that of the positive control (Streptomycin) which is 16.3 mm. The higher zone of inhibition of copper sulfate-embedded paper compare with copper oxide-embedded paper against *S. aureus* is attributed with the higher ability of copper sulfate to be dispersed in aqueous solution due to its higher solubility. Meanwhile, the zone of inhibition against *E. coli* (gram negative) is 8.0 mm, which is a little more than the size of the paper (6 mm). This is far less than the positive control (Streptomycin) which is 16.0 mm.

5. Summary and Conclusion

Copper-based materials (copper sulfate and copper oxide) were embedded on multipurpose paper for antimicrobial applications. Copper sulfate salts were embedded by soaking the paper on copper sulfate solution, while copper oxide was embedded by soaking the copper sulfate-embedded paper in sodium hydroxide solution to undergo series of reactions. After the paper was soaked in copper sulfate solution, its color changes to blue color, indicating the presence of embedded copper sulfate salt. After the paper was embedded in sodium hydroxide solution, the color changes to brown indicating the presence of copper oxide particles. The copper sulfate-embedded paper shows zone of inhibition of 20.3 mm against *S. aureus;* this value is higher than that of the positive control. Meanwhile, the zone of inhibition of copper oxide-embedded paper against *E. coli* and *S. aureus* are 8.0 mm and 13.3 mm, both are lower than the positive control.

Acknowledgement

The author are grateful for the funding support given by the University of the Philippines Los Baños Basic Research Program.

References

[1] Abramova A, Gedanken A, Popov V, Ooi E, Mason T, Joyce E, Beddow J, Perelshtein I, Bayazitov V. 2013. *Mater Lett.* **96** 121-124
[2] Das D, Nath C, Phukon P, and Doluia SK 2012. *Colloid Surface B.* **101** 430-433
[3] Ipeksac T, Kaya F, Kaya C. 2014. *Mater Lett.* **130** 68-70
[4] Sonia S, Jayram ND, Kumar PS, Mangalaraj D, Ponpandian N, and Viswanathan C 2013. *Supperlattice Microst.* **66** 1-9
[5] Zhang Q, Zhang K, Xu D, Yang G, Huang H, Nie F, Liu C, Yang S 2013. *Prog Mater Sci.* **60** 208-337
[6] Horton DJ, Foster H, Binding H, and Scully J. 2015. *Electrochim Acta.* **169** 351-366
[7] Kruk T, Szczepanowicz K, Stefanska J, Socha R 2015. *Colloid Surface B.* **128** 17-22
[8] Mezey R-S, Máthé I, Shova S, Grecu M-N, Roşu T 2015. *Polyhedron.* **102** 684-692
[9] Palza H, Delgado K, Curotto N 2015. *Appl Surf Sci.* **357** 86-90
[10] Shankar S, Rhim J-W 2014. *Mater Lett.* **132** 307-311
[11] Vincent M, Hartemannc P, Engels-Deutsch M 2016. *Int J Hyg Envir Heal.* **219** 585-591
[12] Aarestrup F, Hasman H 2004. *Vet Microbiol.* **100** 83-89
[13] Burrs SL, Bhargava M, Sidhu R, Kiernan-Lewis J, Gomes C, Claussen J, McLamore E 2016. *Biosens & Bioelectron.* **85** 479-487
[14] Ge S, Wu K, Zhang Y, Yan M, Yu J 2016. *Biosens & Bioelectron.* **80** 215-221
[15] Li H, Wang W, Lv Q, Xi G, Bai H, Zhang B 2016. *Electrochem Commun.* **68** 104-107
[16] Oh SY, Kim J, Kim Y 2015. *Mater Lett.* **154** 60-63
[17] Martins N, Freire C, Neto C, Silvestre A, Causio J, Baldic G, Sadocco P, Trindade T 2013. *Colloid Surface A.* **417** 111-119
[18] Youssefa A, Kamelb S, El-Samahya MA 2013. *Carbohyd Polym.* **98** 1166-1172
[19] Li Z, Zhang M, Cheng D, Yang R 2016. *Carbohyd Polym.* **151** 834-840
[20] Ocwieja M, Adamczyk Z, Morga M, Kubiak K 2015. *Adv Colloid Interfac.* **222** 530-563
[21] Raho R, Paladini F, Lombardi FA, Boccarella S, Zunino B, Pollini M 2015. *Mat Sci Eng.* **55** 42-49
[22] Sathishkumar G, Gobinath C, Karpagam K, Hemamalini V, Premkumar K, Sivaramakrishnan S 2012. *Colloid Surface B.* **95** 235-24

IOP Conf. Series: Materials Science and Engineering **201** (2017) 012039 doi:10.1088/1757-899X/201/1/012039

Crack resistanceof glued laminated timber after operational effects

L. Stupishin[1], V. Kabanov[2]and A. Masalov[1]

[1]Urban, Road Construction and Structural Mechanics Department,South-West State University, 305040, 50-let Octyabrya st., 94, Kursk, Russian Federation
2Economics, Management and Politics Department, South-West State University, 305040, Kursk, 50-let Octyabrya st., 94, Russian Federation

lusgsh@yandex.ru

Abstract. The experimental data of crack length growth and crack surface displacements and critical stress intensity factor of glued timber members adhesive joints are presented in this paper.The influence on crack growth of cyclic loading, load duration and atmospheric impacts are considered.

1. Introduction

Various kinds of natural irregularities - knots, slope of grain, etc, are inherent to timber. Except natural irregularities some technological irregularities in timber members are possible - glued joins, notches and so on.

In contemporary design methods for glued laminated timber members those irregularities are taken into account with significant simplifying assumptions. More accurate accounting of irregularities in elements of timber structures in estimation of bearing capacity and durability of glued timber member in operational conditions would improve design effectiveness.

2. Test procedures

In South-West State University some experimental researches to obtain more data about behavior of glued laminated timber and its glue joints under the most severe operatingimpacts were made [1...4]. Large-sized beams were tested to obtain data about crack growth and bearing capacity reduction. Static loading during 540 days and cyclic loading up to 2 million cycles were made on beam specimens with cross-section dimensions 200x200 mm. Those specimens had preexisting cracks of I and II modes.

Small-sized specimens for I mode fracture also were tested for estimation of various types of atmospheric impacts on timber laminated members.

At Fig.1 the large-sized specimen during duration test is presented. Static loading was performed in specially designed lever installations with various levels of loading.

IOP Conf. Series: Materials Science and Engineering **201** (2017) 012039 doi:10.1088/1757-899X/201/1/012039

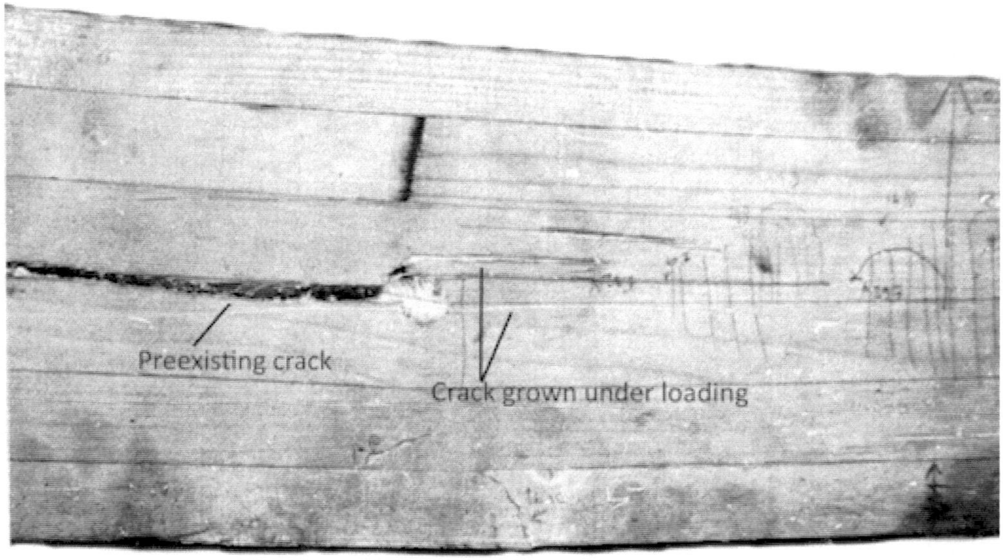

Figure1. Observations of crack growth in duration loading tests

At Fig.2 and Fig.3cyclic test is presented. It was performed also in specially designed installation of inertia-type principle of operation with various levels of loading and load asymmetry. The specimens also had preexisting cracks of I and II modes.

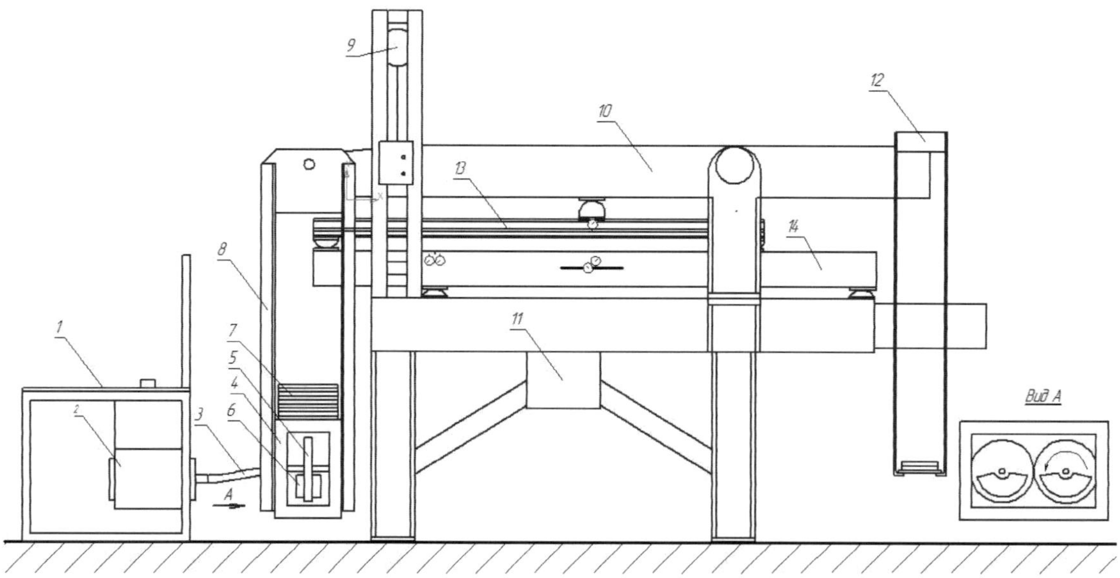

Figure 2. Installation for cyclic loading: 1 - control panel; 2 - DC motor; 3 - flexible shaft; 4 - vibrator body; 5 - vibrator gear wheel; 6 - eccentric load; 7 - loads: 8 - vibrator suspension; 9 - lever-moving winch; 10 - lever; 11 - frame; 12 - counterweight suspension; 13 - dynamometric traverse; 14 - large-sized specimen with dial gauges installed

Figure3. Observations of crack growth and crack face displacement during cyclic loading tests

3. Some results of large-sized specimen tests

Crack length observation during the static loading showed visible crack growth of both modes of cracks was not continuous but had a stepwise character. A synchronism of crack length growth steps and growth of crack surface displacements measured by dial gauges was not registered. At Fig.4 the diagram of shear crack length dependence on duration of loading is presented. At Fig.5 the diagram of shear crack length dependence on number of loading cycles is also presented.

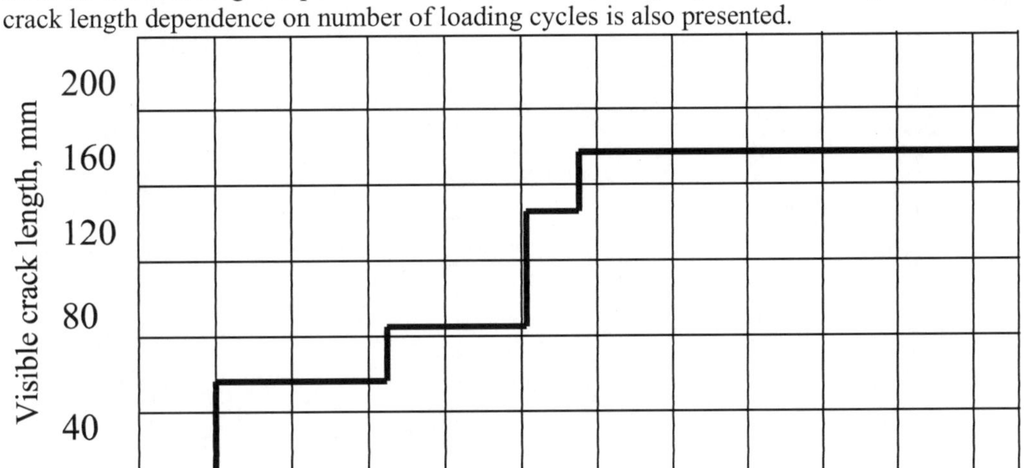

Figure4. Step-wise dependence of shear crack length on duration of loading

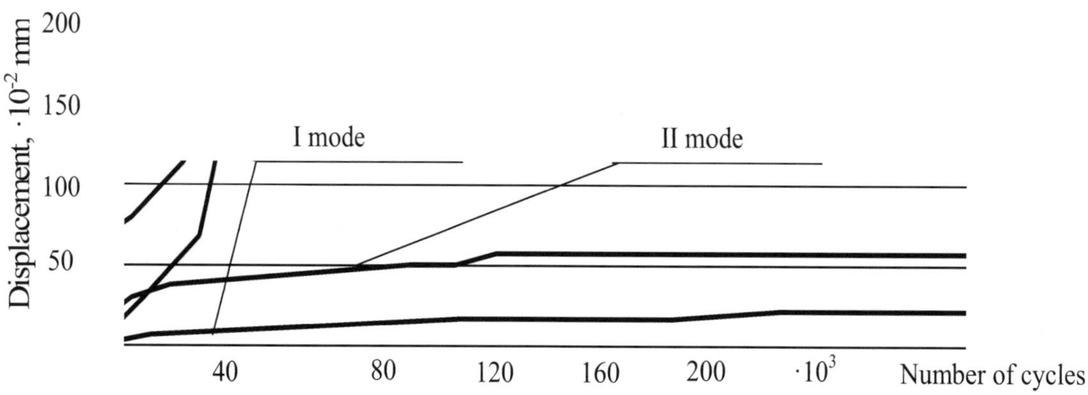

Figure5. Crack faces displacement - number of cycles of loading dependence

4. Some results of small-sized specimen tests

Critical stress intensity factor K1C (CSIF) of timber members adhesive joints on resorcinol glues on small-sized specimens accordingly to Russian standard method was experimentally researched [1] (Fig.6).

Figure 6. Small-sized specimen testing

The following operating impacts were made:

series A - storage at warehouse during 15 years;

series B - atmospheric exposure in subtropics climatic zone - Black Sea area (Sukhumi) during 28 months;

series C - duration static loading during 1 year in normal climatic zone (Kursk) with the following asymmetric loading fatigue tests up to 3.332 millions cycles at asymmetry loading factor R=0.2;

IOP Conf. Series: Materials Science and Engineering **201** (2017) 012039 doi:10.1088/1757-899X/201/1/012039

series D - atmospheric exposure in normal climatic zone during 48 months with the following moistening during 36 months and asymmetric fatigue loading at R=0.2.

Some specimen tests results after service effects are presented in Table 1.

Table 1. Fracture toughness of adhesive joints on FR-100 glue

Series	Average CSIF, K_{1C}, $MPa\cdot m^{1/2}$	Variation
A	0.401	0.18
B	0.433	0.18
C	0.483	0.18
D	0.279	0.25

5. Conclusion

The given results show "non-typical" behavior of large-sized specimens under long-term and cyclic loading. Visible step-wise growth of crack length do not coincides with crack face displacement growth. This phenomenon requiresfurther investigation.

It is important to note that fracture toughness of glued laminated timber is more sensitive to operating effects, than shear strength [2,3,4].

There are more possibilities to estimate failure process using not only values of stresses in timber members but also values of stress intensity factors (SIFs).

References

[1] Kabanov V., Masalov A., and Solianik J. 1996 *The second academic readings RAASN.- Kazan, Russia,* **IV**, 81-83.

[2] Kabanov V., Masalov A. and Solianik J. 1998: *Proc. of the 5th World Conference on Timber Engineering.- Montreux, Switzerland*, **2**, 796-797.

[3] Kabanov V. and Masalov A. 1995 *Proc. of the 5th International Symposium "Wood in engineering structures"*. – Kočovce, Slovakia, 44-50.

[4] Kabanov V. 2001. *Dissertation. Bratislava: SvF STU.*

IOP Conf. Series: Materials Science and Engineering **201** (2017) 012040 doi:10.1088/1757-899X/201/1/012040

Ti/IrO_2/SnO_2 anode for electrochemical degradation of chlorpyrifos in water: optimization and degradation performances

G C Pathiraja*[1,2], M S Wijesingha[3], N Nanayakkara[2,4]

[1]Department of Engineering Technology, Faculty of Technology, University of Ruhuna, Matara, Sri Lanka,
[2]Environmental Engineering/Electrochemistry Research Group, Institute of Fundamental Studies, Hantana Road, Kandy, Sri Lanka,
[3]National Centre for Non Destructive Testing, Sri Lanka Atomic Energy Board, Kelaniya, Sri Lanka,
[4]Department of Civil Engineering, Faculty of Engineering, University of Peradeniya, Sri Lanka.
*gpathiraja@gmail.com

Abstract: Chlorpyrifos, a widely used organophosphate pesticide which can be found in surface water bodies, is harmful for human body. Thus, treating water contaminated with chlorpyrifos is important. In our previous studies, novel Ti/IrO_2–SnO_2 anode was successfully developed for electrochemical degradation of chlorpyrifos in chloride free water. In this study, optimization of previously developed Ti/IrO_2–SnO_2 anode for mineralization of chlorpyrifos was successfully performed through response surface methodology. During the optimization study, two-level factorial design was used to determine the optimal coating solutions concentration for developing the Ti/IrO_2–SnO_2 anode. Cyclic voltammetry and open circuit potential were performed to investigate the electrochemically active surface area and stability of these anodes. The response surface and contour plots show that 0.3 M of [Ir] and 7.5 mM of [Sn] coated electrode has both highest anodic charge and stability. Scanning Electron Microscopic (SEM) images show the evidence of having both compact and porous regions in the surface of the thin film, resulting larger surface area. Within 6 h, the best result for mineralization (55.56%) of chlorpyrifos was obtained with 0.3 M of [Ir] and 7.5 mM of [Sn] coated anode using Total organic Carbon (TOC) analyzer. Therefore, the optimum coating concentration was found as 0.3 M of [Ir] and 7.5 mM of [Sn]. It would require an energy consumption of 6 $kWhm^{-3}$.

1. Introduction

Chlorpyrifos (CP), an organophosphate pesticide, has been used in agricultural pest control and in households as a termiticide. It is commonly used to control insects in agricultural crops such as cotton, grain, fruit, alfalfa, and sugar beets. CP is available in a variety of formulations under many trade names such as Dursban, Lorsban etc. [1,2] Chemical structure of CP is shown in Figure 01. Exposure to CP and its metabolites has been found related to a variety of health effects [3]. Therefore, the use of CP has been restricted in United States (U.S.) and in some European countries. Sri Lanka also took steps to ban CP in 2013, after importing many tons of CP to the country. The significant impacts caused by CP contamination have fostered the need for developing new treatment technologies for its removal.

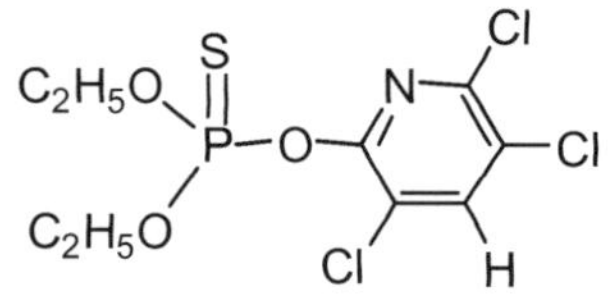

Figure 1. Chemical structure of CP

Dimensionally stable anodes (DSAs) have been widely employed in electrochemical degradation of toxic and non-biodegradable organic pollutants. DSAs typically consist of mixed metal oxide coatings on metal substrates. Titanium is one of the most studied substrate material for DSAs because it has good electrical conductivity, sufficient mechanical strength, and electrochemical inertness [4]. Coatings on substrate are generally produced by a combination of an active catalyst component and a non active secondary component. Coatings are usually prepared by thermal decomposition of precursor solutions [5]. Mainly, the oxidation (degradation) efficiency depends on the properties of the anode materials. The anodic oxidation is useful to directly destroy organic pollutants by reaction with hydroxyl radical ($OH^{\cdot}$) formed at the anode surface from water oxidation [6]:

$$H_2O \rightarrow OH^{\cdot}_{ads} + H^+ + e^- \quad \text{.....(1)}$$

Literature reveals that only few studies are available on the optimization of the anode materials to treat organic contaminants in water. Most of the previous studies on optimization have been used one-factor-at-a-time methodology. However, this approach is traditional, inefficient and time consuming, while fails to consider about possible interactions between factors in a process [7]. Response surface methodology (RSM) is a powerful statistical-based technique, which is able to simultaneously consider several factors at different levels using factorial design of experiments (DOE). Furthermore, it gives a suitable model for the relationship between various factors and the response and thus searching optimum conditions for desirable responses [8,9]. The two most common designs used in RSM are the Central Composite Design (CCD) and the Box–Behnken Design (BBD). The use of CCD is a popular design which provides standard and reliable data, while it can be efficiently used to assess interaction between variables.

Attempts have been made to develop two types of DSA (Ti/IrO_2 and Ti/IrO_2-SnO_2) electrodes for the electrochemical degradation of CP in our previous study [10]. In that study, efficiency in degradation of CP as well as the electrochemical properties of the developed anodes in chloride free electrolytes was carefully investigated. Moreover, it was reported that prepared Ti/IrO_2-SnO_2 anode shows better CP removal efficiency and electrochemical properties. Mechanical dip-coating of titanium plates in precursor solutions was a simple and cost-effective preparation method. In order to improve the efficiency of this anode further, it is recommended to optimize the most effective conditions for the preparation of such dip-coated Ti/IrO_2-SnO_2 anodes for the electrochemical degradation of CP.

The aim of the present work was to optimize and investigate the influence of concentrations of iridium chloride and tin chloride in the dip solution on anode performances. The performances were evaluated by electrochemically active surface area and stability of electrodes. Therefore, in this paper, CCD with five center points of RSM (2 factors and 2 levels) was employed to investigate the effect of coating metal concentration on the electrochemical activity of Ti/IrO_2-SnO_2 anode. The mineralization of CP during the electrolysis for best selected three electrodes, which have better electrochemical properties, was also investigated using Total Organic Carbon (TOC).

2. Materials and methods

All chemicals including commercially available sample of CP (274 mg L^{-1}), Na_2SO_4 anhydrous AR (99%, SDFCL), $IrCl_3.3H_2O$ (Ir -53-56 %, ACRDS Organics), Sn standard solution (Fisher Scientifics), ethanol (99% AR, Fisher Scientifics), isopropanol, $K_2Cr_2O_7$ (99%, Avondale Laboratories), H_2SO_4 (98%, Fisher Scientifics), were of analytical reagent grade. All aqueous solutions were prepared by using deionized (DI) water.

2.1. Development of anode

Titanium plates were used as substrates (Geometrical area ~ 1 cm^2). Prior to the dip coating, substrates were sandblasted followed by a chemical treatment to prepare for a good adhesive metal oxide film on the surface. Firstly, a titanium plates were polished by sand paper with water as lubricant. Secondly, chemical treatment was done using 5% (w/w) oxalic acid solution for 10 min and then 37% (w/w) HCl acid for 5 min. The substrate was dried at 100 ^{0}C until it reaches a constant weight. 0.56 g of $IrCl_3.3H_2O$, 1% of polyethylene glycol (1000 ppm) by Ir concentration was dissolved in 4.6 ml of ethanol. The treated Ti plates were dipped in this solution and then the wet coating surface was air dried in 80 ^{0}C air flow to evaporate solvents. Then electrodes were annealed in an oven at 450 ^{0}C for 10 min. After that, the electrode was washed with distilled water to make the pores in the layer. Then SnO_2 coating solution was prepared by dissolving 0.66 ml of Sn/$SnCl_4$ (1000 ppm) standard solution in 3.3 ml of isopropanol and applied on the washed electrode surface [11]. The wet coating surface was then air dried in 80 ^{0}C air flow to evaporate solvents and calcinated in an oven at 450 ^{0}C for 10 min. This process was repeated until achieving the final coating load of 1 mg/cm^2. After reaching the final coating load, the electrode was postbacked at 500 ^{0}C in muffle furnace for 1 h.

2.2 Design of experiments

In order to optimize the developed anode, a statistical based response surface modeling methodology was used to examine the dependence of anode performance by its electrochemical properties on the concentration of anode materials. CCD design was utilized with two-level factorial design to determine the optimal conditions for Ti/IrO_2–SnO_2 anode preparation. Two main variables are Ir concentration (X_1), Sn concentration (X_2) in the dip solution. These variables were simultaneously changed at highest (+1) and lowest (-1) levels and the range was 0.1 M – 0.3 M and 2.5 mM – 7.5 mM for Ir and Sn concentrations, respectively. A total of 13 experiments were employed in this work including five replicates at the central point. The experimental design used for this work is shown in Table 01.

Table 1. Design of experiments for the optimization of anode

Experiment	Virtual level of factors		Level of factors	
	I	II	[Ir] (moldm^{-3})	[Sn] (*10^{-3}) (moldm^{-3})
1	0.00000	0.00000	0.2	5
2	-1.00000	-1.00000	0.1	2.5
3	1.41421	0.00000	0.3414	5
4	0.00000	1.41421	0.2	8.53
5	-1.41421	0.00000	0.058	5
6	0.00000	-1.41421	0.2	1.46
7	0.00000	0.00000	0.2	5
8	-1.00000	1.00000	0.1	7.5
9	0.00000	0.00000	0.2	5
10	0.00000	0.00000	0.2	5
11	1.00000	-1.00000	0.3	2.5
12	1.00000	1.00000	0.3	7.5
13	0.00000	0.00000	0.2	5

In the case of Ir concentration (-1 = 0.1 M, 0 = 0.2 M, +1 = 0.3 M, 1+α = 1.41421 = 0.3414 M, 1-α = -1.41421 = 0.0586 M), in the case of Sn concentration (-1 = 2.5 mM, 0 = 5 mM, +1 = 7.5 mM, 1+α = 1.41421 = 8.5355 mM, 1-α = -1.41421 = 1.4645 M), and fixed Poly ethylene glycol (PEG) = 0.0100 g were used.

2.3 Electrochemical properties

Electrochemical parameters related to anodic activity were performed for all electrodes by cyclic voltammetry (CV) in the range of voltage scan from –2.5 to 2.5 V at a 0.1 V/s scan rate. CV is an indication of electrochemically active surface area of anodes. All tests were conducted using potentiostat/ galvanostat (Autolab, PGSTAT128N) in a three electrode cell; the electrodes under study were used as the working electrode (WE), titanium plate as the counter electrode (CE) and Ag/AgCl electrode was used as the reference electrode (RE). In order to assess the electrode stability of the prepared anodes, Open Circuit Potential (OCP) was performed. Na_2SO_4 (0.5 M) was used as the electrolyte and experiments were carried out at room temperature (25°C). Triplicate measures for all anodes were prepared in both the cases.

2.4 Mineralization of CP

The three electrodes which have higher electrochemically active surface area and stability out of thirteen electrodes were used to find the mineralization efficiency. The electrodes described above as an anode and a Ti plate as a cathode, were set at a distance of 1.0 cm. The supporting electrolyte was consisted of a 0.5 M of Na_2SO_4 solution. For every electrode, 100 ml of a fresh batch containing 80 ml of 5.48 mg/L of CP solution and 20 ml of Na_2SO_4 was electrolyzed at operating current density of 20 $mAcm^{-2}$. Appropriate amount of samples were collected during 6 hours and Total Organic Carbon (TOC) concentration was measured by using TOC analyzer (multi N/C 2100, Shimadzu) based on the combustion-infrared method. The energy consumption per volume of above water samples were estimated and expressed in $kWh\ m^{-3}$. During the electrolysis, the average cell voltage was taken for the calculation of the energy consumption as follows [10]:

$$\text{Energy consumption} = \frac{U_{cell}\, It}{V\ 3600} \quad \ldots\ldots\ldots (2)$$

where, U_{cell} is the average cell voltage (V), I is the current (A), t is the electrolysis time (s) and V is the volume (L).

3. Results and discussion

3.1 Analysis of electrochemically active surface area

Electrochemically active surface area was evaluated by cyclic voltammetry (CV) for the thirteen electrodes developed, according to the corresponding parameters obtained through CCD. Figure 02 (a) and (b) shows the response surface and contour plots of anodic charge, as a function of [Ir] and [Sn] used to develop the electrode, respectively. As it is obvious from response surface plot, the anodic charge was increased with Ir concentration and Sn concentration. 0.3 M of [Ir] and 7.5 mM of [Sn] coated anode shows the highest anodic charge, possibly due its higher IrO_2 content. Apart from above anode, 0.3414 M of [Ir] and 5 mM of coated anode, 0.058 M of [Ir] and 5 mM of [Sn] coated anode, 0.3 M of [Ir] and 2.5 mM of [Sn] coated anode shows comparatively high anodic charges. Anodic charge is an indication of the electrochemically active sites present at the surface of the coating. These results are in accordance with several previous studies [9,12,13]. They have reported that the surface electrochemistry of the IrO_2 type electrode is governed by the activity of IrO_2, independent from its IrO_2 concentration. In other words, although the concentration of IrO_2 is lower or higher than SnO_2, the electrochemical activity is higher in IrO_2. The anodic charge is closely linked to the probability of $OH^{\cdot}$ production due to high amount of electrochemically active sites. It can be speculated that more electrochemically active sites are present and a large surface area is available in all these anodes.

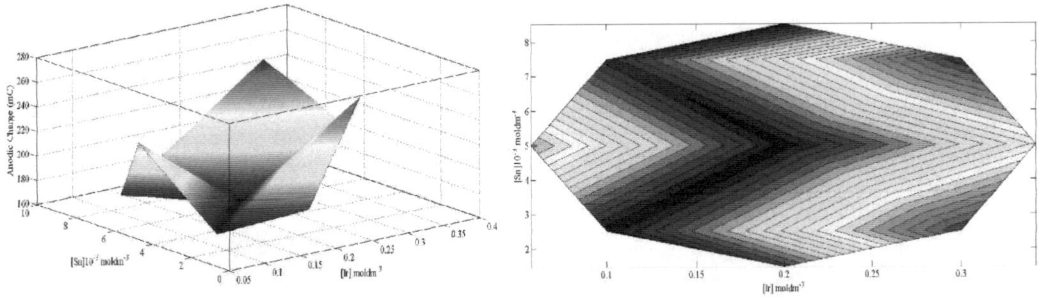

(a) (b)

Figure 2. (a) Response surface plot (b) Contour plot of anodic charge as the function of [Ir] and [Sn] of coated solutions of the electrode

Another reason may be due to the cracked mud or porous surface, leading to help in adsorption of $OH^{\cdot}$ and organic matters, as shown in Figure 03 [14]. Figure 03 shows the SEM micrograph of 0.3 M of [Ir] and 7.5 mM of [Sn] coated Ti/IrO_2–SnO_2 electrode prepared by thermal decomposition method. The coating was heterogeneous with rough morphology. The surface of the thin film was appeared to have both compact and porous regions. Such a feature can be considered advantageous for electrochemical degradation of CP because of the resulting larger surface area.

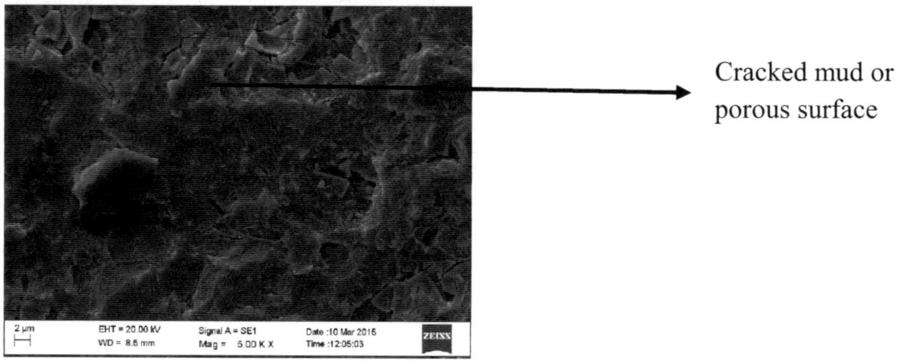

Figure 3. Scanning Electron Microscopic (SEM) images of (1,1) coated anode

3.2 Analysis of anode stability

OCP is a measure of anode stability. The response surface plot and contour plot of OCP for prepared anodes as a function of Ir and Sn concentrations are shown in Figure 04 (a) and (b). Stability was investigated using the OCP values of all anodes immersed in 0.5 M of Na_2SO_4 solution at 25 ^{0}C. The contour plot shows a ''rising ridge'' type, indicating the occurrence of highest stability over a wide experimental region ranging from 0.1 M to 0.3 M of the Ir concentration and 2.5 mM to 7.5 mM of the Sn concentration. It can be seen that OCP at electrode prepared with 0.3 M of [Ir] and 7.5 mM of [Sn] coated anode has the highest stability (-0.64 V), while 0.3414 M of [Ir] and 5 mM of [Sn] coated anode and 0.3 M of [Ir] and 2.5 mM of [Sn] coated anode show comparatively higher stabilities. OCP values of all anodes are negative, since surface of IrO_2 is negatively charged. The presumed reason is that point of zero charge of IrO_2 in Na_2SO_4 solution is less than 1. As an instance, this finding is in agreement with elsewhere, where similar negative values observed in IrO_2 type electrode and dependent with the composition of IrO_2 [15]. Open circuit potential measurements performed were shown that 0.3 M of [Ir] and 7.5 mM of [Sn] coated anode has the highest stability. Therefore, this electrode has both highest anodic charge and stability.

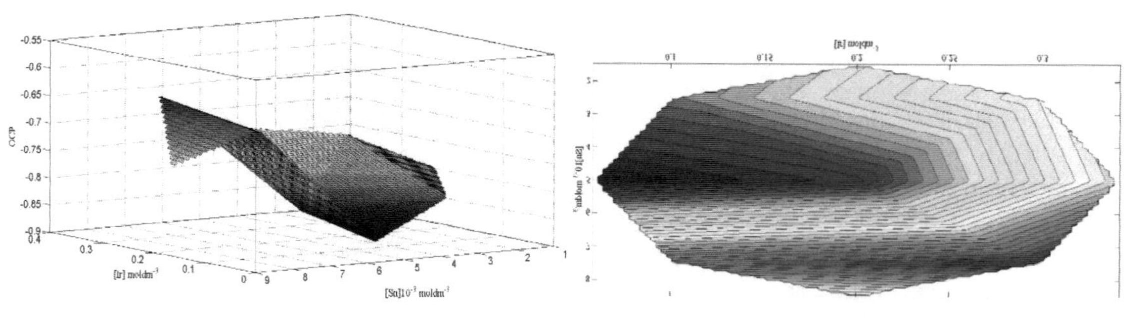

(a) (b)

Figure 4. (a) Response surface plot (b) Contour plot of stability in electrodes as the function of [Ir] and [Sn] of coated solutions of the electrode

3.3 Degradation of CP using optimized anodes

The most common way of estimating the mineralization is to monitor the reduction in TOC. Three anodes with high electrochemically active surface area and stability were selected from above investigations and used in degrading CP. The graph of mineralization efficiencies of selected anodes are shown in Figure 05. The best result was obtained with 0.3 M of [Ir] and 7.5 mM of [Sn] coated anode (55.56%), followed by the 0.3 M of [Ir] and 2.5 mM of [Sn] coated anode (34.29%). Anode coated with 0.3414 M of [Ir] and 5 mM of [Sn] showed an efficiency of 32.09%. Complete CP mineralization was not achieved by any of the anodes during the reaction period of 6 h. It should be noted that the prime objective of this study is to select the best coating concentrations by comparing efficiencies under similar experimental conditions for a specific reaction period. Higher mineralization efficiencies would be obtained at extended reaction durations or at higher current densities. The residual TOC in the solutions after CP removal suggests the formation and accumulation of intermediate organic products, particularly with all the three anodes. The similar phenomena have been reported in literature to remove CP in water [16].

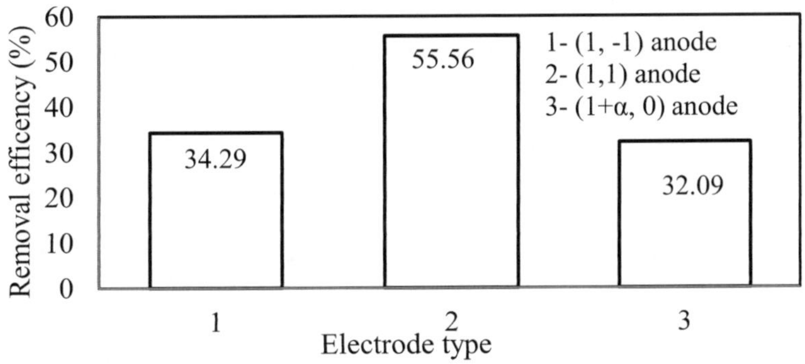

Figure 5. Analysis of mineralization of CP using TOC for (1) (1,-1) coated anode (2) (1,1) coated anode (3) (1+ α,0) coated anode after 6 h of reaction period

Energy consumption is one of the financial facts to consider for a water treatment system. Low energy consumption is essential for any treatment technology, along with high removal efficiency. The energy consumption as a function of TOC removal during electrolysis of above mentioned three electrodes are shown in Table 02, performed under the same condition. The average energy consumption value of 0.3 M of [Ir] and 7.5 mM of [Sn] coated anode (the best anode) was 6 $kWhm^{-3}$. Other two anodes also show approximately low values during the electrolysis time of 6 hours. According to this result, it can be concluded that optimized Ti/IrO_2–SnO_2 anode could be a cost-effective method for CP removal from water.

Table 2. Energy consumption obtained from the electrolysis of 6 hours for different Ti/IrO_2-SnO_2 electrodes

Anode type	Energy consumed ($kWhm^{-3}$)
0.3 M of [Ir], 2.5 mM of [Sn] coated anode	7.2
0.3 M of [Ir], 7.5 mM of [Sn] coated anode	6
0.3414 M of [Ir], 5 mM of [Sn] coated anode	8.4

The optimum coating concentration was found as0.3 M of [Ir] and 7.5 mM of [Sn]. This anode accompanied quantitative TOC of 55.56%, after the electrolysis of 6 hours at 20 $mAcm^{-2}$ of current density. It would require an energy consumption of 6 $kWhm^{-3}$.

4. Conclusions

In this work, Optimization of $Ti/IrO_2–SnO_2$ anode for mineralization of CP is successfully performed by response surface model using statistical experimental design method. Based on the optimization study, two-level factorial design was used to determine the optimal conditions for $Ti/IrO_2–SnO_2$ anode preparation. The factors evaluated included Ir and Sn concentrations. PEG concentration is fixed. Significant influence of fabrication parameters on anodic performance was observed by electrochemical properties and removal efficiency of CP. It can be seen that OCP at electrode prepared with 0.3 M of [Ir] and 7.5 mM of [Sn] coated anode has the highest stability (-0.64 V), This analysis also permits concluding that changes in the IrO_2 and SnO_2 composition of the electrode for its electrochemical properties. The obtained optimum fabrication conditions are 0.3 M of [Ir] and 7.5 mM of [Sn] respectively. The best result for degradation efficiency was obtained with 0.3 M of [Ir] and 7.5 mM of [Sn] coated anode (55.56%) during 6h of electrolysis.

5. References

[1] Murillo R, Sarasa J, Lanao M and Ovelleiro J L 2010 *Wa. Sci. Technol.* **10** 1-6.
[2] Romeh, A A and Hendawi M Y 2013 *Environ. Chem. Lett.* **11** 163-70.
[3] Tang T, Dong J, Ai S, Qiu Y and Han R 2011 *J. Hazard. Mater.* **188** 92-7.
[4] Chen X, Chen G and Yue P L 2003 *Chem. Eng. Sci.* **58** 995 – 1001.
[5] Fockedey E and Lierde A V 2002 *Water Res.* **36** 4169-75.
[6] Belhadj-Tahar N and Savall A 1998 *J. Electrochem. Soc.***145** 3427-34.
[7] Zhang H, Ran X N, Wu X G and Zhang D B 2011 *J. Hazard. Mater.* **188** 261-8.
[8] Hammami S, Oturan N, Bellakhal N, Dachraoui M and Oturan M A 2007 *J. Electroanal. Chem.* **610** 75–84.
[9] Wu J, Zhang H, Oturan N, Wang Y, Chen Lu and Oturan M A 2012 *Chemosphere* **87** 614-20
[10] Pathiraja G C, Jayathilaka P B, Weerakkody C, Karunarathne P and Nanayakkara N 2014 *Curr. Sci.* **107** 219-26.
[11] Tian M, Bakovic L and Chen A 2007 *Electrochim. Acta* **52** 6517–24.
[12] Gargouri B, Gargouri O D, Gargouri B, Trabelsi S K, Abdelhedi R and Bouaziz M **2014** Chemosphere **117** 309–15.
[13] Rossi A and Boodts J F C 2002 *J. Appl. Electrochem.* **32** 735-41.
[14] De Oliveira- Sousa A, Da Silva M A S, Machado S A S, Avaca L A and De Lima Neto P 2000 Electrochim. *Acta* **45** 4467-73.
[15] Alves V A, Da Silva L A and Boodts J F C 1998 *Electrochim. Acta* **44** 1525-34.
[16] Samet Y, Agengui L and Abdelhédi R 2010 *J. Electroanal. Chem.* **650** 152-58.

Acknowledgments

The authors wish to acknowledge National Research Council of Sri Lanka (Grant no. 11-054).

Embedding of polyaniline molecules on adhesive tape using successive ionic layer adsorption and reaction (SILAR) technique

J. K. Pamatmat, A. V. Gillado and M. U. Herrera

Institute of Mathematical Sciences and Physics, University of the Philippines Los Banos, College, Laguna, Philippines 4031

muherrera@up.edu.ph

Abstract. Polyaniline molecules are embedded on adhesive tape using successive ionic layer adsorption and reaction (SILAR) technique. The infrared spectrum shows the existence of molecular vibrational modes associated with the presence of polyaniline molecules on the sample. With the addition of polyaniline molecules, the conductivity of adhesive tape increases. Surface conductivity increases with number of dipping cycle until it reaches a certain value. Beyond this value, surface conductivity begins to decrease. The surface conductivity of the sample is associated with the connectivity of the embedded polyaniline molecules. The connectivity increases as the number of dipping cycle progresses. Meanwhile, the decrease in surface conductivity is attributed to the eroding of existing embedded structure at higher number of dipping cycle.

1. Introduction

Polyaniline [1-3] is an intrinsically conducting polymer. Its conductivity can be increased by protonation or acid doping. On the other hand, its conductivity can be decreased by removing the acid dopant (deprotonation) using base treatment. The color of its conducting state is green while its non-conducting state is blue. Polyaniline has good stability and easy to fabricate. In addition, the monomer used to synthesize polyaniline molecules is relatively cheap. Polyaniline molecules have many different possible applications such as sensors, anti-corrosion agent, absorbent, conductive adhesive, conductive ink, diode, acid-base indicator, memory devices, electrostatic discharged material, erasable circuit board.

Embedding polyaniline molecules on substrates [4-8] (e.g., textile, paper) hybridizes the functionality of polyaniline molecules and the characteristics of the substrate. Double adhesive tape could be turned into stick-on patches by embedding polyaniline molecules. These stick-on patches can be cut to different forms and sizes and can adhere on variety of surfaces. With these stick-on patches, polyaniline layers can be placed on different surfaces for different possible applications.

This study aims to embed polyaniline molecules on adhesive tape using successive ionic layer adsorption and reaction (SILAR) technique. SILAR technique [9-12] is simple, inexpensive, and operational at ambient conditions. This technique employs layer-by-layer deposition of materials on substrate. In SILAR technique, two main steps are done repetitively until the desired outcome is achieved: (1) adsorption of precursors and (2) inducing reaction to produce the material. For the case of embedding polyaniline molecules on adhesive tape, adsorption of aniline-hydrochloric acid molecules is done by dipping the substrate in aniline monomer-hydrochloric solution. Meanwhile, the reaction needed to induce the polymerization of the aniline monomers to polyaniline molecules is done by dipping the substrate in a solution containing the ammonium persulfate (APS) oxidant.

2. Methodology

Polyaniline molecules were embedded on double adhesive tape (Polar Bear® double adhesive tissue tape DS-009) using SILAR technique which is shown in Figure 1. In this technique, newly opened double adhesive tapes were dipped successively in aniline monomer-HCl solution (20 s) and in 0.25M ammonium persulfate solution (10 s) for different number of times (e.g., 10, 20, 30, 40, 50, 60, 70 dipping cycles). The aniline monomer-HCl solution was prepared by mixing 0.8 M HCl and 4 ml aniline monomer (Sigma Aldrich, ≥ 99.5%). Meanwhile, the ammonium persulfate solution was prepared by dissolving ammonium persulfate in distilled water. After the embedding process, the samples were air dried.

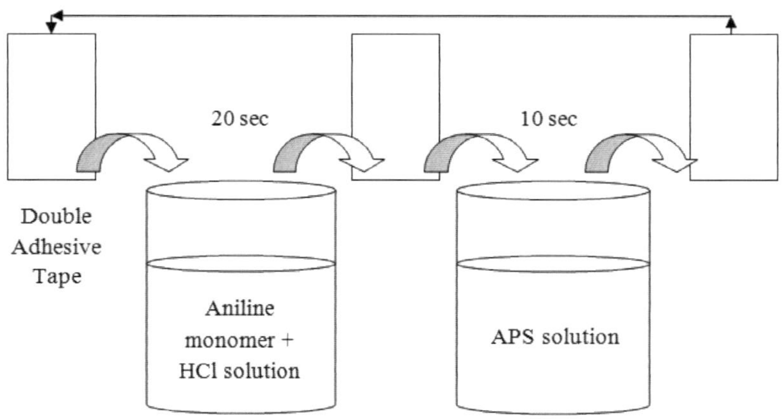

Figure 1. Deposition of polyaniline molecules on double adhesive tape using successive ionic layer adsorption and reaction (SILAR) technique.

The surface conductivities of the samples were determined using four point probe measurement. Meanwhile, the infrared spectra of the sample were determined using Shimadzu IR Prestige 21 Fourier Transform Infrared (FTIR) Spectrometer equipped with Single Reflection Diamond Attenuated Total Reflectance (ATR) accessory. Using the FTIR Spectrometer, the sample was scanned 45 times with a resolution of 4 cm^{-1}.

3. Result and Discussion

We have successfully embedded polyaniline molecules on adhesive tape using SILAR technique. Figure 2 shows the infrared spectrum of the sample while Table 1 presents the peaks of the spectrum with the associated molecular vibrational modes and structures. Infrared spectrum of the sample shows presence of vibrational modes associated with the existence of polyaniline molecules. Polyaniline molecules have aromatic rings which could have C=C stretches and C-H bends. The peaks at 1,604.78 and 1,541.12 cm^{-1} are attributed to the C=C stretches while peaks at 1,062.78, 972.12, 933.55, 869.90, 783.10 and 732.95 $cm^{-1}$are associated to C-H bends. Amine and iminium structures are possible structures in between the aromatic rings. Amine structures could have N-H bends and C-N stretches. The presence of N-H bends is attributed to the peaks at 1,604.78, 783.10 and 732.95 cm^{-1} while the presence of C-N stretches is associated to the peaks at 1,280.73 and 1,230.58cm^{-1}. Iminium structures could have $C=N^+$ stretches and $R-NH^+=R$ stretches. The peak at 1,404.18 cm^{-1} is attributed to the presence of $C=N^+$ stretches while the peak at 1,166.93$cm^{-1}$is attributed to the presence of $R-NH^+=R$ stretches.

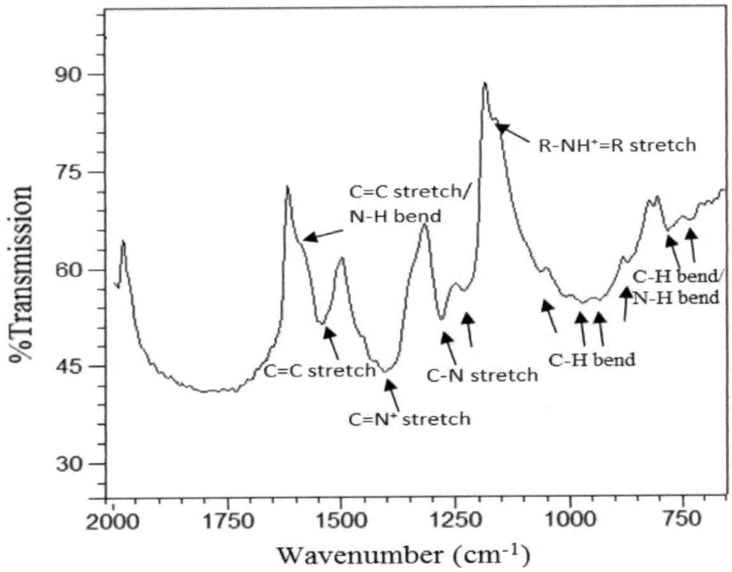

Figure 2. Infrared spectrum of adhesive tape that was embedded with polyaniline molecules using SILAR technique with 50 dipping cycles.

Table 1. Peaks of the infrared spectrum presented in Figure 1 with the associated molecular vibrational modes and structure type

Wavenumber (cm^{-1})	Molecular vibrational modes	Structure Type
1,604.78	C=C stretch/ N-H bend	Aromatic Amine
1,541.12	C=C stretch	Aromatic
1,404.18	$C=N^+$ stretch	Iminium
1,280.73	C-N stretch	Amine
1,230.58		
1,166.93	$R-NH^+=R$ stretch	Iminium
1,062.78		
972.12	C-H bend	Aromatic
933.55		
869.90		
783.10	C-H bend/	Aromatic
732.95	N-H bend	Amine

The surface conductivity of adhesive tape improves with the addition of acid-doped polyaniline molecules. Figure 3 shows the plot of number of dipping cycle with percent increase in surface conductivity of the samples. The plot can be divided into two regions based on the trend of the sample's surface conductivity: (1) the increasing region and (2) the decreasing region. The "increasing region" is from zero to 50 dipping cycle while the "decreasing region" is from 50 dipping cycle and beyond.

The "increasing region", the surface conductivity (y) was fitted exponentially with the number of dipping cycle (x). Fitting the data in the equation $y = A \bullet \exp(k \bullet x)$ yields A = 16,361.72 and k = 0.0897. The increase in surface conductivity is attributed to the increase of connectivity of embedded polyaniline molecules on the adhesive tape. Thus, as the dipping cycle progresses, additional polyaniline molecules is incorporated on the existing structures, thus the embedded polymer's networks becomes more interconnected. The exponential growth of surface conductivity may suggest that the increase in embedded structure's connectivity is possibly dependent on the amount of existing structure.

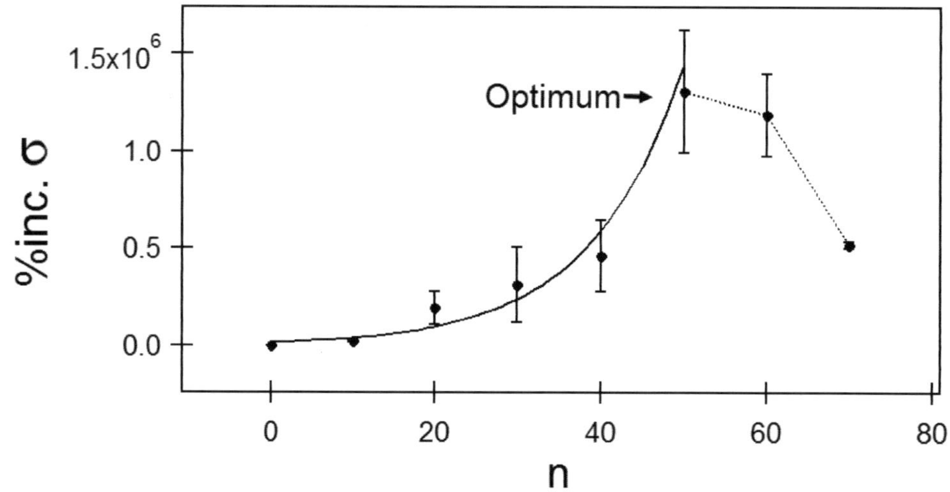

Figure 3. Number of dipping cycle (n) vs. percent increase in surface conductivity (%inc.σ). Note: closed circle represents experimental data points, solid line represents fitted curve at the increasing surface conductivity region while dashed line connects the data points at the decreasing surface conductivity region.

The "decreasing region" shows the surface conductivity decreases with further increase in number of dipping cycle. This region happens at larger number of dipping cycle (e.g., greater than 50). Excessive dipping may have eroded the existing polymeric structure which may have caused the decrease in surface conductivity. Based from the numbers of dipping cycle that were used, the optimum number of dipping cycle to achieve the highest surface conductivity is around 50 times. Beyond this number of cycle, the surface conductivity begins to decrease.

4. Summary and Conclusion

Polyaniline was deposited on adhesive tape using SILAR method with varying dipping cycles. Conductivity measurements show that there is an increase in conductivity with dipping cycle from 0 to 50 dipping cycle. This may be due to the greater connection between the added and the existing polyaniline networks. However, the conductivity started to decrease when the optimum number of dipping cycle (50 dipping cycle) was reached. This may be attributed to the erosion of the existing polymeric structure. The FTIR-ATR spectrum shows the existence of peaks associated with the presence of polyaniline molecules.

References

[1] Bhadra S, Khastgir D, Singha N and Lee J H Progress in preparation, processing and applications of polyaniline 2009 *Prog. Polym. Sci.* **34** 783-810
[2] Stejskal J and Gilbert R G Polyaniline. preparation of a conducting polymer 2002 *Pure Appl. Chem.* **74** 857-67
[3] Dhawan S K, Singh N and Venkatachalam S Shielding behaviour of conducting polymer-coated fabrics in X-band, W-band and radio frequency range 2002 *Synth. Met.* **129** 261-7
[4] Molina J, del Río A I, Bonastre J and Cases F Electrochemical polymerisation of aniline on conducting textiles of polyester covered with polypyrrole/AQSA 2009 *Eur. Polym. J.* **45** 1302-15

[5] Stempien Z, Rybicki T, Rybicki E, Kozanecki M and Szynkowska M I In-situ deposition of polyaniline and polypyrrole electroconductive layers on textile surfaces by the reactive ink-jet printing technique 2015 *Synth. Met.* **202** 49-62

[6] Youssef A M, Mohamed S A, Abdel-Aziz M S, Abdel-Aziz M E, Turky G and Kamel S Biological studies and electrical conductivity of paper sheet based on PANI/ PS/Ag-NPs nanocomposite 2016 *Carbohydr. Polym.* **147** 333-43

[7] Zhou Y, Ding C, Qian X and An X Further improvement of flame retardancy of polyaniline-deposited paper composite using phytic acid as dopant or co-dopant 2015 *Carbohydr. Polym.* **115** 670-6

[8] Treu A, Bardage S, Johansson M and Trey S Fungal durability of polyaniline modified wood and the impact of a low pulsed electric field 2014 *Int. Biodeterior. Biodegrad.* **87** 26-33

[9] Chatterjee S, Saha S and Pal A Formation of all-oxide solar cells in atmospheric condition based on Cu_2O thin-films grown through silar technique 2016 *Sol. Energy Mat. Sol. Cells* **147** 17-26

[10] Gaikwad M A, Suryawanshi M P, Nikam S S, Bhosale C H, Kim J H and Moholkar A V Influence of Zn concentration and dye adsorption time on the photovoltaic performance of M-SILAR deposited ZnO- based dye sensitized solar cells 2016 *J. Photochem. Photobiol.* A: *Chem.* **329** 246-54

[11] Niesen T and De Guire M Review: deposition of ceramic thin films at low temperatures from aqueous solutions 2002 *Solid State Ion.* **151** 61-8

[12] Patil V, Vanalakar S, Patil P and Kim J Fabrication of nanostructured ZnO thin films based NO_2 gas sensor via SILAR technique 2017 *Sens. Actuators* B: *Chem.* **239** 1185-93

IOP Conf. Series: Materials Science and Engineering **201** (2017) 012042 doi:10.1088/1757-899X/201/1/012042

Fabrication of polyaniline-coated Kapok (*Ceiba pentandra*) fibers embedded with copper-based particles

K. E. Arguelles[1], M. U. Herrera[2], C. C. M. Futalan[3], and M. D. L. Balela[1]

[1]Department of Mining, Metallurgical and Materials Engineering, College of Engineering University of the Philippines Diliman, 1101
[2]Institute of Mathematical Sciences and Physics, College of Arts and Sciences, University of the Philippines Los Baños, 4031
[3]AECOM Philippines Consultants Corporation, Fort Legend Towers, Bonifacio Global City, Taguig, Metro Manila, Philippines, 1634

Email: mlbalela1@up.edu.ph

Abstract. Polyaniline-coated kapok (*Ceiba pentandra)* fibers that were embedded with Cu-based particles were fabricated for antimicrobial application. Kapok fibers were coated with polyaniline molecules using oxidative polymerization. The coated fibers were embedded with copper-based particles using soaking method in prepared CuO suspension. X-ray diffraction (XRD) pattern shows presence of Cu and Cu_2O particles on the modified fibers. Scanning electron microscopy (SEM) supports the presence of embedded particles on the modified fibers. The samples showed antimicrobial activity against *Escherichia coli* and *Staphylococcus aureus*.

1. Introduction

Metal-based materials [1-5] such as Ag, Zn, and Cu have been studied for antimicrobial properties. Moreover, copper-based materials have shown potential as bactericidal agent at low cost and ease of synthesis. Copper-based materials in powder form or in film have various possible applications [5-7] such as paints, fabrics, and agriculture.

Kapok (*Ceiba pentandra*) fiber (KF) [9-10] is one of the less given attention agricultural products. The fiber exhibits large hollow lumen, which can be an excellent template for growing polymers. However, one of the drawbacks of the fiber is the hydrophobic character due to its waxy cutin. In such, embedding of particles using aqueous medium requires a hydrophilic substrate such as modified kapok fiber. Conversion of hydrophilic from hydrophobic kapok fibers can be achieved through polymerization of hydrophilic material on the fiber surface.

Polyaniline (PAni) [11-13] is an intrinsically conducting polymer that has been studied for its unique structure, ease of polymerization and applications. The high content of amine and imine groups of PAni makes it an excellent polymer for application of polar environment. Depending on the pH, PAni can exist in different form such as pernigraniline, emeraldine, lecucoemeraldine. PAni has been applied as sorbent [14-15] for heavy metals such as chromium (VI) and copper. The sorbent property of PAni makes it efficient for embedding metal oxides.

IOP Conf. Series: Materials Science and Engineering **201** (2017) 012042 doi:10.1088/1757-899X/201/1/012042

In this study, copper-based particles will be embedded on PAni-coated kapok fiber and tested for its antimicrobial activity against *Escherichia coli* and *Staphylococcus aureus*.

2. Methodology

2.1Materials

Kapok fibers were obtained from Nueva Ecija, Philippines and taken out from their pod. Aniline (99.5%, AR) (95%, AR) was purchased from Loba Cheme. Glacial acetic acid (99.7%, AR) was purchased from Macron Fine Chemical. Sodium Chlorite (80%, AR) and hydrazine monohydrate (98% AR) were from Sigma Aldrich. Ethanol (EtOH) (95%, AR) was from Ajaxfinechem. Hydrochloric acid (HCl, 37%, AR) was from RTC laboratory service and supply house. Ammonium persulfate (98%, AR) was from HIMEDIA. Copper oxide (99% AR) was from Nacalai Tesque. All reagents were used without further purification.

2.2Pretreatment of Kapok fiber

The cutin of kapok fibers was removed to easily adhere the polyaniline molecules on the fiber. About 1.20 g of $NaClO_2$ was dissolved in 100 mL of distilled water in a three-neck flask and 1.90 mL of glacial acetic acid was also added. The mixture was added with 1.50 g of KF and stirred for 60 min at 700 rpm at 90ºC. The treated kapok fibers were washed with distilled water until filtrate reached pH 6-7. The fiber was oven dried at 60 ºC.

2.3Coating of Polyaniline on kapok fibers

Coating of polyaniline on kapok fibers (PAni/KF) was prepared via oxidative polymerization by method of Zheng et al.(2012) [12]. Using 2.0 g of aniline dissolved in 66.0 mL of 1.0 M HCl. The mixture was cooled in an ice bath where 0.40 g of untreated KF was added and stirred for 30 minutes to disperse the fiber. In a separate beaker, 4.0 g of APS was dissolved in 16 mL 1.0 M HCl and cooled in an ice bath. APS mixture was added to the aniline and KF mixture dropwise and stirred for 1 hour. The resulting mixture was left for 16 hours at room temperature. The fibers were filtered then washed with distilled water until the supernatant became colorless. Further washing was done using ethanol. The product was air dried before drying in an oven at 60 ºC.

2.4 Embedding of Copper-based particles on Polyaniline/Kapok fiber

Copper-based particles were embedded by soaking PAni-coated kapok fiber in suspension containing CuO mixture. The mixture was prepared by method of Balela and Amores (2015)[16]. The CuO suspension was prepared by dissolving 4.87 g CuO powder in 42.0 mL deionized water by sonication. In a separate vessel, hydrazine solution was prepared using 3.06 g hydrazine in 42.0 mL deionized water. Around 18.0 g gelatin mixture (10%) was added to CuO and hydrazine mixtures. The pH of the mixtures were adjusted to pH 12 using aqueous NaOH. The mixture was purged with nitrogen for 30 The hydrazine solution was added to CuO mixture slowly while stirring at 500 rpm. The reaction was allowed for 2 hours at 353 K under nitrogen environment. The product was collected by centrifugation then washed with deionized water to remove excess hydrazine.

In a conical tube, 10 mg of PAni/KF was mixed with 10 mL of copper oxide suspension and PAni/KF was thoroughly dispersed. The mixture was stand for 3 hours. The fibers were recovered via filtration and dried. After drying, sonication of the product was done for 10 minutes to remove loosely attached copper oxide. The product was obtained using filtration. Drying was done at room temperature.

2.5 Characterizations
Morphological characterization of kapok and modified kapok fibers was done using Scanning electron microscope JEOL JSM 5310. X-ray diffraction anaylsis was done using SHIMADZU MAXima XRD-7000.

2.6 Antimicrobial activity of copper-based particles-polyaniline/kapok fiber
The test organisms used were *Escherichia coli* (UPCC 1195) and *Staphylococcus aureus* (UPCC 1143). The 24-hr old cultures were prepared as microbial suspensions in 0.1% peptone water. Inoculation was done using swabbing of applicator stick on a pre-poured nutrient agar about 3 mm thick. Even distribution of inoculum was ensured by rotating the plate 60° as the procedure was repeated twice. Samples were applied at three points on the plate. Incubation of NA plates was done at 35°C. Clearing zones were observed after 24 hours.

3. Results and Discussion

PAni-coated kapok fibers embedded with copper-based particles was fabricated. The XRD patterns of the sample are shown in Figure 1. The peaks at diffraction angles (2θ) 36.44°, 61.28 correspond to (111) and (220) of Cu_2O. Peaks at 43.66°, 50.90° and 74.46° correspond to (111), (200) and (220) of Cu. The peak at 22.3° shows typical cellulose (200) plane from the kapok fiber. The XRD profile of the product shows the embedding of Cu and Cu_2O particles on the modified fiber.

SEM images shows the coating PAni and embedding of copper-based particles on KF. Figure 2a-b shows SEM micrographs of kapok fiber while the coated kapok fiber can be seen on Figure 2c-d. The polyanline coating on kapok fiber is not uniform. Some coating aggregated in some sites of the fiber. The rougher surface as compared to the raw KF image indicates PAni coating of KF. Meanwhile, the copper-based particles can be seen on Figure 2 d-e with some agglomeration of the particles. The aggregated materials are suspected to be Cu and Cu_2O. The soaking method employed may have led to agglomeration of the nanoparticle on the surface of modified kapok fiber.

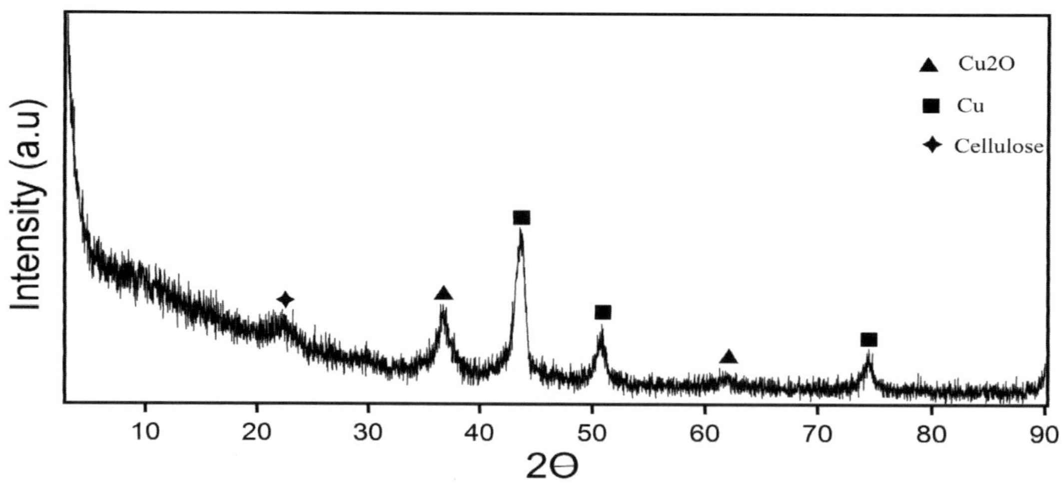

Figure 1. XRD pattern of copper-based particles embedded on modified kapok fiber.

Figure 2. SEM images of a-b) kapok fiber, c-d) PAni/KF, e-f) copper-based particles/PAni/KF.

The copper-based particles/PAni/KF showed antimicrobial activity against *E. coli* and *S. aureus*. Clearing zones were observed around the two samples and one sample with no clearing zone for *E. coli*. The exact antimicrobial of copper particles has not yet been studied. Various mechanism proposed includes disruption enzymatic activity due to release of Reactive Oxygen Species (ROS). The release of ROS leads to production of different radicals which disrupt the cell activity leading to cell death[4] thus, formation of clearing zones. While the zone for *S. aureus* shows the reduced growth instead of clearing zone for the three samples tested as summarized in Table 1. The existence of clearing zone beneath the sample while only having reduced growth of *S. aureus* around the sample suggest that the copper-based particle on modified KF has more effect in contact inhibition where there is an intimate contact of bacteria to sample.

Table 1. Antimicrobial activity of Copper-based particles/PAni/KF on *Escherichia coli* (Gram-negative) and *Staphylococcus aureus* (Gram-positive).

Test organism	**Clearing zone present (+) or absent (-)**			**Growth beneath the sample**	**Antimicrobial activity**
	1	2	3		
Escherichia coli (Gram-negative)	+	+	-	none	Present
Staphylococcus aureus (Gram-postive)	reduced	reduced	reduced	none	Present

4. Summary and conclusion

In this study, copper-based particles were embedded on polyaniline-coated kapok fibers. The fibers were coated with polyaniline using oxidative polymerization. Embedding of copper-based particles was done using soaking method in a prepared copper oxide suspension. Embedding of Cu and Cu_2O particles were shown by XRD profile and SEM images. The product was subjected to agar disc diffusion test. Clearing zones were observed against *E. coli* while reduced growths were observed against *S. aureus*. Results showed the potential of copper-based embedded PAni-coated KF as an antimicrobial material.

5. Acknowledgement

This work was supported by a grant from UP System Emerging Inter-Disciplinary Research Program (OVPAA-EIDR C06-032).

References

[1] Dizaj, S., Lotfipour, F., Barzegar-Jalali, M., Zarrintan, M., Adibkia, K. 2014 *Materials Science & Engineering C* **44** 78-284
[2] Chernousova, S. and Epple, M. 2013 Angew. Chem. Int. Ed. **52** 1636–1653
[3] Noimark, S., Weiner, J., Noor, N., Allan, E., Williams, C. K., Shaffer, M. S. P., Parkin, I. P. 2015 Adv. Funct. Mater. **25** 1367–1373
[4] Meghana, S., Kabra, P., Chakraboty, S., Padmavathy, N. 2015 *RSC advances* **5** 12293-12299
[5] Datu E and Balela MD 2016 *Key Materials Engineering* **705** 163-167
[6] Bellotti, N., Romagnoli, R., Quinteroc, C., Domínguez-Wong, C., Ruiz, F., Deyá, C. 2015 *Progress in Organic Coatings* **85** 33-40
[7] Bajpai, S. K., Bajpai, M. and Sharma, L. *2012 J. Appl. Polym. Sci.* **126** E319–E326
[8] Kanhed, P, Birla, S., Gaikwad, S., Aniket Gade, A., Seabra, A., Rubilar, O., Duran, N., Rai, M. 2014 *Materials Letters* **115** 13-17
[9] Wang F., Zheng, Y., Zhu., Y., Wang A. 2016 *Frontier in Environmental Science* **4** 1-12
[10] Wang, J., Zheng, Y., Wang, A. 2012 *Ind. Crop. Prod.* **40** 178–184
[11] Shinde, S. and Kerr, J. 2014 *International Journal of Innovative Research in Science, Engineering and Technology* **3** 16570-16576
[12] Zheng, Y., Lin, Y., Wang, A. 2012 *Industrial and Engineering Chemistry Research* **51** 10079-10087
[13] Stejskal, J., Kratochvil, P., Jenkins, A. 1997 *Polymer* **2** 367-369
[14] Kanwal, F., Rehman, R., Manhmud, T., Anwar, J., Ilyas, R. 2012 *Journal Chilean Chemistry Society* **57** 1058-1063
[15] Jiang, N., Xu, Y., Dai, Y., Luo, W., Dai, L 2012 *Journal of Hazardous Materials* **215-216** 17-24
[16] Balela, M and Amores, K. 2016 *Science Diliman* **27** 10-20

Carbon monoxide gas sensing using zinc oxide deposited by successive ionic layer adhesion and reaction

E. A. Florido, N. A. C. Dagaas

Institute of Mathematical Sciences and Physics, University of the Philippines Los Banos 4031, Laguna, Philippines

eaflorido@up.edu.ph

Abstract. This study was aimed to determine the carbon monoxide (CO) gas sensing capability of zinc oxide (ZnO) film fabricated by successive ionic layer adsorption and reaction (SILAR) on glass substrate. Films consisting of a mixture of flower-like clusters of ZnO nanorods and nanowires were observed using scanning electron microscopy (SEM). Current-voltage characterization of the samples showed an average resistivity of 13.0 Ω-m. Carbon monoxide gas was synthesized by mixing the required amount of formic acid and excess sulfuric acid to produce CO gas concentrations of 100, 200, 300, 400, and 500 parts per million (ppm) v/v with five trials for each concentration. Two sets of data were obtained. One set consisted of the voltage response of the single film sensor while the other set were obtained from the double film sensor. The voltage response for the single film sensor and the double film sensor showed an average sensitivity of 0.0038 volts per ppm and 0.0024 volts per ppm, respectively. The concentration the single film can detect with a 2V output is 526 ppm while the double film sensor can detect up to 833 ppm with a 2V output. This shows that using the double film sensor is advantageous compared to single film sensor, because of its higher concentration range due to the larger surface area for the gas to interact. Moreover, the measured average resistance for the single film sensor was 10 MΩ while for the double film sensor the average resistance was 5 MΩ.

1. Introduction

Carbon monoxide (CO) gas is colourless, odourless, highly toxic and extremely dangerous. Natural sources account for 40% of the CO in the environment while 40% is the result of human activities. It is produced by the incomplete combustion of fuels and exhaust emissions [1]. CO molecules bind with the hemoglobin to form a new molecule, carboxyhemoglobin (COhgb), which reduces the ability of the blood to carry oxygen into the body which is deadly at high levels [2]. It can cause death if the concentration inhaled by a person is 35ppm for over 8 hours ascribed by the United States National Institute of Occupational Safety and Health (NIOSH). CO can exist at very low concentrations which requires extremely sensitive sensors for its monitoring. Different types of CO poisoning include acute, chronic, and occult poisoning. Acute poisoning involves one time large exposure to CO gas. Chronic poisoning involves repeated exposure to low levels of the gas and the effects are manifested at a later time. Occult poisoning can involve either acute or chronic poisoning but the diagnosis is missed or attributed to other causes [1].

Zinc oxide (ZnO) is an n-type metal oxide semiconductor (MOS) that has been widely developed as gas sensors for many years because of their chemical sensitivity to different adsorbed gases, high chemical stability, amenability to doping, non-toxicity, affordability and relative simplicity. It has a wide range of electronic, chemical and physical characteristics and has become a well-known commercial sensor because of the sensitivity of its properties to variations in its chemical environment [3,4]. Molecules can interact with the semiconductor surface and affect its surface properties such as resistance, conductivity and surface potential. In ambient air, O_2 molecules get adsorbed on ZnO surface and bind to its vacancy sites, extracting or trapping electrons from the conduction band (E_c), then conduct these electrons to the surface and stay strongly attached as charged oxygen anions. The trapped electrons are no longer accessible for conductivity in the solid, resulting to an increase in its resistance and the formation of a space-charge layer. Reducing gases such as CO react with the adsorbed oxygen resulting to its desorption as CO_2 with a consequent release of electrons and increase in conductivity of the semiconductor [5].

Different methods of depositing semiconductor films are in use which include metalorganic chemical vapor deposition (MOCVD), molecular beam epitaxy (MBE), liquid phase epitaxy (LPE), sol-gel processing, spray pyrolysis, electrophoretic deposition (EPD) [6] and successive ionic layer adhesion and reaction (SILAR). SILAR involves successive dipping of a substrate in precursor solutions such as sodium zincate (Na_2ZnO_2) prepared by mixing zinc sulfate ($ZnSO_4 .7H_2O$) with excess sodium hydroxide in aqueous solution. This results to deposition of Na_2ZnO_2 on the substrate surface. When the substrate is next dipped in hot water bath, the sodium zincate is converted to ZnO is subsequently produced. The process is then repeated several times [7,8].

2. Methodology

2.1. Film fabrication

Zinc oxide film was deposited on a pre-cleaned 9mm by 25mm glass substrate. Two sets of sensors were fabricated. Single-film sensor was fabricated by clipping two glass substrates together to allow film deposition on only one side of the substrate. Zinc oxide was deposited on both sides of the glass substrate to produce a double-film sensor.

The precursor solution was prepared by mixing 15 ml of 0.1M of $ZnSO_4$ with 20 ml of 1M of NaOH to produce sodium zincate (Na_2ZnO_2). The substrates were dipped alternately in the sodium zincate solution and a 95^0C hot water bath. Dipping was repeated 200 times with a dipping time of two seconds. The deposited film was air dried for one hour and then annealed at 300^0C for 30 minutes. After annealing silver electrodes were printed one centimetre apart at each end of the film surface.

2.2. Film characterization

The morphology of the film was examined using scanning electron microscopy (SEM) using a JEOL 5300 scanning electron microscope. Four-point probe was used to examine the electrical property of the deposited film. In addition, the actual resistance of the film for the sensor circuit was determined by the Wheatstone bridge method.

2.3. Gas sensing

The gas sensing circuit consisted of a Wheatsone bridge using the film sensor as one arm of the bridge. The bridge was supplied with 5V from a DC power supply. The sensor was placed inside an airtight container of known volume. Voltage output of the bridge was obtained using a digital multimeter while data logging was done with the aid of a video recorder.

The voltage response of the ZnO film was determined using CO gas concentrations of 100, 200, 300, 400, and 500 ppm v/v. To generate the required concentration of CO gas in the chamber, the corresponding amounts of formic acid and sulphuric acid were mixed in a beaker inside the gas chamber using a magnetic stirrer. Five trials were conducted per concentration. An excess of sulfuric acid was used in the ratio of 6:1 to ensure complete consumption of the formic acid to produce the required stoichiometric concentration of CO gas. Gas response data were obtained using single film and double film ZnO sensor.

3. Data and Results

Scanning electron microscope (SEM) image revealed the nanostructure of the zinc oxide film as shown in Figure 1. The structure consisted of rod-like structures as well as nanowires. The large surface area offered by the nanorods and nanowires is beneficial to the function of the film sensor since gas sensing is a surface phenomenon. The nanorods obtained were not exactly hexagonal which could be attributed to impurities retained during the dipping process and the low annealing temperature of 300^{0}C. Studies have shown that increasing the annealing temperature of deposited zinc oxide film increases its crystal quality [9, 10, 11].

Figure 1. Scanning electron microscope (SEM) image of the zinc oxide film showing nanorods and nanowires at 10k (a) and 20k magnification.

The average resistivity of the films determined by the four-point probe method was 13.0 Ω-m while the actual resistance of the film between the electrodes measured using a Wheatstone bridge was 10 MΩ.

Table 1 shows the average voltage response of the single-film and double film sensors for each of the five different concentrations conducted at five trials each. The voltage response is the difference between the voltage output of the film before and after exposure to the CO gas.

Figure 2 shows the graph of voltage response of the film sensor with concentration. Linear regression analysis for the single film sensor showed an average sensitivity of 0.0038 volt per ppm with an R^2 value of 0.997. This value can be used to determine the operating range of the film sensor. Using the regression equation, the maximum concentration the film can detect for a maximum output of 5 volts obtained is 1320ppm.

Table 1. Voltage response of the single and double films at different concentrations.

Concentration (ppm)	Single-film voltage response (V)	Double-film voltage response (V)
100	0.402	0.574
200	0.695	0.631
300	1.138	0.919
400	1.493	1.260
500	1.903	1.472

For the double film sensor (Figure 3) the sensitivity of the film was 0.0024 volt per ppm with an R^2 value of 0.963. From the regression equation it can be shown that the sensor is capable of detecting up to 1980 ppm with a 5 volt output which is higher than that of the single film sensor. This shows that using the double film sensor is more advantageous compared to single film sensor because of its higher concentration range due to the larger surface area for the gas to interact with. Moreover, the measured average resistance for the one-sided film was 10 MΩ while for the two-sided-film the average resistance was 5 MΩ.

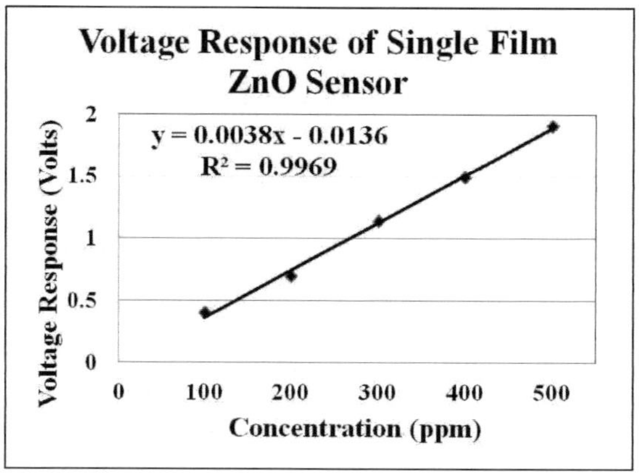

Figure 2. Voltage response of single film sensor with CO gas concentration.

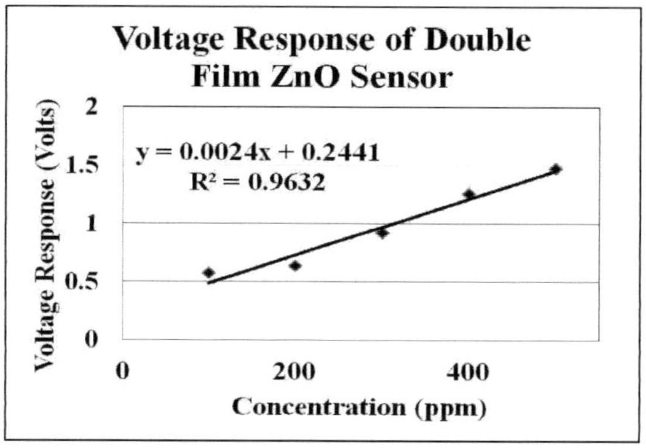

Figure 3. Voltage response of double film sensor with CO gas concentration.

4. Summary and Conclusion

Zinc oxide film was fabricated by successive ionic layer adhesion and reaction on precleaned glass substrates using sodium zincate solution as precursor solution. The process involved 200 alternate dippings in the precursor solution and a 95^0C hot water bath and annealing of the deposited films at 300^0C. Two sets of data were obtained. One set consisted of the voltage response of the single film sensor while the other set were obtained from the double film sensor. The voltage response for the single film sensor and the double film sensor showed an average sensitivity of 0.0038 volts per ppm and 0.0024 volts per ppm, respectively. The concentration the single film can detect with a 2V output is 526 ppm while the double film sensor can detect up to 833 ppm with a 2V output. This shows that using the double film sensor is advantageous compared to single film sensor, because of its higher concentration range due to the larger surface area for the gas to interact. Moreover, the measured average resistance for the single film sensor was 10 MΩ while for the double film sensor the average resistance was 5 MΩ.

Acknowledgements

The authors wish to thank the Philippine Council for Industry, Energy and Emerging Technology Research and Development (PCIEERD) for the project funds, and the Institute of Mathematical

Sciences and Physics (IMSP) University of the Philippines Los Banos (UPLB) as implementing agency of the research project.

References

[1] Wright J 2002 Chronic and occult carbon monoxide poisoning: we don't know what we're missing *Emerg Med J.* **19** 386–390
[2] Meredith T and Vale A 1988 Carbon onoxide Poisoning *British Med. J.* 77-79
[3] Weng J ,et al. 2004 *Thin Solid Films* **478** 25
[4] Wellings J, Chaure N, Heavens S, and Dharmadas I 2008 *Thin Solid Films* **516** 3893
[5] Chakraborty S, Sen A and Maiti H S 2006, *Sensors and Actuators B* **115** 610–613
[6] Hsu C, Chen K, Tsai T and Hsueh T 2013 *Sensors and Actuators B.* **182** 190
[7] Chatterjee A P, Mitra P and Mukhopadhyay A K 1999 *J. of Mat. Sci.* **34** 4225
[8] Mitra P and Mukhopadhyay A K 2007 *Bull. Pol. Ac. Tech.* **55(3)** 281
[9] Chin H and Chao L 2013 The effect of thermal annealing processes on structural and photoluminescence of zinc oxide thin film *J. of Nanomat.* **2013** 8 pages
[10] Ezema F I and Nwankwa U O A 2010 Effect of annealing temperature on the structural and optical properties of al-doped zinc oxide (ZnO) crystals *J. of Optoelec. and Biomed. Mat.* **1** 1677-173
[11] Lu L and Wong M 2014 The resisitivity of zinc oxide under different annealing configurations and its impact on the leakage characteristics of zinc oxide thin film transistors 2014 *IEEE Transactions on Electron Devices* **61(4)** 1077-1084

IOP Conf. Series: Materials Science and Engineering **201** (2017) 012044 doi:10.1088/1757-899X/201/1/012044

Effect of cryogenic temperatures on the failure strain and surface morphology of glass fiber

Wang Qingwei[1,a], Chen Jian[2,b], Luo Lida[1,c],Ning Wei[2,d]

1 State Key Laboratories for Modification of Chemical Fiber and Polymer Materials, Donghua University, Shanghai, 201620, P.R.china;
2 Engineering Research Center of Advanced Glasses Manufacturing Technology, MOE, Donghua University, Shanghai 201620, China

[a]wqwq888@dhu.edu.cn; [b]Chenjian@dhu.edu.cn; [c]lld@dhu.edu.cn;
[d]ningwei@dhu.edu.cn,

Abstract: Glass fiber used as reinforced composite presented excellent properties and was used widely in various environment including cryogenic temperature. The failure strain and surface morphology of glass fiber dipped in the liquid nitrogen and room temperature were measured in the study. Through the comparison of two different environment, results showed the glass fiber present better failure strains compared to those at room temperature, and there are no visible defect in the surface of glass fiber after immerged into the liquid nitrogen. The reduction of failure strain in the room temperature would owe dominantly to the relative humidity, but the temperature just exercised less influence on it.

1. Introduction

E glass fibers is one of most common glass fibers, which is widely used as the reinforced materials in different engineering industries due to their high specific strength to weight ratio, chemical and corrosion resistance and ease of manufacturing into complicated forms[1]. With the development of the aircraft and aerospace engineering industry [2-3], Glass fiber is used to manufacture the structural components such as wings, fuselages and cryogenic fuel tanks due to their higher stiffness/weight and strength/weight ratios compared to conventional metallic materials[4]. In the low earth orbit environment, where spacecrafts and satellites are located, the materials used for making their structures are always subject to extreme temperatures in the range of -170°C to 200°C[5]. Therefore, carrying out continuous research and development to optimize the mechanical properties of the glass fibers in cryogenic condition is undoubtedly essential.

Depending on the temperature level, glass fibers may behave more ductile or more brittle in affecting their performance of service. In literature, there are a few studies interested in the strength of glass fibers at low temperatures. Pukh, et al [6] used a three-point bending technique to measure the strength of a variety of compositions prepared as 100-150μm fibers, under liquid nitrogen, and reported intrinsic strengths of 12.0 GPa for polymer coated silica glass. Tang, et al[7] studied the inert strength and inert fatigue of glass fiber in liquid nitrogen. Results showed the calculated failure stress were 5.1 to 5.2 $\pm$ 0.1 GPa in inert conditions. Bulent ,et al studied the impact behavior of laminated glass-epoxy composites at several temperatures (20 °C, -20 °C and -60 °C).

These studies concentrated only on the properties of glass-fiber reinforced composites, and paid less attention on the failure strain or surface morphology of glass fiber in the cryogenic temperatures. The aim of this study was to investigate the low temperature effects on failure strain and surface morphology of glass fiber.

2. Experimental procedures

2.1 Material and fiber pulling

Commercial E-glass marbles which composition in weight percent was: 61% SiO_2, 13.5% Al_2O_3, 20.5 CaO, 4.6% MgO,0.2% alkali oxide were melted in platinum crucibles at 1 550 °C for 8 hours prior to fiber drawing. The crucible containing molten glass is transferred to another furnace set to the temperature at 1300 °C.

Glass melts were fiberized using a top-drawing system as Fig.1. Fibers are wound onto a rotating cage which is designed to prevent fiber overlap and damage each other. The fiber diameter is controlled by the pulling temperature and the rotating speed of the cage. All fibers of 100 to 200μm diameter are prepared. These pristine glass fibers are put into the cabinet to age immediately. The cabinet with ambient air was kept at constant temperature at 50°C and constant relative humidity at 80%.

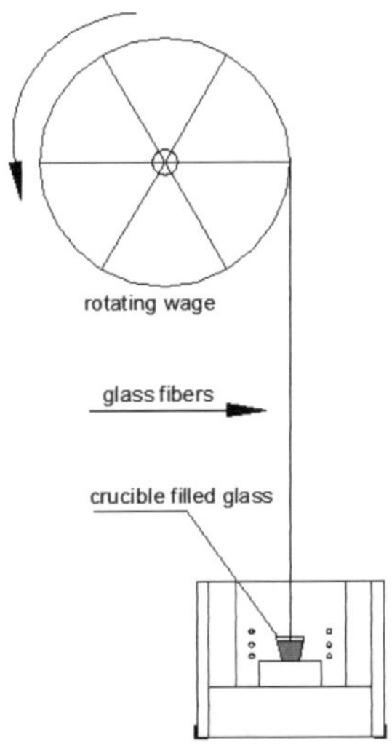

Figure.1 schematic map of the process of top-drawing glass fibers

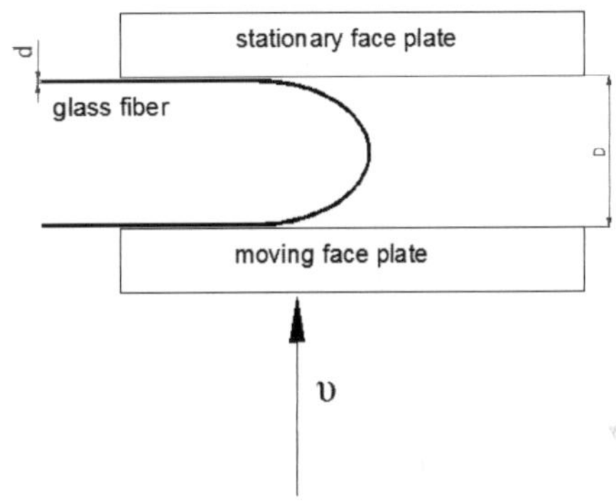

Figure.2 schematic map of two-point bend testing

2.2 Two-point bending failure measurement

All the failure strain is measured by Two-point bend (TPB) technique. The TPB technique was used as early as 1980 to testing glass fibers [6], and it is convenient for studying the mechanical properties of glass fibers. The TPB test does not require excessive handling of the samples, instead those tested for the TPB are freshly-drawn fibers with the original surfaces. Moreover, the most crucial factor is the TPB test of glass fiber can be implemented in liquid nitrogen circumstances just as figure 2.

Once the glass fiber is broken owing to the moving of faceplate, the failure is detected by an acoustic sensor and the shift of faceplate stops simultaneous. The failure strain (ε) is calculated from the faceplate gap distance at failure (D) and the fiber diameter (d). the equation is $\varepsilon = 1.198d/(D-d)$[6] .

In the article, the two-point fiber bending system is fabricated by TNL Tool and Technology, LLC (Parnell, IA). There are two testing circumstances used in the experiments: the one is in the ambient air at room temperature (22±1°C) at a relative humidity of 50±1%, the other is to dip in the liquid nitrogen(at -196°C). Each time at least 20 fibers are tested respectively.

2.3 Morphology characterization

The surface features of glass fibers would be characterized by the Optical Microscope (Hirox KH-700, Japan) and Scanning Electron Microscope (FEI Helios NanoLab 600 FIB).

3. Results

3.1 Effects on the failure strain

The Weibull distributions of the failure strain in ambient air for glass fibers aging for different days measured with the face plate speed of 4000 μm/s were shown in Fig.3. Results showed the failure strain of fibers was decreased with the aging time. The failure strains of most pristine fibers were 5.8%~6.0%, but it had decreased to 4.6~5.1% in 28 days later. It could also be observed that the failure strain became broader in the distribution.

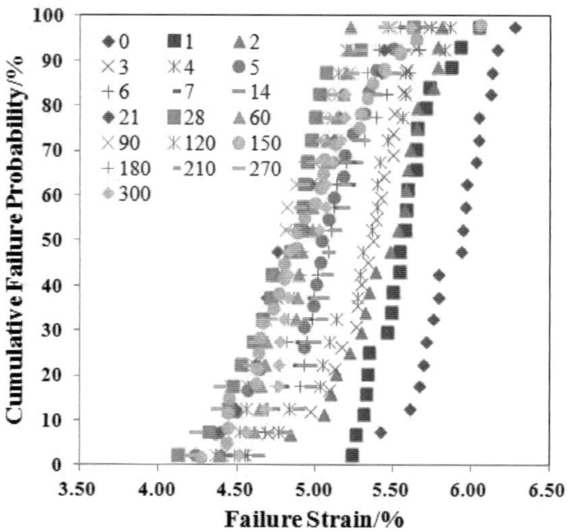

Figure.3. Weibull plots of the failure strains in air environment (at 22 °C with 50% relatively humidity) for glass fibers aging at 50°C with 80% relative humility for different days.

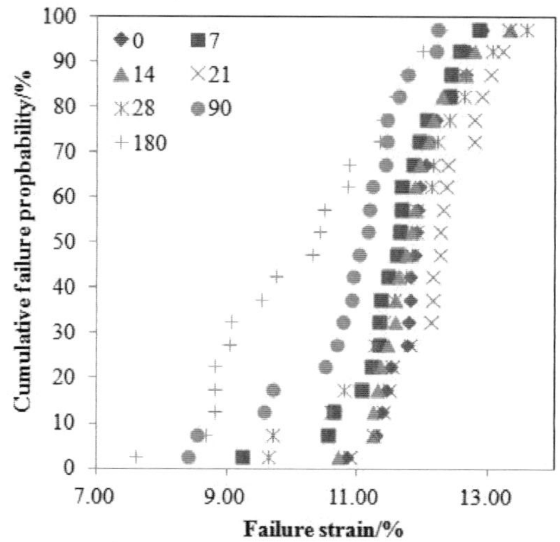

Figure. 4. Weibull plots of the failure strains in the liquid nitrogen(at -196 °C) condition for the glass fiber aging at 50 °C with 80% relative humility for different days.

Fig.4 showed the failure strain distributions in liquid nitrogen condition for glass fibers aging for different days measured with the face plate speed of 4000 μm/s. it was found the failure strains in inert condition were significant greater than in air condition, although it also descended with the aging time.

Fig.5 was the curve of the average values of failure strains. Besides above-mentioned changes, it was seen the failure strain after aging for a week did not decrease until two months later, and the descent trend was similar for two different circumstances.

3.2 Surface morphology

3.2.1 Optical pictures. Fig.6 showed the optical micrographs of glass fiber treated by liquid nitrogen. The surface of glass fiber was very neat and smooth except some dots. There are no any other changes in the surface.

3.2.2 SEM Micrographs. The SEM imagines of glass fiber which had been immerged into liquid nitrogen for 30 seconds were showed as Fig.7. Nothing except something like waterlogging was observed. It seemed that the cryogenic temperatures could not result in any perceptible defects.

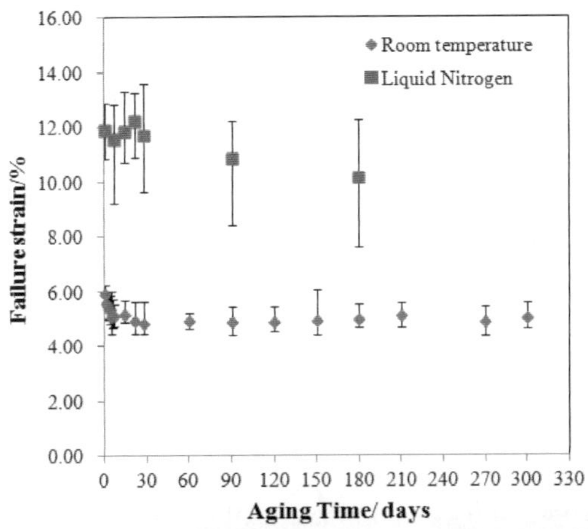

Figure.5 Effects of the aging time on the failure strain of glass fibers measured in the room temperature(at 22 °C with 50% relatively humidity) /liquid nitrogen condition(at -196 °C). The aging condition is at 50°C with 80% relative humility.

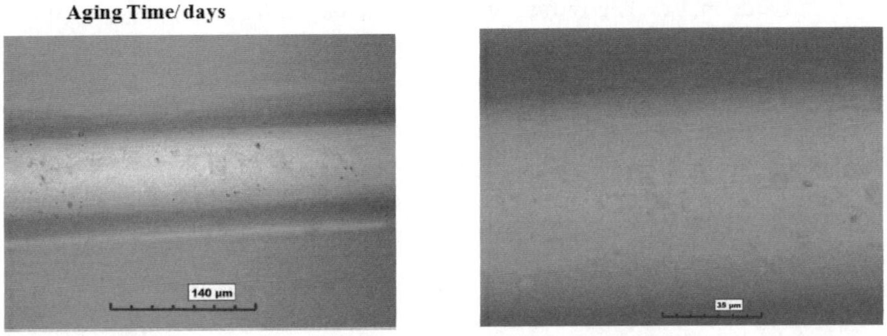

Figure. 6 Optical pictures of pristine E glass fibers immerged into liquid nitrogen for 30s (a: *500, b: *2000)

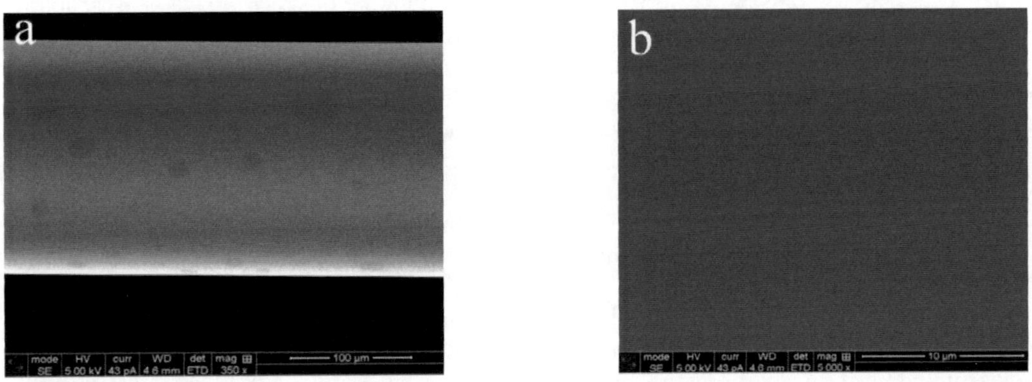

Figure.7 SEM imaging of glass fibers digging into the liquid nitrogen for 30s (a: *350, b: * 5000)

4. Discussion

It was clearly observed from Figure 3 and Figure 4 that there were two differences. One was the failure strains in the liquid nitrogen were as higher as nearly two times than those in room temperature, the other is the deviation of the failure stain in liquid nitrogen is also significantly broader.

According to the Griffith theory [8], failure strain (ε) can be calculated from the length of surface flaws (c) and the surface free energy (γ) just as the equation (1):

$$\varepsilon f = (2\gamma / E\pi c)1/2 \qquad (1)$$

Where E is the elastic modulus.

As for the failure strain, it was known in the room temperature environment there must be certain relative humidity. Because the relative humidity would sharply boost the propagation of the crack, it paid a great influence on the failure strain of fibers[9,10]. Duncan[11,12] et al presented the humidity dependence of failure strains usually was divided into three stages. Because we were not able to measure failure strains below ~0.1% RH, all the measurement was done in controllable 50% relative humidity.

Moreover, As for the Gibbs surface free energy, it can be calculated with the equation (2):

$$\Delta G=\Delta H-T\Delta S \qquad (2)$$

Where ΔH is the enthalpy, as we know, H=U+PV, U is internal energy, it is dependent on the vibrating velocity of molecule. As solid material, H of glass would change little wherever in air or LN. The entropy ΔS depends on the temperature and volume. ΔS increases when the temperature increases.

Therefore, when the glass fiber was immersed into the LN, the temperature varied with a difference of 218°C, ΔS must be decrease. The reduction of both temperature and entropy would result in the limited growth of ΔG, and it would contribute to the broaden of the failure stain.

5. Conclusion

In the paper the two measure environments was studied. It was found the failure strain and the error of value presented significant differences. Through discussing on the morphology of surface of fibers, there was nothing except something like waterloggings. It was inferred the reduction of failure strain in room temperature would be attributed mainly to the relative humidity since its sharp influence on the propagation of the crack. The temperature does have certain effect on the surface energy but the change was very slight.

References

[1] Lezzi PJ, Seaman JH, Tomozawa M. Strengthening of E-glass fibers by surface stress relaxation, J Non-Crys Solids, 402 (2014)116-127.

[2] Cormier L, Joncas S,Nijssen RPL. Effects of low temperature on the mechanical properties of glass fibre-epoxy composites: static tension, compression, R=0.1 and R =-1 fatigue of +/-45 degrees laminates, Wind Energy,19(2016) 1023-1041

[3] Lezzi PJ, Seaman JH, Tomozawa M. Strengthening of E-glass fibers by surface stress relaxation, Journal of Non-Crystalline Solids, 402 (2014):116-127.

[4] Ma H, Jia Z,Lau K, Leng J,,Hui D. Impact properties of glass fiber/epoxy composites at cryogenic environment. Composites, Part B. Engineering,92(2016) 210-217

[5] Liu X, Cheng L, Zhang L, Dong N, Wu S, Meng Z. Tensile properties and damage volution in a 3D C/SiC composite at cryogenic temperatures. Mater Sci Eng A,528(2011)7524-7528.

[6] Pukh VP, Baikova LG, Kireenko MF, Tikhonova LV, Kazannikova TP, Sinani AB, Atomic Structue and Strength of Inorganic Glasses, Phys. Solid State, 47 (2005) 850-855.

[7] Tang Z, Lower NP, Gupta P K, Kurkjian CR, Brow RK. Using the two-point bend technique to determine failure stress of pristine glass fibers. J Non-Crys Solids, 428 (2015) 98-104

[8] Bradt RC. Applying fractography and fracture mechanics to the energy and mass of crack growth for glass in the mirror region. J. Euro. Ceram. Soc., ,34(2014) 3255-3262.

[9] Wang Q., Brow R., Li H. et al. Effect of aging on the mechanical properties and surface morphology of E glass fiber without sizing, J Mater Sci, 51(2016)2404-2410

[10] Wang Q., Brow R., Li H. et al. Effects of Aging Conditions on E-Glass Failure Strain and Dynamic Fatigue. Eu J Glass Sci Tech:A, Eur. J. Glass Sci. Technol. A, 57 (2016) 58-61

[11] Duncan W J, France PW, Graig S P. The Effect of Environment on the Strength of Optical Fiber, in: C. R. Kurkjian (Ed.), Strength of Inorganic Glass, Plenum, 1985, 309-28.

[12] Tang Z. Two-point bend studies of glass fibers, Missouri University of Science and Technology , Ph D. Theisi, 2011.

Effect of Al on the mechanical properties and corrosion resistance of Pb-Al alloy

LU Zhicheng[a], LIU Zhenlin[a], LI Yongliang[b], WU Dan[b], WANG Fuming[b]

[a]China Electric Power Research Institute, Beijing, 100192, China,
[b]University of Science and Technology Beijing, Beijing 100083, China,

E-mail:wangfuming@metall.ustb.edu.cn

Abstract. A set of binary Pb-Al alloys with different Al contents were designed in this work. The mechanical properties and corrosion resistance of Pb-Al alloys were investigated with help of tensile test, Charpy V-notch impact test and salt spray corrosion test (SSCT). And the microstructure was observed by optical microscopy. The results showed that microstructure of all alloys were twin structure, and the twin structure was gradually refined with the increase of Al content. Al dissolved in matrix could significantly improve the tensile strength, impact energy and corrosion resistance. However, a higher content of Al would harm the mechanical properties and corrosion resistance. It may be due to the heterogeneous precipitation of Al rich phase.
Key Words: Pb-Al alloy, mechanical property, corrosion resistance, salt spray corrosion test, microstructure.

1. Introduction

Electrical equipment with porcelain bushings is the most basic electrical components in power system. However, ceramics is prone to fragility fractures due to the resonance in earthquake. The practice has proved that installing dampers can enhance the seismic capacity of electric porcelain equipment by changing the free vibration characteristic and increasing the damping[1]. Traditionally, dampers are made of rubber, but rubber is easy to ageing and its service life is short[2, 3]. Compared with rubber dampers, metal dampers have the advantages of long service life, stable performance, etc. Metal dampers are mainly made of lead or mild steel, and the plastic deformation of the metal is used to absorb the earthquake energy[4-6]. Lead has excellent toughness and plasticity at room temperature and low temperature, and can absorb a large amount of energy during the process of deformation. While lead has good plasticity, the low tensile and compressive strength limits its applications. In order to apply lead better in industry, it is necessary to improve the strength of lead. On the other hand, metal dampers are usually used in the wild field. Under the comprehensive effects of atmospheric oxygen, Cl^-, HCO_3^- and SO_4^{2-} ions in the long-term service process, lead work pieces are corroded seriously. It causes serious damage to the matrix and reduces the reliability of work pieces[7]. Therefore, it is needed to improve the corrosion resistance of lead in the natural environment on the basis of improving the strength, so as to prolong their service life.

In theory, the strength of metals can be improved by grain size strengthening, cold deformation dislocation strengthening, solid solution strengthening, aging precipitation strengthening, phase transformation strengthening, etc[8]. However, the strengthening effect of grain refinement is actually ineffective in lead due to the fact that the creep strain rate of fine-grain lead is larger than that of coarse-grain lead. Dislocation enhancement can only play a role below the recrystallization temperature. The room temperature is in the range of recovery and recrystallization temperature of lead, so it is generally unable to make use of cold deformation to strengthen the lead. Aging hardening is a strengthening way for many alloys. But the enhancing effects will gradually disappear because of the coarsening behavior of the precipitates in lead alloys[7]. Based on the above reasons, lead alloys were strengthened by solid solution strengthening in this paper. Moreover, adding a small amount of alloying elements is beneficial to the improvement of corrosion resistance[9].

The study of lead based binary alloy shows that the solid solution strengthening ratio, ds/dc(s: strength, c: concentration), increasing in order as follows: thallium, bismuth, tin, cadmium, antimony, lithium, arsenic, calcium, copper and barium[7]. However, thallium is a highly toxic element[10], so it is not suitable for large scale application in industry. In the melting process of bismuth and tin, oxidization burning loss is quite serious on account of the low melting point[11], and thus reduces the yielding rate. The aging strengthening effect of lead-cadmium alloy is weak and short. Therefore, Sb is commonly used to strengthen lead alloys in industry[12]. But Sb and its oxides are toxic and pollute environment[13]. Research shows that alloying lead with aluminum would reduce weight and increase tensile strength[14]. However, these studies mainly focused on the higher Al content[15], and Al mainly exists in the form of precipitates. There is little research about the effect of minor Al on the performance of lead. On the other hand, corrosion resistance of lead alloys has been studied widely, but most of them are concentrated on the battery material[16-19]. The study on corrosion resistance of Pb-Al binary alloy in natural environment is very few.

In the present work, minor Al is added to improve the performance of lead alloys. Pb-Al alloy is melted in vacuum induction furnace and cooled by using forced fast cooling technology. The influence of the concentration gradient of Al on the mechanical properties of Pb-Al alloy was studied by mechanical property test and microstructure observation, meanwhile, the influence of the concentration gradient of Al on the corrosion resistance was studied by artificial atmosphere-neutral salt spray corrosion test.

2. Experimental

The chemical compositions of different Pb-Al alloys are shown in Table 1. In order to prevent the high temperature oxidation of alloying elements and avoid the environmental pollution caused by the volatilization of lead, the melting process is carried out in a vacuum induction furnace. The melting alloy are cast into 10kg ingots under the protection of argon. In the casting process, the mold is cooled by flowing water in order to accelerate the solidification of the liquid alloy, thereby preventing the gravity segregation and grain boundary segregation of Al.

Table 1. The chemical compositions of Pb-Al alloys

Specimens	Al / wt%	Pb / wt%
Pure lead	-	99.999
Pb-0.05Al	0.05	Bal.
Pb-0.10Al	0.10	Bal.
Pb-0.15Al	0.15	Bal.

2.1. Mechanical properties test

The ingots were forged into 17mm cross-sectional bars at room temperature, and then were processed into a variety of test specimens. Standard samples were used for tensile and impact tests. The impact test temperatures were 20°C, -20°C, and -40°C. Breaking the tensile test sample along the axial direction by using the electric spark line cutting, and metallographic specimen was prepared by cold setting in order to avoid that the microstructure of the alloy is affected by high temperature and high pressure. Sample was eroded by a mixed solution (glycerin: glacial acetic acid: hydrogen peroxide; 4: 1: 1), and the microstructure was observed by optical microscopy.

2.2. Salt spray corrosion test
The ingot was rolled into a sheet of 1 mm in thickness by cold rolling mill, and samples with a size of 50 mm×50 mm were cut from the sheet. The sample surface was mechanically-polished and washed in an ultrasonic bath of acetone. The neutral salt spray atmosphere (5%NaCl water solution) was chosen as artificial fog in this experiment. Salt spray corrosion was carried out in salt spray test chamber, with the temperature controlled at 25°C± 1°C. Non-test surface of the sample was wrapped by silica gel so as to improve the accuracy of weight loss measurement per unit area. The treated samples were placed in four corners of the salt spray test chamber. The surface morphology changes were observed regularly. After the test, the samples were soaked in hydrochloride aqueous solution (volume ratio, 1:1; ρ=1.18g/mL), adding CH12N4 as a corrosion inhibitor, to remove corrosion products. The samples were cleaned with water and acetone at room temperature. After rapid drying, measured sample weight and calculated the quality loss per unit area. The morphology of etched surface was observed by optical microscope and the effect of Al element content on corrosion morphology was analyzed.

3. Result and discussion

3.1. Microstructure
At room temperature, lead and lead alloys consist of the alpha phase (Pb) and the terminal solid solution based on other elements. The microstructure of lead alloys with different Al contents is shown in Figure 1. There was no significant difference between the alloys with different Al. The Microstructures of all alloys were twin structure, and the twin structure was gradually refined with the increase of Al content. Research shows that the grains of Pb alloys were refined with 0.01-0.08% aluminum addition[20]. As aluminum content increased to 0.15%, the lead alloy grains were further refined in the present work. Moreover, there were obvious precipitates in grains of Pb-0.15%Al alloy. The solubility of Al in Pb was lower than 0.10% at room temperature[21]. Therefore, the addition of excess aluminum may produce Al-rich phase.

Figure 1. The microstructure of Pb-Al alloys: (a)Pb; (b)Pb-0.05%Al; (c)Pb-0.10%Al; (d)Pb-0.15%Al

3.2. Mechanical property
The mechanical properties of pure lead and Pb-Al alloys are shown in Table 2

Table 2. Mechanical properties of Pb-Al alloys

Specimen	R_m / MPa	A / %	Z / %	A_{KV}/ J		
				20°C	-20°C	-40°C
Pure lead	13.2	75.3	100	18.0	18.3	18.5
Pb-0.05%Al	17.9	73.7	100	19.5	20.2	20.2
Pb-0.10%Al	15.5	78.2	100	17.4	18.7	19.2
Pb-0.15%Al	16.5	80.5	100	18.8	19.0	19.5

3.2.1. Tensile property. The mechanical properties of lead alloys changed obviously with the increase of the content of Al. The tensile strength of Pb-0.05%Al alloy was the highest, which is 17.9 MPa. The strength of lead alloys appeared to fluctuate with aluminium content increasing. But it showed the enhancement effect overall. Compared with pure Pb, the strength of Pb-0.05%Al, Pb-

0.10%Al and Pb-0.15%Al alloys increased by 4.7 MPa, 2.3 MPa and 3.3 MPa, respectively. It can be seen that the strengthening effect of Al element on Pb was obvious.

Section shrinkage rates of pure lead and three lead alloys were all 100%. Although the elongation of Pb-0.05%Al alloy decreased slightly compared with others, but it still exceed 70%. The result showed that the addition of Al element has little influence on the plasticity of the alloy, while increasing the strength of the lead alloy.

3.2.2. Impact toughness. As can be seen in Table 2, impact energy of Pb and Pb-Al alloys were all improved with the decrease of the temperature. Compared with other alloys, Pb-0.05%Al alloy had the highest impact toughness. The impact energy was decreased with Al content was further increased to 0.10% and 0.15%, but the low temperature impact energy was slightly higher than that of Pb. It showed that the low temperature impact toughness of Pb-Al alloy can be improved by adding Al element.

The main factors affecting the impact toughness are grain size, microstructure, and macro-fiber flow and so on. And the grain size is the main influencing factor. The smaller the grain size of the metal material, the higher the impact energy, and vice versa reduced. The recrystallization temperature of Pb and Pb-Al alloys is below zero. Above the critical point of recrystallization temperature, the grain of Pb-Al alloy has a tendency to grow up with the increase of temperature. However, coarse grains can lead to the reduction of impact energy. Therefore, the impact energy that is measured in the experiment will be decreased with the increase of the temperature. Pb-Al alloys were not broken after impact test at different temperatures. This shows that Pb-Al alloys had better stability at low temperatures.

3.3. Salt spray corrosion test

3.3.1. Macroscopic morphological. Figures 2 and 3 are the macro photos of the samples after the salt spray corrosion. As shown in Figure 2, the surface of the sample was lost and the colour gradually faded after 2h corrosion. From the colour of the corrosion film, the surface of pure lead was the most uneven. It showed that the thickness of the product is not uniform, and accompanied with drops phenomenon. Compared with pure lead, the surface color of Pb-0.05%Al and Pb-0.10%Al alloy were more uniform. When further increasing the Al content reached 0.15%, the surface of the sample was relatively rough. But there was no obvious change in the color of the corrosion film at different locations. The thickness and composition of the product are the main reasons for the change of the surface color, so it can be concluded that the corrosion resistance of Pb-0.05%Al and Pb-0.10%Al alloy is slightly different from that of lead. However, due to the limitations of the experimental conditions and objectives, the specific product did not carry out quantitative analysis and further research is needed to determine it.

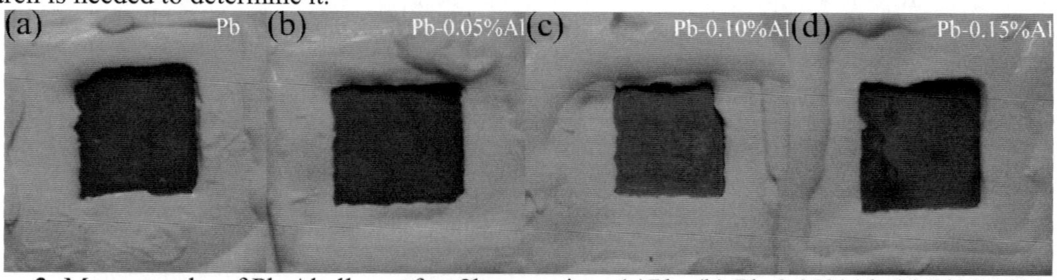

Figure 2. Macrographs of Pb-Al alloys after 2h corrosion: (a)Pb; (b) Pb-0.05%Al; (c) Pb-0.10%Al; (d) Pb-0.15%Al

Macrographs of Pb-Al alloys after 96h corrosion are shown in Figure 3. From figure 3(a) and 3(d), we can see that the colored corrosion film was formed on the surface of Pb and Pb-0.15%Al alloy. And there existed a certain amount of corrosion groove on the substrate of the two alloys. Compared with the sample after 96h corrosion, the surface roughness of Pb-0.15%Al alloy was deeper and the surface became dim after 96h corrosion. Although there was a certain product film off on the surface of Pb-0.5%Al and Pb-0.10%Al alloy, the color change was not obvious. The result showed that the product film was still relatively uniform and the binding force of the membrane and the substrate was strong after long time corrosion.

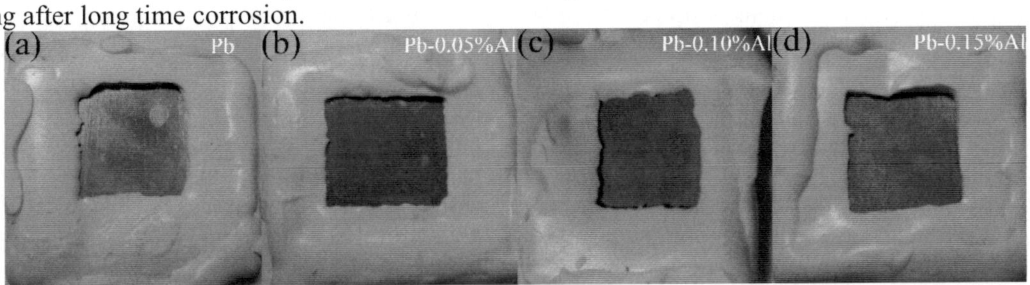

Figure 3. Macrographs of Pb-Al alloys after 96h corrosion: (a)Pb; (b) Pb-0.05%Al; (c) Pb-0.10%Al; (d) Pb-0.15%Al

3.3.2. Mass loss. Mass loss of Pb and Pb-Al alloys after 96h corrosion is shown in Table 3. As can be seen from Table 3, the values of mass loss per unit area of Pb and Pb-0.15%Al alloy unit area were 12.5mg/cm^2 and 10.5g/cm^2, respectively; as for Pb-0.05%Al and Pb-0.10%Al alloys, they were 5.7mg/cm^2 and 4.3mg/cm^2, respectively. Compared with Pb and Pb-0.15%Al alloy, Pb-0.05%Al and Pb-0.10%Al alloys have higher corrosion resistance.

Table 3. Mass loss of Pb-Al alloys after 96h corrosion

Specimen	Pre-test quality/mg·cm^{-2}	Post-test quality/mg·cm^{-2}	Mass loss/ mg·cm^{-2}
Pure lead	3940.8	3928.2	12.5
Pb-0.05%Al	4994.9	4989.2	5.7
Pb-0.10%Al	4504.4	4500.1	4.3
Pb-0.15%Al	4940.2	4929.7	10.5

Compared with Pb, the mass loss of Pb-0.05%Al and Pb-0.10%Al alloy decreased by 6.8mg/cm^2 and 8.2 mg/cm^2 after 96h corrosion, respectively; as for Pb-0.15%Al, by contrast, it only decreased by 2mg/cm^2. The result shows that the corrosion resistance of Pb-0.05%Al and Pb-0.10%Al is more than twice that of Pb.

3.3.3. Microstructure. After 96h corrosion, there were a large number of ditch corrosion and a certain number of corrosion pits on Pb surface, and its size is larger, as shown in Figure 4(a). Compared with Pb, the number of ditch corrosion and corrosion pits of Pb-0.05%Al and Pb-0.10%Al were significantly less, as shown in Figure 4(b), 4(c). However, when the Al content is further increased to 0.15%, their numbers were significantly increased, as shown in Figure 4(d). By comparing the microstructure changes and mass loss, it is found that the variable tendency of the weight loss and the number of ditch corrosion and corrosion pits on the surface of Pb alloy is consistent. When the number and size of ditch corrosion and corrosion pits were increased, the weight loss became larger, and vice versa.

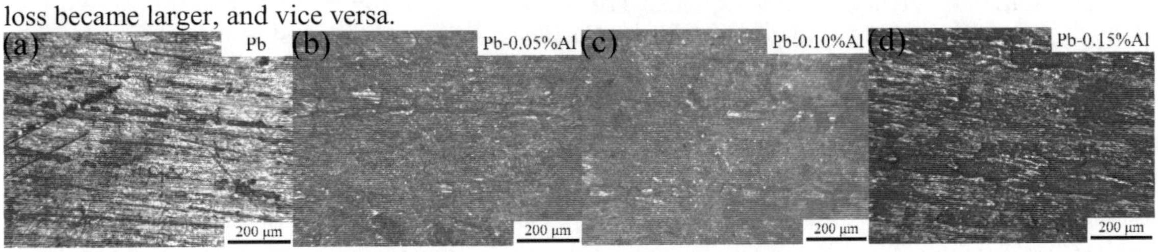

Figure 4. Micrograph of Pb-Al alloys after 96h of corrosion: (a)Pb; (b)Pb-0.05Al; (c)Pb-0.10Al; (d)Pb-0.15Al

3.3.4. Mechanism of salt spray corrosion. Fmoreas been carried out to study the effect of alloying elements on the corrosion resistance of lead[16]. Lead based solid solution is formed by adding a small amount of alloying element, such as tin, antimony, bismuth, aluminium. As stated above, not only can it improve the mechanical properties of the alloy, but also help to improve the corrosion resistance[9]. However, adding excessive alloy elements may harm the corrosion resistance of lead alloys[22].

The addition of excess aluminium may produce Al-rich phase, as shown in figure 1(d). And it enables the formation of a local micro cell between the lead base and the Al-rich phase. Anodic lead dissolution under the action of positive and negative ions in aqueous solution, and a certain amount of oxidation products was generated on the surface. Eventually, ditch corrosion and corrosion pits were formed. Although pure water has no effect on lead, dissolved matter in water may lead to surface erosion of lead. Corrosion in aqueous solution occurs in an electrochemical process. Lead atoms enter the solution with a metal cation state or transition into solid state compounds at the anode. In the neutral salt solution, the cathodic reaction is the reduction of dissolved oxygen.

In the neutral salt solution, Cl^{-1} in water is not directly involved in chemical reactions. But it accelerates the corrosion of the lead matrix in the whole process of the reaction. In the case of a certain concentration of Cl^{-1}, the higher the content of Al in the alloy, the more intermetallic compounds are formed. And then the galvanic reaction of the alloy is stronger. Finally, the weight loss is increased in the process of salt spray corrosion.

4. Conclusions

(1) The microstructures of lead and all lead alloys were twin structure, and the twin structure was gradually refined with the increase in Al content. And there were obvious precipitates in grains of Pb-0.15%Al alloy.

(2) The strengths of lead alloys were significantly improved after adding Al element. Compared with pure lead, the tensile strength of Pb-0.05Al alloy was increased by 31%, reaching 17.9MPa. And the addition of Al had little influence on the plasticity.

(3) The addition of Al could significantly improve the low temperature impact performance of lead, especially in Pb-0.05%Al alloy. The impact energies of Pb and Pb-Al alloys were increased with the decrease in temperature, and there was no low temperature embrittlement phenomenon.

(4) The corrosion resistances of Pb-0.05%Al and Pb-0.10%Al alloys were significantly increased compared with pure lead. However, the mass loss of Pb-0.15%Al alloy was similar to pure lead. It indicated that the solid solution of aluminium in lead lattice can effectively improve the corrosion resistance of lead alloys. But excessive Al could easily form an Al-rich phase which leads to the local micro cell effect and accelerate the corrosion of Pb matrix.

References

[1] Li Y, Li X and Liu X 2005 *J. Acta Seismol Sin* **18** 89
[2] Amin A and Alam M, Okui Y 2002 *J. Mech Mater* **34** 75
[3] Gu H, Itoh Y 2010 *J Adv. Struct.Eng.* **13** 1105
[4] Robinson W H and Greenbank L R 1976 *J. Earthquake Eng Struct Dyn* **4** 251
[5] Ming-hsiang S, Wen-pei S and Cheer-germ G 2004 *J. J Zhejiang Univ: Sci* **5** 326
[6] Dargush G F, Soong T T 1995 *J. Earthquake Spectra* **11** 545
[7] Hofmann W 1970 Lead and Lead Alloys *Lead and Lead alloys*(Berlin Heidelberg:Springer) chapter B pp 25-320
[8] Arzt E 1998 *J. Acta mater* **46** 5611
[9] Ivanov I, Stefanov Y, Noncheva Z, Petrova M and Dobrev T 2000 *J. Hydrometallurgy* **57** 109
[10] Karbowska B, Zembrzuski W, Jakubowska M and Wojtkowiak T, Pasieczna A and Lukaszewski Z 2014 *J.J. Geochem. Explor.* **143** 127
[11] Weeks J R 1971 *J. Nucl Eng Des* **15** 363
[12] Roumina R, Raeisinia B and Mahmudi R 2004 *J. Scripta Mater* **51** 497

[13] Klein J, Dorge S, Trouvé G, Venditti D and Durécu S 2009 *J. J. Hazard. Mater.* **166** 585
[14] Cole J F and Goodwin F E 1990 *J. JOM* **42** 41
[15] Zhu M, Che X Z, Li Z X, Lai JKL and Qi M 1998 *J. J Mater Sci* **33** 5873
[16] Clancy M, Bettles C J, Stuart A and Birbilis N 2013 *J. Hydrometallurgy* **131-132** 144
[17] Zhang W, Li A, Chen H, Lan B, Pan K,Zhang T, Fang M, Liu S and Zhang W 2012 *J. J Power Sources* **203** 145
[18] Tunnicliffe M and Mohammadi F 2012 *J. J Electrochem Soc* **159** C170
[19] Osório W R, Rosa D M, Garcia A 2008 *J. J Power Sources* **175** 595
[20] Bouirden L, Hilger J P and Hertz J 1991 *J. J Power Sources* **33** 27
[21] Gabrisch H, Kjeldgaard L, Johnson E and Dahmen U 2001 *J. Acta mater* **49** 4259
[22] Albert L, Chabrol A, Torcheux L, Steyer Ph and Hilger J P 1997 *J. J Power Sources* **67** 257

Effects of hydrogen on microstructure and hot deformability of directionally solidified Ti–44Al–6Nb–1Cr alloy

T. F. Ma, R. R. Chen, J. J. Guo, H. S. Ding, Y. Q. Su and H. Z. Fu

School of Materials Science and Engineering, Harbin Institute of Technology, Harbin, 150001, China

ruirunchen@hit.edu.cn

Abstract. Effect of hydrogen on microstructure and the hot flow behavior of the directionally solidified Ti–44Al–6Nb–1Cr (at.%) alloy was studied. Ti–44Al–6Nb–1Cr alloys were prepared by vacuum induction melting (as-cast alloy) and cold crucible directional solidification (CCDS). Effects of hydrogen on the hot deformation behavior were tested on Gleeble-1500D at 1373 K and 1473 K under strain rate of 0.01 s^{-1}. Results show that the volume fraction of B2 phase is increased after CCDS due to the solute distribution. The residual B2 phase in the hydrogenated alloy is more than that in unhydrogenated alloy because hydrogen is beta stabilizing element and stabilizes the B2 phase. The peak stress of the CCDS alloy is higher than that of as-cast alloy at the same deformation condition due to the directional solidified microstructure. Hydrogen can decrease peak stress of the as-cast alloy and CCDS alloy significantly, which is decreased by about 46% and 43% respectively. This is attributed to the hydrogen-promoted dynamic recrystallization and hydrogen-increased the volume fraction of B2 phase.

1. Introduction

TiAl alloys have attractive properties, such as low density, high specific strength, and excellent oxidation resistance, which can be applied in aerospace area and automotive filed at high temperature range of 700–950 °C [1–3]. The mechanical property of TiAl alloys are improved significantly by directional solidification, the average room temperature tensile ductility and failure strength of the Ti–45Al–8Nb single crystals is 6.9% and 978 MPa respectively [4]. The small size of the single crystal samples restricts the industrial applications. CCDS can be used to prepare TiAl ingots with industry size efficiently, such as the Ti–44Al–6Nb–1Cr–2V alloy with size of Φ 19 mm×300mm is directional solidified successfully [5]. Although the mechanical property of CCDS TiAl alloys is lower than the TiAl single crystals, it is higher than the as-cast alloy. The tensile ductility and yield strength of CCDS Ti–47Al–2Cr–2Nb alloy is 5% and 530 MPa respectively [6]. Hydrogen is considered to be a positive element in titanium alloy and Ti_3Al-based alloy during hot deformation [7, 8]. Hydrogen can decrease the flow stress of Ti600 alloy by about 70% after hydrogenation with 0.3 wt.% H, the peak stress of Ti–25Al–l0Nb–3V–1Mo alloy is decreased by 30% after hydrogenation with 0.2 wt.% H [8, 9]. Recently, hydrogen has been introduced in TiAl alloy, and the positive effect of hydrogen is also observed, the flow stress of Ti–47Al alloy and Ti–45Al–5Nb–0.8Mo–0.3Y alloy is decreased by 33% and by 26% respectively after hydrogenation [10, 11]. While, the effect of hydrogen on the CCDS TiAl alloy has never reported. The different microstructures of CCDS TiAl alloy and as-cast alloys result in different hot deformation behavior, thus effect of hydrogen on microstructure and hot deformation behavior needs to be studied deeply. In present study, the as-cast and CCDS Ti–44Al–6Nb–1Cr alloys are hydrogenated at 1473 K and 1623 K under hydrogen pressure of 0.1 MPa, the effect of hydrogen on microstructural evolution is observed and analyzed. The hot deformation behavior of as-cast and CCDS Ti–44Al–6Nb–1Cr alloys was tested on Gleeble-1500D at 1373 K and 1473 K under strain rate of 0.01 s^{-1}, and the deformation mechanism is discussed.

2. Experimental Methods

The material used in this study was Ti–44Al–6Nb–1Cr alloy, which was prepared by vacuum induction melting. The raw materials are titanium sponge with purity of 99.98%, Al-50%Nb master alloy, Cr particles. The melt was cast in the steel mould which was preheated to 573 K. The ingot size was about Φ 100 mm × 120 mm. The CCDS experiments had been described in detail in reference [5, 6]. Cylindrical specimens (8 mm in diameter and 12 mm in height) were cut from the center of the ingot and the CCDS ingot using wire-electrode cutting. The specimens were heated at 1473 K and 1623 K in a vacuum furnace under hydrogen atmosphere, and soaked for 30 min and 60 min then furnace cooled to room temperature. The hydrogen content was determined by weighting the specimens before and after hydrogen absorption using a high precise electronic balance with accuracy scale accurate to 10^{-5} g, and hydrogen content was calculated in mass. The isothermal heat treatment was carried out under argon atmosphere for comparison in order to eliminate the effect of heat treatment.

The hot deformation was carried out on Gleeble-1500D thermal simulation machine at 1373 K and 1473 K under strain rate of 0.01 s^{-1}. The engineering strain was 60% (≈0.8 true strain). The specimens were immediately water-quenched after hot compression to retain the deformed microstructures. The microstructures before and after hot deformation were observed by Quanta 200FEG scanning electron microscopy (SEM). For SEM observations, the specimens were ground by sandpapers from P200 to P2000, polished with the Cr_2O_3 solution, and then slightly etched by a Kroll solution (10 vol% HNO_3 +10 vol% HF+80 vol% H_2O).

3. Results and discussion

3.1. Original microstructure

Figure 1 shows the microstructures of as-cast and CCDS Ti–44Al–6Nb–1Cr alloys. The as-cast microstructure consists of α_2/γ lamellar, and mixture of block γ phase and B2 phase which is commonly seen in TiAl alloys contained high content of beta stabilizing element [12]. The B2 phases are mainly distributed at the grain boundary and inter-dendrite. The volume fraction of B2 phase is increased after CCDS, this is because of the solute distribution during the solidification. The β phase is rich of beta stabilizing element Nb, Cr, the α phase nucleates at the boundary of β phase and discharges the beta stabilizing elements resulting in β-segregation. The B2 phase is the ordering of β phase. The mixture of block γ phase and B2 phase disappears after CCDS, and the B2 phase is distributed in the inter-dendrite. The B2 phase along the fix direction because the distribution of beta elements is related to the heat flux direction. Similar microstructure is also observed in the Ti–44Al–8Nb alloy using Bridgeman method [13].

3.2. Effect of hydrogen on microstructure

The microstructures of as-cast and CCDS Ti–44Al–6Nb–1Cr alloys after isothermal heat treatment and hydrogenation at different temperatures are shown in Figure 2. It can be seen that the B2 phase dissolves partly after isothermal heat treatment and hydrogenation at 1473 K. The residual B2 in the CCDS alloy is more than that in the as-cast alloy. The residual B2 phase in the hydrogenated alloy is more compared with that in unhydrogenated alloy for both as-cast alloy and CCDS alloy. This is because hydrogen is beta stabilizing element and stabilizes the B2 phase. Although the B2 phase is detrimental to the room-temperature mechanical property, it contributes to hot deformation. The B2 phase dissolves completely after treatment at 1623 K, and the directional solidification microstructure is destroyed, and fully lamellar structure is obtained. The lamellar structure is formed by the sequence of $\alpha \rightarrow \alpha+\gamma \rightarrow \alpha_2+\gamma$ during continuous cooling from the single α phase field according to S. Zghal [14].

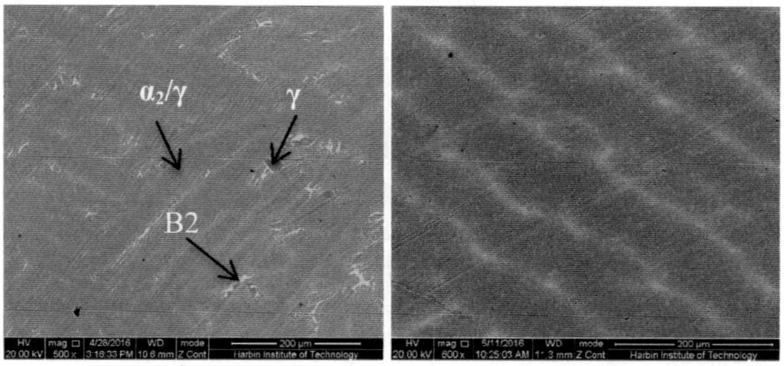

(a) (b)

Figure 1. The as-cast alloy (a) and CCDS alloy (b).

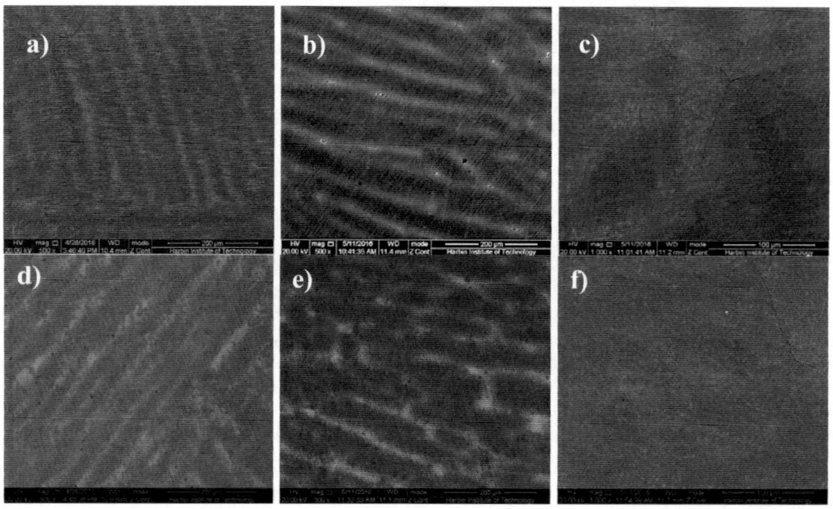

Figure 2. Microstructures of as-cast and CCDS Ti–44Al–6Nb–1Cr alloys after isothermal heat treatment and hydrogenation: (a)as-cast alloy at1473 K (Ar gas); (b)CCDS alloy at 1473 K (Ar gas); (c)CCDS alloy at 1623 K (Ar gas); (d)as-castalloy at 1473 K (H_2); (e)CCDS alloy at 1473 K (H_2); (f)CCDS alloy at 1623 K (H_2).

3.3. Effect of hydrogen on hot deformation behavior

The true stress-strain curves of the as-cast and CCDS Ti–44Al–6Nb–1Cr alloys hydrogenated with different hydrogen content deformed at different temperatures under strain rate of s^{-1} are shown in Figure 3. The curves show the same shape indicating that the deformation behavior is not changed. The curves present the dynamic recrystallization (DRX) softening behavior which is commonly seen in TiAl alloys [15, 16]. The peak stress is increased after CCDS, the peak stress of the as-cast alloy and CCDS alloy deformed at 1473 K is 197 MPa and 350 MPa respectively. This is attributed to the directional solidified microstructure. The peak stresses of the as-cast alloy and CCDS alloy can be decreased significantly by hydrogen, which decreases with increasing of hydrogen content. The maximum decrease of the peak stress for the as-cast alloy and CCDS alloy deformed at 1473 K is 46% and 43% respectively. Hydrogen can decrease the flow stress of titanium alloys, Ti_3Al-based alloy and TiAl alloys by promoting the DRX and dislocation motion [9, 10]. The mechanism of hydrogen-induced the decrease of peak stress in CCDS alloy is discussed below.

IOP Conf. Series: Materials Science and Engineering **201** (2017) 012046 doi:10.1088/1757-899X/201/1/012046

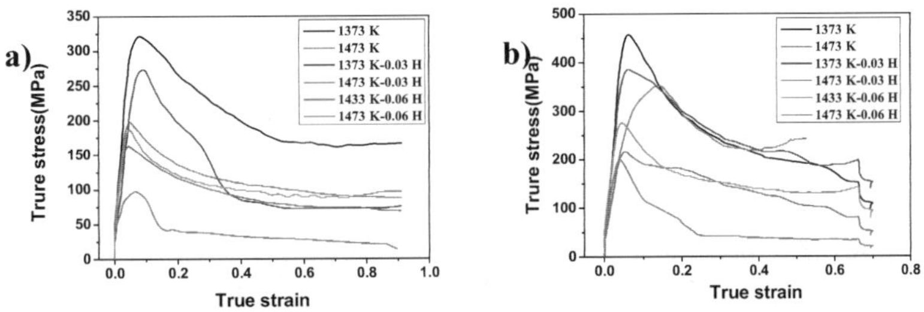

Figure 3. The true stress-strain curves of the as-cast and CCDS Ti–44Al–6Nb–1Cr alloys hydrogenated with different hydrogen content at different temperatures under strain rate of s^{-1}: (a) the effect of hydrogen content on as-cast alloy; (b) the effect of hydrogen content on CCDS alloy.

3.4. Effect of hydrogen on deformation microstructure

Effects of hydrogen on the deformation microstructure of the as-cast and CCDS Ti–44Al–6Nb–1Cr alloys deformed at 1373 K and 1473 K are show in Figure 4. The deformation microstructures consist of residual lamellar and DRX grains. When the deformation temperature is low (1373 K), the bending and elongating of lamellar occur and the mixture of B2 phase and γ phase at the grain boundary coordinates deformation. The DRX occurs at the grain boundary due to the high dislocation density and deformation energy. However, obvious DRX occurs in the hydrogenated alloy suggesting that hydrogen promotes the DRX. The volume fraction of B2 phase is increased after hydrogenation, which contributes to the decrease of flow stress. The volume fraction of DRX increases with increasing of temperature. The complete DRX is formed in the hydrogenated alloy deformed at 1473 K. By comparing the deformation microstructure, it can be seen that hydrogen decreases the deformation temperature by about 50 K. The DRX in CCDS alloys is not as obvious as the as-cast alloys, suggesting that the CCDS microstructure is more stable than the as-cast alloys. As a result, higher flow stress is obtained in CCDS alloys, which agrees well with the true stress-strain curves. The fibrous microstructures consisting of residual lamellar and DRX grains are observed in the CCDS alloy. The DXR is more obvious in the hydrogenated alloy compared with that of unhydrogenated alloy. It reveals that hydrogen can promote the DRX of the CCDS alloy.

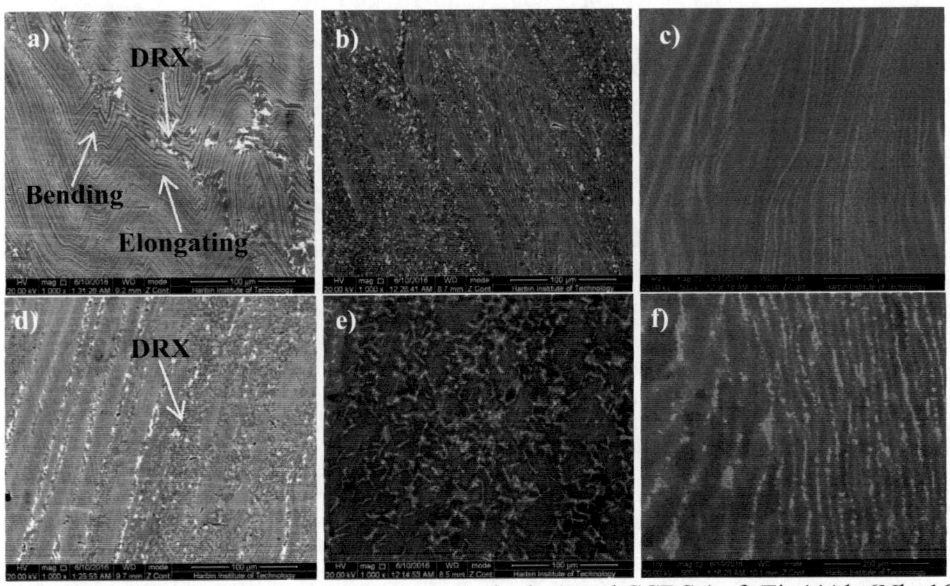

Figure 4. Deformation microstructure of as-cast (a, b, d, e) and CCDS (c, f) Ti–44Al–6Nb–1Cr alloys after hydrogenation with 0.06 wt.% H at different temperatures: (a) 1373 K (0 H); (b) 1473 K (0 H); (c) 1473 K (0 H); (d)1373 K (0.06 H); (e)1473 K (0.06 H); (f)1473 K (0.06 H).

4. Summary

(1) The volume fraction of B2 phase is increased after CCDS due to solute distribution. Hydrogen stabilizes the B2 phase during hydrogenation at 1473 K because hydrogen is beta stabilizing element.
(2) The flow stress of the CCDS alloy is higher than the as-cast alloys because of the directional solidified microstructure. The peak stresses of the as-cast alloy and CCDS alloy are decreased by 46 % and 43% after hydrogenation respectively.
(3) The microstructure of the CCDS alloy is more stable than the as-cast alloy resulting in higher flow stress. Hydrogen can promote the dynamic recrystallization of the CCDS alloy as well as the as-cast alloy.

Acknowledgement

This work was supported by National Natural Science Foundation of China (51274076) and The National Science Fund for Distinguished Young Scholars (NSFC51425402).

References

[1] Kothari K, Radhakrishnan R and Wereley N M 2012 *Prog. Aerosp. Sci.* **55** 1
[2] Clemens H and Mayer S 2013 *Adv. Eng. Mater.* **15** 191
[3] Jiao Z H, Yu H C, Su Y J, Kong F T and Chen Y Y 2016 *Mater. Mech. Eng.* **38** 50
[4] Chen G, Peng Y, Zheng G, Qi Z, Wang M, Yu H, Dong C and Liu C T 2016 *Nat. Mater.* **15** 876
[5] Dong S L, Chen R R, Guo J J, Ding H S, Su Y Q. and Fu H Z 2015 *Mater. Des.* **84** 118
[6] Nie G, Ding H S, Chen R R, Guo J J and Fu H Z 2012 *Mater. Des.* **39** 350
[7] Froes F H, Senkov O N and Qazi J I 2004 *Int. Mater. Rev.* **49** 227
[8] Zhang Y, Zhang S Q and Tao C 1997 *Int. J. Hydrogen Energ.* **22** 125
[9] Zhao J W, Ding H, Hou H L and Li Z Q 2010 *J. Alloy. Compd.* **491** 673
[10] Liu X W, Su Y Q, Luo L S, Liu J, Guo J J and Fu H Z 2010 *Int. J. Hydrogen Energ.* **35** 13322
[11] Wen D S, Zong Y Y, Xu W C, Shan D B and Guo B 2014 *Int. J. Hydrogen Energ.* **39** 17404
[12] Niu H Z, Chen Y Y, Xiao S L, Xu L J 2012 *Intermetallics* **31** 225
[13] Liu G H, Li X Z, Su Y Q, Chen R R, Guo J J and Fu H Z 2012 *T. Nonferr. Metal. Soc.* **22** 1342
[14] Zghal S, Thomas M, Naka S, Finel A and Couret A 2005 *Acta. Mater.* **53** 2653
[15]Schwaighofer E, Clemens H, Lindemann J, Stark A and Mayer S 2014 *Mater. Sci. Eng. A* **614** 297
[16] Jiang H, Zeng S, Zhao A, Ding X and Dong P H 2016 *Mater. Sci. Eng. A* **661** 160

IOP Conf. Series: Materials Science and Engineering **201** (2017) 012047 doi:10.1088/1757-899X/201/1/012047

Microstructural evolution of two binary β-titanium alloys during cold deformation

Saad Mohamed Ebied[a], Mohamed Abdel-Hady Gepreel and Atef Hamada

Materials Science and Engineering Department, Egypt-Japan University of Science and Technology, Borg El Arab, Alexandria 21934, Egypt.

[a]saad.ebied@ejust.edu.eg

Abstract. The deformation mechanism that reflects on the cold workability, of two beta-Ti alloys with chemical compositions of Ti-14Cr and Ti-17Mo in wt. % are studied in this work. The present alloys were solution treated at 900 °C for 1.8 ks. Followed by cold rolling with different reduction ratio from 5% to 30%. The microstructural evolution during cold deformation was followed using optical microscopy, X-ray diffraction and micro hardness. It was observed that the hardness and work hardening rate are affected significantly by the amount of cold deformation for the present Ti-alloys. It is apparent from the intensive microstructural observations that the deformation twinning (TWIP effect) is the dominant deformation mechanism in Ti-17Mo alloy at room temperature. However, the formation of slip bands and extended shear bands (SBs) were observed to be the deformation-induced microstructural features in Ti-14Cr alloy. The XRD analyses also revealed that the stress/strain induced martensitic transformation (TRIP effect) has not activated in the deformed Ti-alloys despite high deformation of 30%.

1. Introduction

Titanium alloys are utilized in many industrial applications including aerospace, biomedical, structural and automotive industries [1,2]. Most β-titanium alloys have unique properties for example high specific strength, acceptable cold workability, moderate ductility and high fatigue performance. It is well known that titanium alloys with β-*bcc* structure have better mechanical properties than those titanium alloys with α-*HCP* or α+β structure, since; the work-hardening capacity of β-Ti alloys is significantly higher [3]. Furthermore, the β-Ti alloys are heat treatable alloys, consequently, mechanical and physical properties could be enhanced with deformation [4].

In the recent years, beta titanium alloys were undergone development to enhance the mechanical properties such as tensile strength and also the formability property [5,6]. The deformation mode in beta titanium alloys could be significantly affected by the alloying elements, as reported in the open literature, e.g. Gouda et al. [7] and Abbasi et al. [8]. The hot workability of Ti alloys has been studied extensively to determine the activation energy of hot deformation and other deformation parameters to use them in the hot forming processes, such as rolling. However, the hot deformation processes have disadvantages such as high energy consumption, oxidation problems and the formation of α-case. Furthermore, a non-uniform temperature distribution can be occurred as a result of the contact between the alloy and the roller. Consequently, non-uniform plastic deformation and anisotropy in mechanical properties [9,10].

Hanada and Izumi [11,12] studied the deformation mode of a series of β-Ti alloys with different alloying elements by uniaxial tensile tests at ambient temperature with low strain rate of 5.2×10^{-4} s^{-1}. They found that the deformation mode depends on the β-phase stability. Twinning appears in the low stability β-phase alloys while with increasing the β-phase stability the deformation mode changes to slipping. they also investigated the effect of Cr content on the deformation mechanism and found that there are three types of deformation modes depending on the Cr content. Zhou and Luo [13] studied the microstructure and the mechanical properties in cold rolling of Ti-(10-20) wt.% Mo and found that the Mo-content control the microstructure and the mechanical properties of the alloys. Zhao and Niinomi [14] investigated the deformation mechanism and relate it to the Mo-content on the mechanical properties. In another work, Zhao et al. [15] studied the optimization of Cr content in series of binary titanium chromium beta titanium alloys and found that in ST and 10%CR Ti-14Cr alloy comprise β-phase. Nakai et al. [16] studied the effect of cold rolling on the young's modulus of Ti-xCr alloy by using free resonance technique and found that binary beta titanium alloys with low Cr content up to 12%wt exhibit deformation induced omega phase transformation in cold deformation. Wang et al. [17] investigated the microstructures and tensile properties of a series of Ti-Mo alloys after solution treatment and found that the martensite transformation becomes difficult with increasing the stability of the β-phase.

To the best of our knowledge, there is no a published work on β -Ti alloys concern the effect of the different reduction ratio on the deformation mode and the microstructural evolution. In the present work, we will study the deformation mechanism dependence of the reduction ratio through following the deformed microstructures in two β -Ti alloys, Ti-14Cr and Ti-17Mo with almost the same β-phase stability. Moreover, the work-hardening rate of the studied alloys was evaluated from the micro hardness tests and correlated with the deformation mode during cold working.

2. Experimental Procedure

Two Ti-alloys with nominal chemical compositions of Ti-17 wt. % Mo and Ti-14 wt. %Cr are used in this study. These compositions were selected to have the same level of stability of the β-phase in both alloys using the Mo_{eq} empirical approach. The Mo equivalent, Mo_{eq}, is a measure of the power of different alloying elements to stabilize β-phase in Ti-alloys and calculated as the sum of the weighted averages of the alloying elements (wt. %) present in the alloy as follow: $Mo_{eq} = Mo+0.67V+0.44W+0.28Nb+0.22Ta+1.25Cr-1Al$ [18]. The alloys were melted using an electric arc furnace (ARCAST 200, USA) with single electrode under high-purity argon atmosphere. This furnace is equipped with an electromagnetic stirrer for insuring the composition homogeneity and prevents alloying segregation. Subsequently, the cast ingots were homogenized at 1373 k for 28.8 ks in a tube furnace under high-purity argon atmosphere then quenched in ice water [16]. The oxide layer was mechanically removed after every step of the process. The alloys were hot rolled at 1123 k with total reduction ratio of 40% to remove the as cast structure. The alloys were then solution treated at 1173k for 1.8 ks in argon atmosphere. After ST condition, the alloys were cut into small samples for the cold deformation process. Four reduction ratios of 5, 10, 20 and 30 % were applied to both alloys by using Durston rolling mill (FSM 130, U.K).

X-ray diffraction (Model-6100, Shimadzu, Japan) was employed to detect the present phases with cold deformation. XRD was operated at 30 kV and 30 mA at a speed of 0.5 degrees/min. The microstructural evolution was followed using the optical microscopy (Model-Olympus GX71). Shimadzu micro-hardness tester was used to measure the micro-hardness with testing load 5N and the time of indentation is 15 seconds. Each specimen was measured 5 times and the average value was plotted.

3. Results and Discussion

3.1 Ti-14Cr alloy

The XRD patterns of the cold deformed structures along with the solution treated structure of the Ti-14Cr alloy are shown in figure 1. It can be observed that all peaks are indexed for the β-*bcc* phase. It means that the 14 wt.% Cr seems enough to stabilize the β-*bcc* structure against stress-induced martensitic formation during cold deformation. The β-phase seems predominant phase in the ST 14 wt.%

Cr alloy, however, small fractions of other phases such as martensites and ω-phase may be existed but its quantity may be very small to be detected by XRD technique and other techniques like TEM is needed for further phases characterization. According to figure 1, there is no observed stress induced phase transformation during cold rolling up to 30% . Which confirms that 14% Cr-content is sufficient to stabilize β–phase as a predominant phase in the solution treated condition and even in cold deformed condition as planned in the design stage, in good agreement with the previous results reporting that the least single β-phase composition is at around 10 % Cr-content[15]. These results were confirmed by the microstructure investigation. Typical microscopic images for solution treated and each case of cold rolling process are shown in figure 2. All conditions show single β-phase with no sign for any other phases which confirmed by XRD results. Solution treated samples show large equiaxed grains with average grain size of 350 μm (as measured using the linear-intercept method, ASTM E112) without any precipitates, as shown in figure 2(a). After 5% reduction ratio, there is no change in the microstructure because no mechanical twin or slip band appear in figure 2(b). With increasing the reduction ratio up to 10% small amount of slip bands (SBs) start to appear in grain labeled (A) in figure 2(c). It is reasonable to assume that the present Ti-14Cr alloy has high stacking fault energy, thereby cross slip of dislocations is induced and leading to the rearrangement of dislocations into slip bands. As the deformation proceeds to 20% reduction, extended shear bands were formed, as displayed in figure 2(d). Similarly, with additional higher reductions up to 30%, the fraction and the thickness of the extended shear bands increased, figure 2(e). The micro-hardness measurements of the Ti-Cr alloy was carried out after cold deformation to different reduction ratio. Figure 3(a) show the micro hardness data as a function of the reduction ratio. It can be seen that the solution treated structure, as expected, shows a lowest hardness value of 290 Hv. The hardness values increase with increasing reduction ratio (RR). This can be explained to be a result of the strain hardening effect which occurs by the plastic deformation which confirmed by the existence of the shear band on the microstructure [19].

.

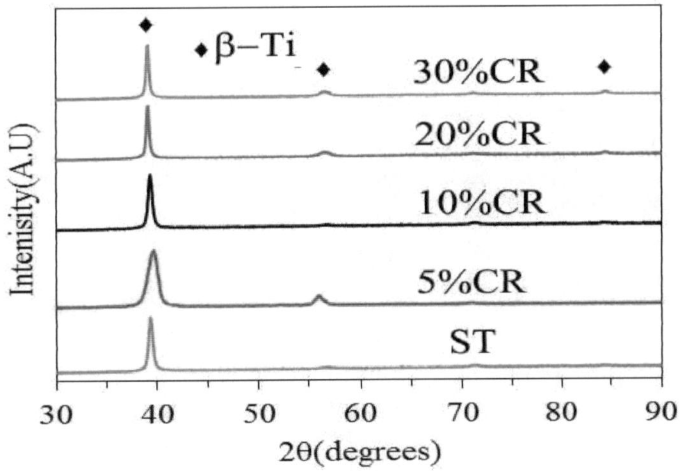

Figure 1. XRD patterns of the solution treated, ST, and 5, 10, 20 and 30% cold rolled, CR, Ti-14Cr alloy.

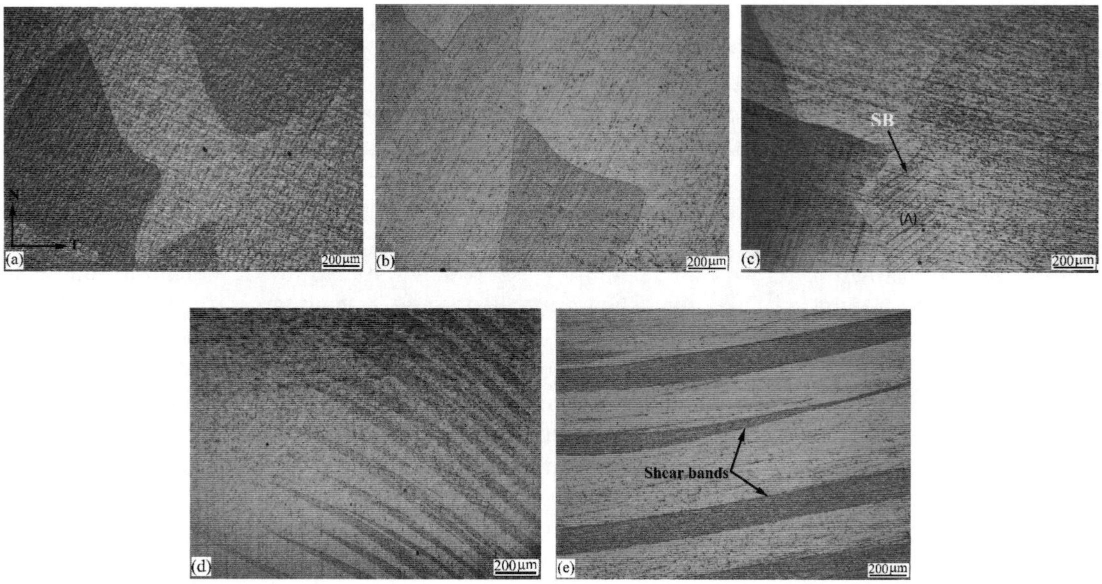

Figure 2. Optical microstructure of Ti–14Cr alloy in solution treatment (ST) and various cold rolled conditions (CR): (a) ST, (b) 5% CR, (c) 10% CR, (d) 20%CR, (e) 30% CR.

The hardness change with the reduction ratio has been drawn in figure 3(b) to reveal the hardness change sensitivity with reduction ratio (∂ΔH/∂RR). Different behavior through the reduction range can be seen.

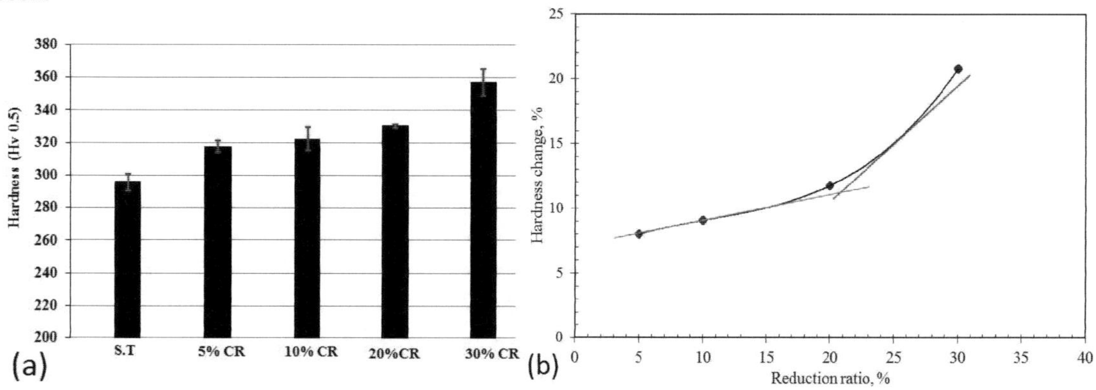

Figure 3. (a) Micro-hardness of theTi-14Cr alloy under different conditions, and (b) The change in hardness by cold rolling reduction ratio (%) with respect to solution treated condition of Ti-14Cr alloy.

The Hardness changing rate (∂ΔH/∂RR) dependence is better depicted from drawing tangent straight lines (green-red lines) for the hardness change plot. Two straight lines may fit two different reduction ranges, one is in the reduction range (5-20 CR) and other above 20 CR reduction ratio. The straight lines would have different slopes, and show more powerfully that the deformation mechanism has been changed starting from 20 CR reduction ratio by shear band formation as observed in the optical microstructure, *see* figure 2 (d).

3.2 Ti-17Mo alloy

It is well known that Mo is a strong beta stabilizer and 15%Mo is sufficient to retain β-phase at room temperature [18]. Microstructure and XRD results prove that the microstructures are consisted mainly from only β–phase in both the solution treated and cold rolled conditions. As shown in figure 4. There is

IOP Conf. Series: Materials Science and Engineering **201** (2017) 012047 doi:10.1088/1757-899X/201/1/012047

no sign for stress induced phase transformation as confirmed by the XRD results. In 17% wt Mo-alloy, other phases such as ω-phase may be existed but its amount is very small to be detected by XRD technique, as discussed above. These results were confirmed by the microstructure investigation. Typical microscopic images are shown in figure 5. The alloy in all conditions show single β-phase with no sign for any other phases which confirmed by XRD results. Solution treated sample shows large equiaxed grains with average grain size of 450 μm, without any precipitates as shown in figure 5 (a). After 5%CR the mechanical twin starts to appear with almost one direction in every grain, as shown in figure 5(b). The thickness of the twin is reduced and its number increases with introducing more plastic deformation as shown in figure 5(c). However, the twin is in two directions in two grains. For 20% CR, the twin become thinner and denser and the twin become in more than two directions on the same grain. The formation of sub twin starts to form at 20%CR and increases in the 30%CR condition as shown in figure 5(d,e) .

Figure 6 shows the hardness variation of the cold deformed Ti-Mo with the reduction ratio. It is clear that the hardness is noticeably affected by changing the amount of deformation. The hardness is significantly increased at reduction ratio 20 pct. It can be observed that the hardness of the cold deformed structure with 30 %CR is close to 400 Hv. Similarly, the hardness change with the reduction ratio ($\partial \Delta H/\partial RR$) of the Ti-17Mo alloy has been plotted in figure 6(b). It is clear that the hardness change slope in the CR range (10 to 20 %) is significant high (red line). This is attributed to the formation of twins network, see figure 5(d). This twin network strengthens the structure by enhancing the dynamic Hall-Petch dependence on the grain size. Therefore, the work hardening of the deformed structures beyond the 20 %CR increases with proceeding of the deformation straining. However, in the third CR range (20 to 30 %), the hardness change slope is very small comparing to the slope in the second CR regime, as shown in figure 6(b)

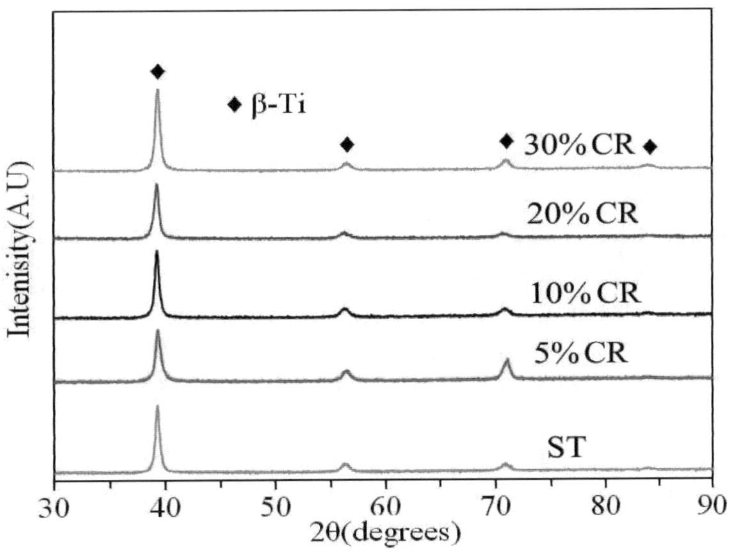

Figure 4. XRD patterns of the ST and CR conditions of Ti-17Mo alloy.

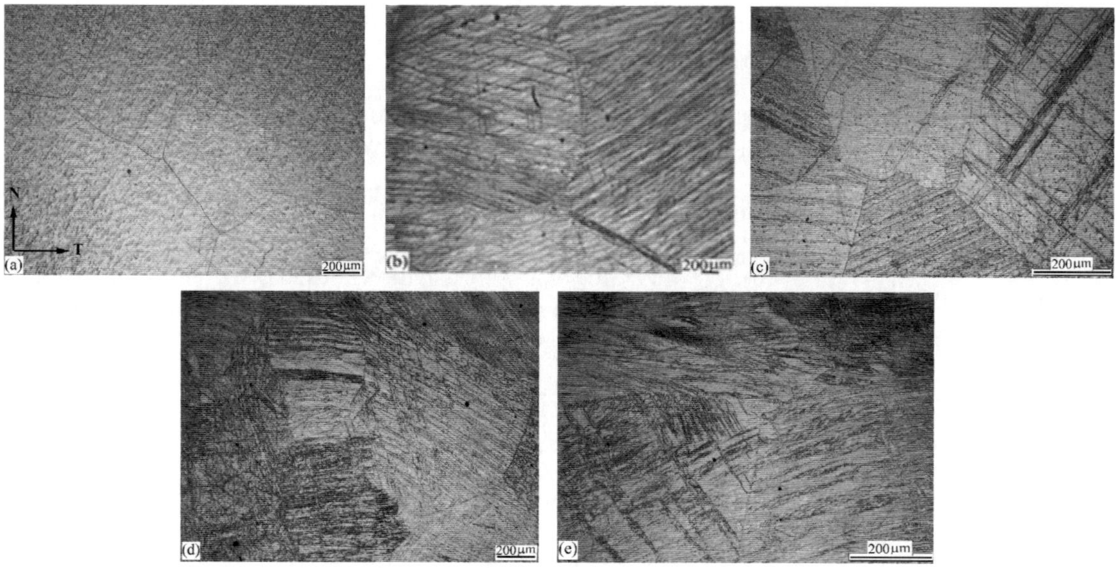

Figure 5. Optical microstructure of Ti–14Cr alloy in solution treatment (ST) and various cold rolled conditions (CR): (a) ST, (b) 5% CR, (c) 10% CR, (d) 20%CR, (e) 30% CR.

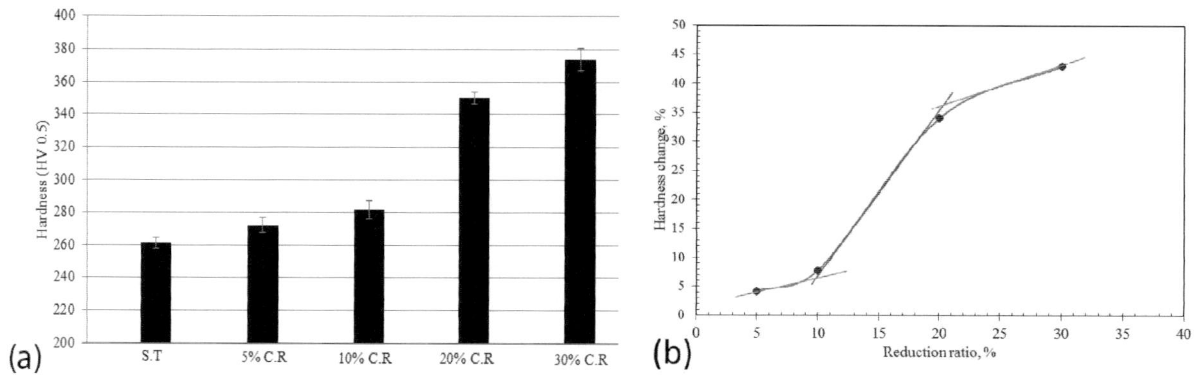

Figure 6. (a)Micro-hardness of solution treated and cold rolled conditions of theTi-17Mo alloy and (b) Micro-hardness change with cold rolling reduction ratio (%) with respect to solution treated.

4. Summary

The effect of the cold deformation with different reduction ratio on the microstructural evolution of Ti-14Cr and Ti-17Mo alloys with β-phase structure has been studied. The results allow drawing the following conclusions:

(1) The deformation mode in the Ti-14Cr alloy changes from deformation band to shear band with cold rolling at and above 20%CR.

(2) The deformation mode in the Ti-17Mo alloy is mechanical twinning (TWIP).

(3) The effect of cold deformation reduction ratio on the micro-hardness change rate in the studied β-Ti alloys can be attributed to the change in the deformation mechanism.

(4) However, the two alloys under this study have almost same level of β-phase stability; they possess different modes of deformation and therefore different level of cold deformability. Further investigations are needed to understand why this happens.

References

[1] M.J. Donachie, *Titanium*: a technical guide, ASM international, 2000.
[2] G. Lütjering, J.C. Williams, *Titanium*, Springer, 2003.
[3] O. Ivasishin, P. Markovsky, Y.V. Matviychuk, S. Semiatin, C. Ward, S. Fox, A comparative study of the mechanical properties of high-strength β-titanium alloys, *Journal of Alloys and Compounds* **457** (2008) 296-309.
[4] H. Matsumoto, S. Watanabe, S. Hanada, Microstructures and mechanical properties of metastable β TiNbSn alloys cold rolled and heat treated, *Journal of Alloys and Compounds* **439** (2007) 146-155.
[5] H. Oikawa, K. Nishimura, M. Cui, High-temperature deformation of polycrystalline beta titanium, *Scripta metallurgica* **19** (1985) 825-828.
[6] N. Nagasako, K. Nishino, T. Saito, C. Iwamoto, Y. Ikuhara, and T. Sakuma, Origin for "Super" Properties in GUM METAL, Ti-2003 Science and Technology; Proceedings of 10th World Conference on Titanium, G. Lutjering and J. Albrecht, Ed., Jul 13-18, 2003 (Hamburg), Wiley, (2004)1527-1534
[7] Mohamed Kamal Gouda, Koichi Nakamura, Mohamed Abdel-Hady Gepreel, Effect of Mn-Content on the deformation behavior of binary Ti-Mn alloys, *Key Engineering Materials*,**705**(2016)214-218.
[8] Seyed Mehdi Abbasi, Maryam Morakkbati, Amir Hossein Sheikhali,, Hot Deformation Behavior of Beta Titanium Ti-13V-11Cr-3Al Alloy, *Metall and Mat Trans* , **45A**(2014) 5201-5211.
[9] G. Welsch, R. Boyer, E. Collings, Materials properties handbook: *titanium alloys*, ASM international, 1993.
[10] Joseph D. Beal, Rodney Boyer, and Daniel Sanders, The Boeing Company Forming of Titanium and Titanium Alloys, ASM Handbook, Volume **14B**(2006)656-669.
[11] S.Hanada and O.Izumi, studied Correlation of Tensile Properties, Deformation Modes, and Phase Stability in Commercial β-Phase Titanium Alloys,*Metallurgical Transaction*, **18A** (1987)265-271.
[12] S.Hanada and O.Izumi,v Deformation behaviour of retained β phase in β -eutectoid Ti-Cr alloys, *journal of material science* **21** (1986) 4131-4139.
[13] Ying-Long Zhou and Dong-Mei Luo, Microstructures and mechanical properties of Ti–Mo alloys cold-rolled and heat treated, materials *characterization* **62** (2011) 931–937.
[14] Xingfeng Zhao , Mitsuo Niinomi , Masaaki Nakai , Junko Hieda, Beta type Ti–Mo alloys with changeable Young's modulus for spinal fixation applications, *Acta Biomaterialia* **8** (2012) 1990–1997.
[15] Xingfeng Zhao , Mitsuo Niinomi , Masaaki Nakai , Junko Hieda , Takuya Ishimoto ,Takayoshi Nakano. Optimization of Cr content of metastable β-type Ti–Cr alloys with changeable Young's modulus for spinal fixation applications, *Acta Biomaterialia* **8** (2012) 2392–2400.
[16] Masaaki Nakai , Mitsuo Niinomi , Xingfeng Zhao , Xiaoli Zhao , Self-adjustment of Young's modulus in biomedical titanium alloys during orthopaedic operation, *Materials Letters* **65** (2011) 688–690 .
[17] C. H. Wang, C. D. Yang, M. Liu, X. Li, P. F. Hu, A. M. Russell, and G. H. Cao, Martensitic microstructures and mechanical properties of as-quenched metastable b-type Ti–Mo alloys, *J Mater Sci*, **51** (2016) 6886–6896.
[18] Nilson T.C. Oliveira , Giorgia Aleixo , Rubens Caramb, Antonio . Guastaldi, Development of Ti–Mo alloys for biomedical applications:Microstructure and electrochemical characterization, *Materials Science and Engineering*, **A 452–453** (2007) 727–731.
[19] R. PrakashR Kolli , William J. Joost and Seeramamurthy Ankem, Phase Stability and Stress-Induced Transformations in BetaTitanium Alloys, *JOM*, **67**(2015), 1273–1280.
[20] Dengke Yang, Yang An, Pavel Cizek, Peter Hodgson, Development of adiabatic shear band in cold-rolled titanium, *Materials Science and Engineering*, **A 528** (2011) 3990–3997.

IOP Conf. Series: Materials Science and Engineering **201** (2017) 012048 doi:10.1088/1757-899X/201/1/012048

Fabrication of zinc oxide-embedded kapok (*Ceiba pentandra*) paper

E S Mones[1], M D L Balela[1], C C M Futalan[2], R D Manalo[3], and M U Herrera[4]

[1]Department of Mining, Metallurgical, and Materials Engineering, College of Engineering, University of the Philippines Diliman, Quezon City, Philippines 1101
[2]AECOM Philippines Consultants Corporation, Fort Legend Towers, Bonifacio Global City, Taguig, Metro Manila, Philippines 1634
[3]Department of Forest Products and Paper Science, College of Forestry, University of the Philippines Los Baños, Laguna, Philippines 4031
[4]Institute of Mathematical Sciences and Physics, College of Arts and Sciences, University of the Philippines Los Baños, Laguna, Philippines 4031

Email: muherrera@up.edu.ph

Abstract. Zinc oxide-embedded kapok (*Ceiba pentandra*) papers with antimicrobial property were fabricated. Fabrication of papers from kapok fibers was done using chloroform and sodium chlorite treatments. Meanwhile, embedding of zinc oxide particles on the fabricated kapok papers was done using an *in situ* method. The said method involved soaking the paper in zinc acetate and sodium hydroxide solutions. The method also involved heat treatment of the sample to fasten the formation of zinc oxide particles. X-ray diffraction (XRD) measurement shows the presence of synthesized ZnO particles on the cellulose fibers while scanning electron microscopy (SEM) revealed the flake-like morphology of the embedded ZnO particles. Agar diffusion test shows that the samples have higher antimicrobial effect against *Staphylococcus aureus* than *Escherichia coli*.

1. Introduction

Metal oxides [1-8] have attracted attention in the inhibition of bacterial growth. Studies were done to modify and improve the properties of different materials by inclusion of metal oxides [5,9,10]. Embedding of metal oxides on templates or substrates (e.g., metal oxide-glass, metal oxide-paper) can lead to the enhancement of the properties.

Zinc oxide (ZnO) [6-9,11-15] is an n-type oxide semiconductor that has an energy gap of 3.37 eV. Its physical and chemical properties make it useful in diodes, UV sensors, thin films, antimicrobial applications, and photocatalysis.

Paper [5,9], a regularly-used material, is flexible and portable. Different metal oxides were embedded on paper for various applications. Papers are commonly made of fibers from woods. In the view of conserving woods for environmental protection, finding alternative sources of fibers can be explored. Kapok fibers [16] were studied for its potential in papermaking. Kapok fibers are from the fruits of kapok trees (*Ceiba pentandra*). They have cotton-like appearance and have light-yellowish color. They are fluffy, less dense, and have microtubular structures. Due to its hydrophobicity, kapok fibers are used in oil removal applications. This study aims to modify the properties of handmade kapok paper by embedding ZnO particles. Papers have different applications such as template for printing and decorative items, food and material packaging. Embedding of ZnO in paper opens many possible applications wherein the desirable properties of both ZnO and paper are taken advantage.

2. Methodology

2.1. Fabrication of Kapok (Ceiba pentandra) papers

2.1.1. Chloroform treatment. Twelve grams (12g) of Kapok fibers from San Jose del Monte, Bulacan, Philippines and 1.2 L chloroform ($CHCl_3$, J.T. Baker Chemical Co.) were mixed and stirred constantly for 4 h. The fibers were filtered using Buchner funnel while being supported by a rotary vacuum pump (ROTARY VACUUM PUMP SA18 032616 SHIMADZU Corporation) to quicken the filtration. The fibers were washed with methanol (CH_3OH, Fisher ChemAlert) and then air dried for at least 12 h.

2.1.2 Sodium chlorite treatment. Thirty millilitres (30mL) of stock solution consisting of 60 mL acetic acid (CH_3COOH, MACRON Fine Chemicals), sodium hydroxide solution (1.3 g NaOH per 1 water) and 2 g of dry weight chloroform-treated kapok were mixed in 250 mL Erlenmeyer flask. The Erlenmeyer flasks containing the mixture were placed in hot water bath at 75°C. Three milliliters (3 mL) of 20% sodium chlorite ($NaClO_2$, SIGMA-ALDRICH) solution was added to the mixture and the 250 mL Erlenmeyer flask was swirled afterwards. Another 3 mL of sodium chlorite solution was added after 0.5, 1.0, 1.75, and 2.5 h. After 2.5 h, the mixture was maintained in hot bath for 45 min. The mixture was removed from hot bath and then filtered with the aid of rotary vacuum pump. The sodium chlorite-treated kapok was cleaned using 60 mL chilled distilled water with simultaneous filtration of filtrate. It was further cleaned using four 75 mL 1% acetic acid solution and two 30 mL methanol and then air dried for at least 36 h.

2.1.3 Kapok paper production. One gram (1g) of sodium chlorite-treated kapok was broken down into small pieces and soaked in 299 mL distilled water for 24 h. The slurry was osterized using a commercial blender (Waring Commercial) for 5 min. The slurry underwent filtration in with the aid of rotary vacuum pump. After filtration, the sodium chlorite-treated kapok was slightly air dried for 5 to 10 min. The sample was flattened into sheets using rolling pin and then air dried for at least 24 h.

2.2 Embedding of zinc oxide

Zinc oxide particles were embedded in kapok paper via *in situ* method. Kapok papers with 1 mm diameter were immersed in 0.75M zinc acetate ($(CH_3{\cdot}COO)_2Zn{\cdot}H_2O$, TECHNO PHARMCHEM HARYANA, 99.5%) solution for 1 hr. The papers were then immersed in 0.5M sodium hydroxide (NaOH, MACRON Fine Chemicals, AR) solution. Afterwards, the papers were heated for 1 min at 50°C on a hot plate (CORNING PC-420D).

2.3 Characterizations

X-ray diffraction measurement was done to determine the composition of the deposited material on the paper. The said measurement was done using XRD SHIMADZU MAXima XRD-7000. Meanwhile, the surface morphology of the sample was observed using scanning electron microscope (SEM, JEOL Ltd., JSM-5310).

Agar disc diffusion was employed to determine the antimicrobial indices of the samples against *Escherichia coli* and *Staphylococcus aureus.* The samples have diameter of 10 mm. Microbial suspensions were prepared from 24-hour old culture of the bacteria *Escherichia coli* (UPCC 1195) and *Staphylococcus aureus* (UPCC 1143). Pre-poured Nutrient Agar (NA), about 3 mm thick, was inoculated with the respective microbial suspension by swabbing the agar surface. The NA plates were incubated at 35°C and observed after 24 h.

3. Results and Discussion

Zinc oxide (ZnO) was successfully embedded on the kapok paper. Figure 1 shows the XRD pattern of ZnO-embedded kapok paper. The diffraction peaks at 15.8° and 23.1° correspond to (101) and (002) planes of the cellulose fibers (Kapok paper). On the other hand, the diffraction peaks at 31.1°, 32.9°, 36°, and 47.8° correspond to (100), (002), (101), and (102) planes, respectively, of hexagonal wurtzite ZnO.

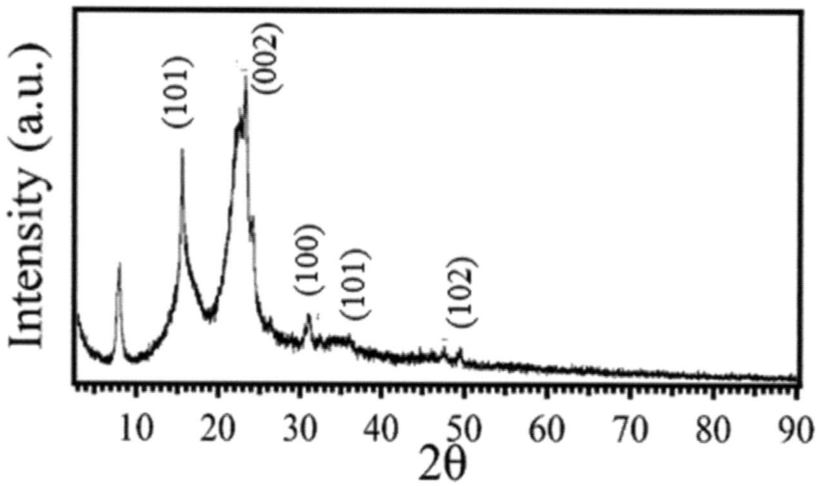

Figure 1. XRD pattern of ZnO-embedded kapok paper synthesized via in situ method

Figures 2a and 2b show the SEM images of kapok paper while Figures 2c and 2d show the SEM images of ZnO on kapok paper synthesized using *in situ* method. As can be observed in Figure 2c and 2d, ZnO with flake-like morphology were distributed on the surface of the paper. In the *in-situ* method, zinc acetate salts were dissolved in aqueous solutions into ions. The smallness of ions enabled them to penetrate into the pores of the paper. Soaking the treated paper in sodium hydroxide solution enabled the zinc ions to transform to zinc hydroxide. With heating, zinc hydroxide transformed into zinc oxide particles.

Figure 3a and 3b show the ZnO-embedded kapok paper subjected to antimicrobial test against *Escherichia coli* (gram-negative) and *Staphylococcus aureus* (gram-positive), respectively. The clearing zones were measured in millimeters and the average diameters of the clearing zones were calculated. The antimicrobial indexes (AI) were computed using the following formula:

$$AI = \frac{Diameter\ of\ clearing\ zone - Diameter\ of\ sample}{\text{Diameter of sample}} \quad (1)$$

Table 1 shows the clearing zones and the AI of ZnO-embedded kapok papers for each test organisms. For zinc-based materials, one possible antibacterial action is the generation of reactive oxygen species (ROS). It is possible that the antimicrobial action is more evident in *S. aureus* because this type of organisms has thin cell wall.

Figure 2. SEM images of kapok papers at (a) x1000 and (b) x5000 magnifications and flake-like ZnO at (c) x1000 and (d) x5000 magnifications

Table 1. Average clearing zones and antimicrobial indices of ZnO-embedded kapok papers tested against *E. coli* and *S. aureus*

Test organism	Average clearing zones (mm)	Antimicrobial index (AI)
Escherichia coli (gram negative)	10.67	0.1
Staphylococcus aureus (gram positive)	15.33	0.5

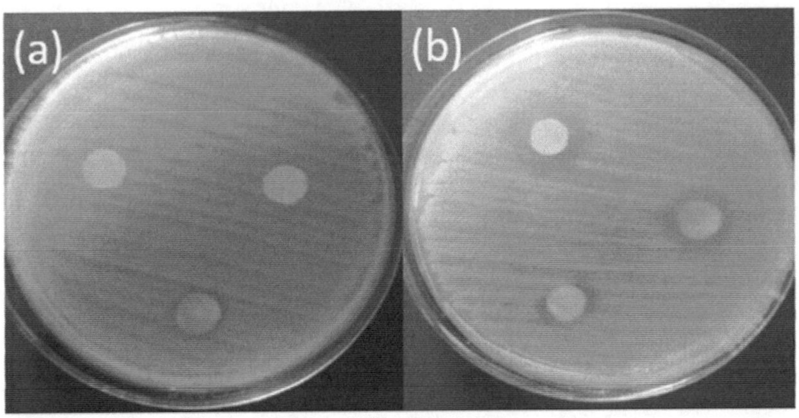

Figure 3. ZnO-embedded kapok papers in agar disc diffusion test against (a) *E. coli* and (b) *S. aureus*

4. Summary and Conclusion

ZnO-embedded kapok papers were successfully fabricated. Kapok papers were fabricated from Kapok fibers using chloroform and sodium chlorite treatments. Meanwhile, zinc oxide particles were embedded in the fabricated kapok paper using *in situ* technique. XRD measurements show presence of

ZnO particles on kapok paper. Moreover, SEM shows the flake-like morphology of the embedded ZnO. In addition, the antimicrobial activity of the ZnO-embedded paper showed higher antimicrobial index against gram-positive *S. aureus* than gram-negative *E. coli.*

5. Acknowledgements
This work was supported by a grant from UP System Emerging Inter-Disciplinary Research Program (OVPAA-EIDR C06-032).

References
[1] Dizaj S M, Lotfipour F, Barzegar-Jalali M, Zarrintan M H and Adibkia K 2014 *Materials Science and Engineering C* **44** 278-284
[2] Azam A, Ahmed A S, Oves M, Khan M S, Habib S S and Memic A 2012 *International Journal of Nanomedicine* **7** 6003-6009
[3] Katwal R, Kaur H, Sharma G, Naushad M and Pathania D 2015 *Journal of Industrial and Engineering Chemistry* **31** 173-184
[4] Ren G, Hu D, Cheng E W C, Vargas-Reus M A, Reip P and Allaker R P 2009 *International Journal of Antimicrobial Agents* **33** 587-590
[5] Booshehri A Y, Wang R and Xu R 2015 *Chemical Engineering Journal* **262** 999-1008
[6] Thati V, Roy A S, Ambika Prasad M V N, Shivannar C T and Gaddad S M 2010 *J Biosci Tech* **1** 64-69
[7] Wang C, Liu L-L, Zhang A-T, Xie P, Lu J-J and Zou X-T 2012 *African Journal of Biotechnology* **11(44)** 10248-10254
[8] Ann L C, Mahmud S, Bakhori S K M, Sirelkhatim A, Mohamad D, Hasan H, Seeni A and Rahman R A 2014 *Ceramics International* **40** 2993-3001
[9] Manekkhathodi A, Lu M-Y, Wang C W and Chen L-J 2010 *Adv. Mater* **22** 4059-4063
[10] Longo A, Wang X L, Ruotolo A, Peluso A, Carotenuto G and Lortz R 2012 *J Nanopart Res* **14** 1314
[11] Mones E S, Gillado A V and Herrera M U 2016 *Key Engineering Materials* **705** 186-189
[12] Bai S, Wu W, Qin Y, Cui N, Bayerl D J and Wang X 2011 *Adv. Funct. Mater.* **XX** 1-6
[13] Tahar N B H , Tahar R B H, Salah A B and Savall A 2008 *J. Am. Ceram. Soc.* **91** 846-851
[14] Kaur J, Bansal S and Singhal S 2013 *Physica B* **416** 33-38
[15] Shen W, Li Z, Wang H, Liu Y, Guo Q and Zhang Y 2008 *Journal of Hazardous Materials* **152** 172-175
[16] Chaiarrekij S, Apirakchaiskul A, Suvarnakich K and Kiatkamjornwong S 2011 *BioResources* **7(1)** 475-488

IOP Conf. Series: Materials Science and Engineering **201** (2017) 012049 doi:10.1088/1757-899X/201/1/012049

Antibacterial activity of copper-based particles synthesized using an electroless deposition technique

M. J. G. Biguerras[1], M. U. Herrera[2], C. C. Futalan[3] and M. D. L. Balela[1]

[1]Department of Mining, Metallurgical and Materials Engineering, College of Engineering, University of the Philippines Diliman, 1101
[2]Institute of Mathematical Sciences and Physics, College of Arts and Sciences, University of the Philippines, Los Baños, 4031
[3]AECOM Philippines Consultants Corporation, Fort Legend Towers, Bonifacio Global City, Taguig, Metro Manila, Philippines

Email: mdbalela1@up.edu.ph

Abstract. Copper-based particles were synthesized using an electroless deposition technique. The said synthesis was done in an aqueous solution by reducing copper oxide powders using hydrazine. In this technique, gelatin was used as protective agent. X-ray Diffraction (XRD) measurement shows that the synthesized sample is composed of cuprous oxide (Cu_2O) and copper (Cu) particles. Scanning Electron Microscopy (SEM) shows the morphology of the synthesized copper-based particles. Antimicrobial test shows that the number of *Escherichia coli* organisms reduced to 62.06% after 2 minutes of contact with the sample. Likewise, SEM micrographs of the *Escherichia coli* organisms show that the said organism underwent morphological changes in the presence of the synthesized copper-based particles.

1. Introduction

Copper [1-6] has antibacterial activity against gram-positive and gram-negative bacteria. Copper particles can be incorporated with different systems such as in wastewater treatment to perform its antimicrobial functions. The antimicrobial activity of copper is a function of contact area with microorganisms. Close interaction of copper particles with cells can inhibit the cell's normal function, which could eventually lead to cell death.

The synthesis of metallic copper [7-11] was a challenging task because it can rapidly oxidize in air or in aqueous media. To inhibit or minimize oxidation, copper metals are protected with capping agents or stabilizers. In a case wherein the capping agents do not completely inhibit the formation of oxides, it is important to note that Cu oxide [12-13] also has antimicrobial properties.

The aim of this research is to synthesize copper-based particles using an electroless deposition technique. In this technique, copper oxide powders were reduced using hydrazine in an aqueous solution. This technique used gelatin as protective agent or stabilizer. The antimicrobial activities of the synthesized copper-based particles against *Escherichia coli* were tested.

2. Experimental

2.1. Materials

Copper (II) oxide powders (Nacalai tesque) were used as copper source and lab grade gelatin (Nacalai Tesque) was used as protective agent. Ninety eight weight percent (98 wt%) hydrazine monohydrate ($N_2H_4 \cdot H_2O$, Sigma Aldrich Inc.) and sodium hydroxide pellets (NaOH, Macron Fine Chem) were also used in the synthesis of the copper-based particles.

2.2. Synthesis of Cu-based particles by Electroless deposition

The synthesis of Cu-based particles was prepared according to Balela and Amores [14]. The protective and reducing agent were prepared separately and both were added with 10 wt% gelatin. The pH of reaction mixture was adjusted to 12 by dropwise addition of 1.5 M of NaOH. The total solution was allowed to react for 2 h at 353 K under continuous nitrogen gas purging to prevent oxidation. The Cu-based particles were collected and washed with deionized water.

2.3. Characterization

The morphology of the synthesized Cu-based particles was determined using Scanning Electron Microscopy (JEOL JSM 5310) while the identification of phases present was done using X-ray Diffraction (Shimadzu XRD-7000).

2.4. Antibacterial activity

Contact kill test was used to determine the amount of *Escherichia coli* organisms that remains after the copper-based particles were brought into contact with the test organism. Meanwhile, the effects of copper-based materials on the cell morphology of *Escherichia coli* organisms were observed using Scanning Electron Microscopy (SEM).

3. Results and Discussion

Figure 1 shows the X-ray diffraction pattern of the sample stored for 2 weeks in aqueous solution. Both peaks of cuprous oxide (Cu_2O) and copper (Cu) particles were present in the XRD pattern, suggesting that some of the copper particles oxidized to Cu_2O.

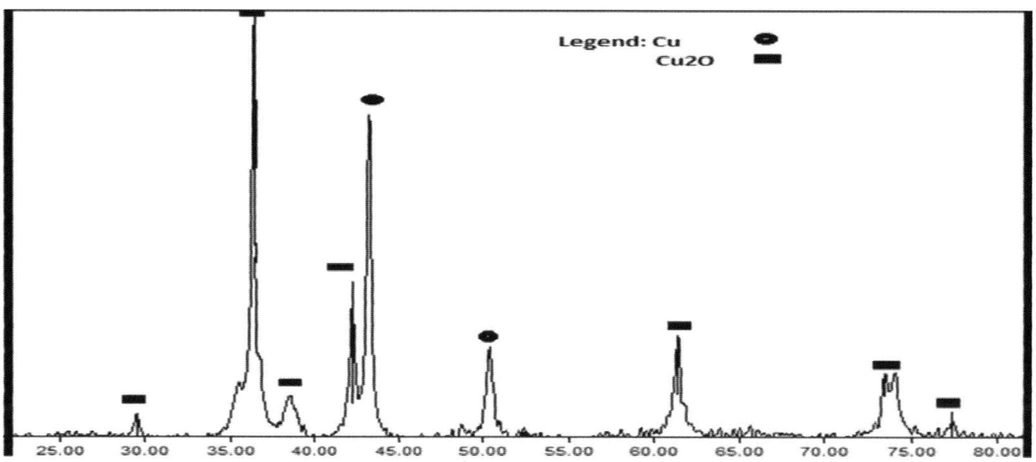

Figure 1. XRD pattern of Cu-based particles after two weeks of storage in aqueous solution

Figure 2 shows the SEM image of synthesized copper-based particles, which was stored in aqueous solution for two weeks. The figure shows that these particles have relatively uniform distribution. The particles appear aggregated but this is just a typical behaviour when they are taken out of solution. This could be also an indication of oxidation of copper to Cu_2O.

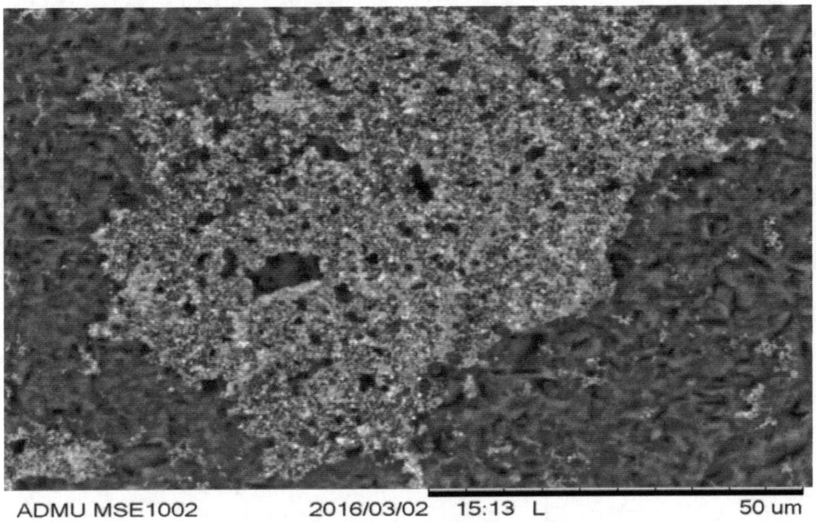

Figure 2. SEM image of Cu-based particles after two weeks of storage in aqueous solution

Table 1 show the result of Contact kill test of Cu-based particles against *Escherichia coli.* The initial count of bacterial samples was 2.3 x 10^6 (CFU/ml) and were reduced to 8.6 x 10^5CFU/ml (62.06%) after two minutes of contact time with the synthesized copper-based particles. The interaction of Cu to outer and inner bacterial membrane might be the possible effect of inhibition of the bacterial samples.

Table 1. Summarize result of Contact kill test against *Escherichia coli.*

Type of Organism	Initial Count	Final Count
Escherichia coli	2.3 x 10^6 CFU/ml	8.6x 10^5 CFU/ml

Figure 3a shows the SEM images of *Escherichia coli* incubated in the absence of copper-based particles. Meanwhile, Figure 3b and Figure 3c show the SEM images of *Escherichia coli* treated with copper-based particles in the presence and absence of light, respectively. Without the copper-based particles, the average size of the test organism is 329 nm. This average size reduces to 152 nm when copper-based particles were introduced in the presence of light. On the other hand, when copper-based particles were introduced in the absence of light, the average size of the test organism reduces to 90 nm.

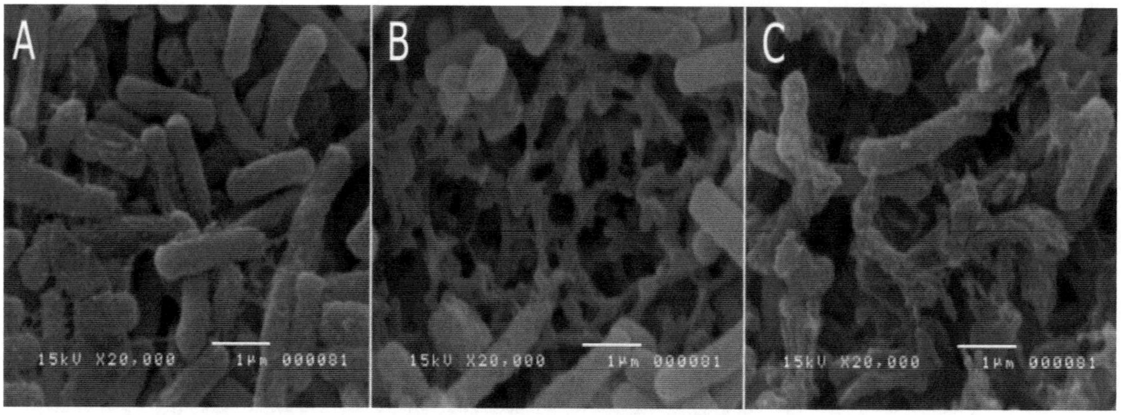

Figure 3. SEM images of (a) *Escherichia coli* organisms and SEM image of *Escherichia coli* organisms treated with the synthesized Cu-based particles in the (b) presence of light and (c) absence of light.

4. Summary and Conclusion

Copper-based particles were synthesized using an electroless deposition technique. The synthesis was done in aqueous solution using copper oxide as copper source, hydrazine as reducing agent and gelatin as protective agent. X-ray diffraction shows that the sample is composed of copper particles and Cu_2O. Scanning electron microscopy shows that the copper-based particles were aggregated. The synthesized copper-based particles show antibacterial activity against *Escherichia coli.* In such, after 2 minutes of contact with the sample, the number of *Escherichia coli* organisms reduced to 62.06%. Furthermore, SEM micrographs of *Escherichia coli* organisms also show that morphological changes happened on the said organisms in the presences of the synthesized copper-based particles.

Acknowledgment

This work was supported by a grant from UP System Emerging Inter-Disciplinary Research Program (OVPAA-EIDR C06-032).

References

[1] Meyer T, Ramlall J, Thu Phyo, and Gadura N 2015 *International Journal of Biological, Agricultural, Food and Biotechnological Engineering* **9** 274-27
[2] Chatterjee, A K, Chakraborty R, and Basu T 2014 *Nanotechnology* **25**
[3] Vincent M, Hartemann P and Engels-Deutsch 2016 *International Journal of Hygiene and Environmental Heath* **219** 585-591
[4] Sawant A R, Raut R R, Patil T D and Malwade R R 2014 International *Journal of Chemical and Physical Sciences* **3** 77-81
[5] Betancourt-Galindo R, Reyes-Rodriguez P Y, Puente-Urbina B A, Avila-Orta C A, Rodríguez-Fernández O S, Cadenas-Pliego G, Lira-Saldivar R H, and García-Cerda L A 2014 *Journal of Nanomaterials*
[6] Ramyadevi J, Jeyasubramanian K, Marikani A, Rajakumar G, and Rahuman A A *Material Letters* **71** 114–116
[7] Mott D, Galkowski J, Wang L, Luo J and Zhong C J 2007 *American Chemical Society* **23** 5740-

5745

[8] Musa A, Ahmad M, Hussein, M Z, Saiman M I and Sani H A 2016 *Nanoscale Research Letters* **11:438**

[9] Zhang D and Yang H 2013 *Physica B* **415** 44-48

[10] Park B K, Jeong S, Kim D, Moon J, Lim S and Kim J S 2007 *Journal of Colloid and Interface Science* **311** 417-424

[11] Datu E and Balela MD 2016 *Key Materials Engineering* **705** 163-167

[12] Meghana S, Kabra P, Chakraborty S and Padmavathy N 2015 *The Royal Society of Chemistry* **5** 12293-12299

[13] Hans M, Erbe A, Mathews S, Chen Y and Solioz M 2013 *American Chemical Society* **29** 16160-16166

[14] Balela MD and Amores K. 2016. *Advanced Materials Research* **1131** 255-259

IOP Conf. Series: Materials Science and Engineering **201** (2017) 012050 doi:10.1088/1757-899X/201/1/012050

Facile synthesis of cupric hydroxide and cupric oxide on copper foil for potential electrochemical applications

E M A Espejo[1], M D L Balela[2]

Sustainable Electronic Materials Group, Department of Mining, Metallurgical, and Materials Engineering, University of the Philippines Diliman, Quezon City 1101, Philippines

[1]erikamaeespejo@gmail.com, [2]mlbalela1@up.edu.ph

Abstract. This paper presents the synthesis of cupric hydroxide [$Cu(OH)_2$] and cupric oxide (CuO) *via* a simple *in situ* oxidation of copper (Cu) foil in alkaline solution for electrode materials in pseudocapacitors. $Cu(OH)_2$ and $Cu(OH)_2$/CuO film grown on Cu substrate offer binderless and additive-free active materials directly in contact with the current collector. This approach allows more efficient interaction of electrolyte and the active material in terms of charge and mass exchange for higher capacitance. The Cu-based nanomaterials have uniform morphology, and are needle-like and flower-like in structure for $Cu(OH)_2$ and $Cu(OH)_2$/CuO, respectively. The fabricated nanostructures exhibit a specific capacitance in the range of 400–2000 F/g.

1. Introduction

Research on efficient, clean, sustainable, and renewable energy sources has been on the rise due to carbon emission, exhaustion of fossil fuels, and increase in environmental pollution [1–8]. However, clean energy research can only progress with the simultaneous development of efficient energy storage systems.

Supercapacitors are energy devices which have high power density, operating safety, environment benignity, fast charging/discharging rate, long-term cycle stability, high energy density, faster response time and near-infinite life cycle [1–8]. They are classified into two types, namely electric double-layer capacitors (EDLCs) and pseudocapacitors. In EDLCs, charge/energy storage takes place through ion adsorption/desorption at the electrode–electrolyte interface [1]. The process is non-Faradaic; no electron transfer takes place across the electrode interface and the storage of electric charge and energy is electrostatic. However, EDLCs have limited specific capacitance due to their double-layer charge storage mechanism [2].

Pseudocapacitors, on the other hand, store energy by Faradaic battery-type oxidation–reduction reactions leading to their pseudocapacitive behavior wherein the electrical charge can be built up *via* electron transfer that produces the changes in the chemical or oxidization state of the electrochemically active materials [1, 2].

Transition metal hydroxides and oxides have been explored as electrode materials of pseudocapacitors to increase the capacitance due to their nanostructures which can store charges for long periods of time without appreciable leakage, making them suitable for electrochemical applications [1, 3]. In particular, metal hydroxides and oxides have received significant attention due to their cost-effective characteristics. Cupric hydroxides and oxides are promising materials for pseudocapacitors in view of their abundance, low cost, environmental friendly nature and high specific capacitance of 100–800 F/g [2, 3, 4, 5, 6, 9]. In addition, copper compounds can exhibit numerous nanostructures which is advantageous for energy storage [4].

Herein, the morphology and properties of $Cu(OH)_2$ and $Cu(OH)_2/CuO$ nanostructures fabricated on Cu foil *via* a simple, binderless, and additive-free liquid–solid reaction are presented. Oxidation was performed by immersing the Cu foil in $NaOH/(NH_4)_2S_2O_8$ aqueous solution. Such a facile approach to growing various oxide nanostructures using water has been applied to ZnO, resulting in hierarchical structures of nanorods, nanotubes and nanoflowers [10]. The effect of reaction time on the morphology, phase composition and crystal structure of the resulting Cu compound was investigated by scanning electron microscopy (SEM) and X-ray diffraction (XRD). Additionally, the electrochemical properties was studied by cyclic voltammetry.

2. Experimental

All chemicals used were of analytical grade and used as received without further purification. In a typical experiment, 1 × 1 cm^2 pieces of copper foil were ultrasonically cleaned in acetone, ethanol, and deionized water consecutively for 10 min. The copper substrates were then immersed in 1 M HCl solution to remove other oxides and impurities. The pre-cleaned Cu foils were then dried in air and nail enamel was used to cover one side of the substrates. To fabricate the copper compounds, the substrates were immersed in a solution containing 12 mL of 10 M NaOH, 6 mL of 1 M $(NH_4)_2S_2O_8$, and 27 mL of deionized water for 30 to 60 min. The substrates were then rinsed with ethanol and deionized water, followed by air drying. The morphology of the Cu products was investigated in a scanning electron microscope (SEM, JEOL Ltd., JSM-5310). The phase composition and crystal structure were also examined by X-ray diffraction (Shimadzu MAXima XRD-7000, Cu K radiation, λ = 0.1542 nm, 30 mA). Cyclic voltammetry was performed using the oxidized Cu substrate, Pt sheet, and silver/silver chloride (Ag/AgCl in 3.5 M KCl, Horiba) as the working, counter and reference electrode, respectively. Measurement was performed in 6 M potassium hydroxide (KOH) aqueous solution at scan rate of 2 mV.

3. Results and discussion

Figure 1 shows the optical images of the Cu foil before and after oxidation in $NaOH/(NH_4)_2S_2O_8$ aqueous solution for 30–60 min. The pre-cleaned Cu foil appeared darker brown in color compared to the sample after cleaning in HCl, which can be attributed to surface oxides and other impurities. After 30 min of reaction in $NaOH/(NH_4)_2S_2O_8$ aqueous solution, the Cu foil changed from rose gold to blue, which possibly indicates the formation of Cu compound on the surface. Prolonged reaction to 60 min resulted in the blackening of some areas of the blue Cu film. This suggests the presence of a different Cu compound.

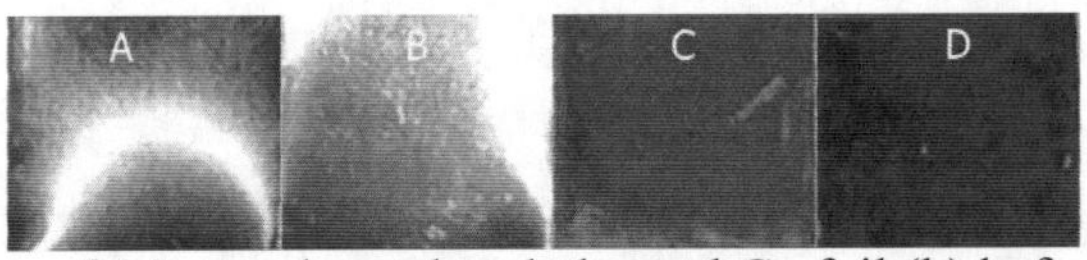

Figure 1. Optical images of (a) pre-cleaned and cleaned Cu foil (b) before and after oxidation in $NaOH/(NH_4)_2S_2O_8$ aqueous solution for (c) 30 and (d) 60 min at room temperature.

Figure 2 shows the corresponding XRD patterns of cleaned Cu foil after immersion in $NaOH/(NH_4)_2S_2O_8$ aqueous solution for 30–60 min. Three peaks at 2θ values of 43.54°, 50.66°, and 74.48° corresponding to the (111), (200), and (220) planes of Cu were observed in both Cu products

(JCPDS file no. 04–0836). In figure 2b, the diffraction peaks at the 2θ values of 16.92°, 24.1°, 34.34°, 36.12°, 38.78°, 40.02°, and 53.58° can be indexed to the (020), (021), (002), (111), (022), (130), and (150) planes of orthorhombic $Cu(OH)_2$, respectively (JCPDS file no. 80-0656). The diffraction peaks corresponding to the (111) and (200) planes of CuO were observed at 2θ values of 35.74° and 38.84° in figure 2a (JCPDS file no. 80-1916). The peaks corresponding to the (130) and (150) planes of $Cu(OH)_2$ were also observed in figure 2b at 2θ values of 40.46° and 53.6°, respectively. This shows that the change in color from blue to black of the Cu film observed in figure 1 could be due to the formation of CuO when the reaction was prolonged to 60 min.

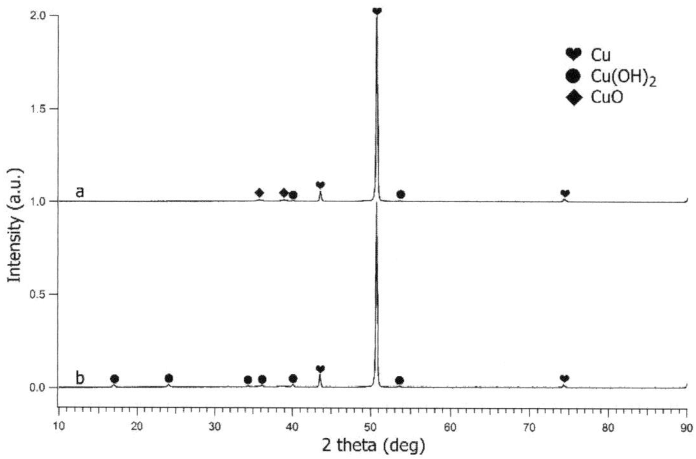

Figure 2. XRD patterns of cleaned Cu foil after immersion in $NaOH/(NH_4)_2S_2O_8$ aqueous solution for (a) 60 and (b) 30 min at room temperature.

Figure 3 shows the SEM images of the Cu products after oxidation in $NaOH/(NH_4)_2S_2O_8$ aqueous solution at room temperature. The $Cu(OH)_2$ film formed after 30 min of reaction exhibits a needle-like structure with a typical length of about 5 μm and width of 200–300 nm as shown in figures 3a–c. On the other hand, microflowers with a diameter of about 3–5 μm mixed together with $Cu(OH)_2$ needles are seen in figures 3d–f. The microflowers possibly correspond to CuO, which explains the observed change in color of some areas of the film from blue to black and the appearance of CuO peaks in the XRD pattern as shown in figure 2a.

Figure 3. SEM images of cleaned Cu foil after immersion in $NaOH/(NH_4)_2S_2O_8$ aqueous solution for (a–c) 30 and (d–f) 60 min at room temperature.

The formation of $Cu(OH)_2$ and $Cu(OH)_2/CuO$ on Cu foil by oxidation of Cu foil in alkaline solution can be understood from the equations below. At 30 min, the growth of $Cu(OH)_2$ on Cu foil is due to the redox reaction between Cu and persulfate ($S_2O_8^{2-}$) forming Cu^{2+} and sulfate (SO_4^{2-}) as shown in equation (1). Cu^{2+} and OH^- then generate μ4–OH bridges, which accompany the nucleation process and formation of $Cu(OH)_2$ [7].

$$2Cu + 8NaOH + 4(NH_4)_2S_2O_8 \rightarrow 2Cu(OH)_2 + 4Na_2SO_4 + 4NH_3\uparrow + 4H_2O \quad (1)$$

With increasing reaction time, some of the $Cu(OH)_2$ needles are split into nanopieces of CuO sheets and partially form the flower-like structure [7]. Equation (2) shows that CuO are produced due to the hydrogen linkage being unstable under oxidative conditions resulting in removal of water.

$$Cu(OH)_2 \rightarrow CuO + H_2O \quad (2)$$

The change in color of the Cu foil at 60 min reaction time and the mix of $Cu(OH)_2$ needles and CuO flowers in the SEM images in figures 3d–f can be explained by equation (3) wherein the some of the $Cu(OH)_2$ produced at 30 min by equation (1) are converted to CuO by equation (2).

$$2Cu + 8NaOH + 4(NH_4)_2S_2O_8 \rightarrow Cu(OH)_2 + CuO + 4Na_2SO_4 + 4NH_3\uparrow + 5H_2O \quad (3)$$

Transition metal oxides/hydroxides usually have relatively low power density and poor cycling stability, and the low power density can be ascribed to the poor electric conductivity of metal oxides/hydroxides, which confines electron transfer rates. However, combining metal oxides/hydroxides with conductive materials and developing porous nanostructures can accelerate the reaction kinetics and provide more ion adsorption or active sites for charge transfer reactions as well as shorten the diffusion and transfer pathways of electrolyte ions [11]. The specific capacitance values of some oxide/hydroxide pseudocapacitors have been close to or higher than their theoretical Faradic capacitances, *e.g.* >1000 F/g [3]. Figure 4 shows the CV profiles of the Cu products at a scan rate of 2 mV/s. The specific capacitance of $Cu(OH)_2$ and $Cu(OH)_2/CuO$ are 2136 F/g and 422 F/g, respectively based on equation (4):

$$C_s = \frac{1}{mv(E_2 - E_1)} \int_{E_2}^{E_2} i(E)dE \quad (4)$$

where C_s is the specific capacitance, m is the mass loading, v is the scan rate, and E_2 to E_1 is the potential window. Results show that the Cu product obtained after 30 min reaction time displayed higher capacitance.

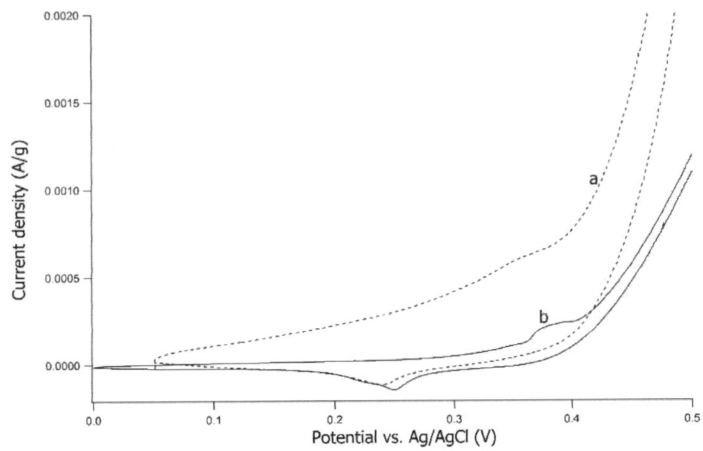

Figure 4. CV curves of the cleaned Cu foil after immersion in $NaOH/(NH_4)_2S_2O_8$ aqueous solution for (a) 30 and (b) 60 min at room temperature.

The cross needle-like architecture of $Cu(OH)_2$ grown on Cu foil provides larger electrode–electrolyte interface, thus good electrical conductivity, and have effective electrolyte-accessible channels for ion transportation and efficient redox reaction compared to CuO flowers [8]. Moreover, the needles are interconnected with each other and create pores and crevices which make a large surface area for easy diffusion of the electrolyte onto the area [9]. These resulted in the high capacitive performance observed from $Cu(OH)_2$.

2D and 3D solids can also have possible applications in pseudocapacitors as their layered structure permits the ionic or neutral species in the electrolyte to intercalate and interact with it [4]. The increase in granular size from needles to flowers may increase the active surface area and hence improve the material's electrochemical properties [12]. However, since one-dimensional morphologies like needles and rods possess the property of directional charge transport, the needle-like $Cu(OH)_2$ displays lower charge transfer resistance and smaller IR drop than the flower-like CuO which greatly enhanced its capacitive performance [2].

Energy stored in a pseudocapacitor is affected by the surface area, structure, particle size, and crystallinity of the material [4]. Therefore, utilizing nanostructured materials is of great importance in this field [8]. The fabricated needle-like $Cu(OH)_2$ and flower-like CuO mixed with $Cu(OH)_2$ microneedles on Cu foil showed high capacitive performance which presents them as potential electrode materials for pseudocapacitors.

4. Conclusion

In summary, needle-like $Cu(OH)_2$ and flower-like CuO mixed with $Cu(OH)_2$ microneedles were successfully fabricated on the surface of Cu foil by a facile oxidation reaction in alkaline solution. The synthesis was a cost-effective one-step method wherein binderless and additive-free materials were obtained. The synthesized copper compounds have unique nanostructures which enable easy interaction with the electrolyte resulting in a specific capacitance of 2136 F/g and 422 F/g for $Cu(OH)_2$ and $CuO/Cu(OH)_2$, respectively.

5. Acknowledgments

This work is supported by the National Research Council of the Philippines under the project "Electrochemical Investigation of the Pseudocapacitive Properties of Ni/NiO Nanowire Nonwoven Electrode Formed Under External Magnetic Field." Project start date: 1 June 2016.

References

[1] Yin B, Zhang S, Zheng X, Qu F and Wu X 2014 *Nano-Micro Lett.* **6** 340–6
[2] Xu P, Ye K, Du M, Lui J, Cheng K, Yin J, Wang G and Cao D 2015 *RSC Adv.* **5** 36656–64
[3] Chen K, Song S, Li K and Xue D 2013 *CrystEngComm* **15** 10367–73
[4] Nwanya A, Obi D, Ozoemena K, Osuji R, Awada C, Ruediger A, Maaza M, Rosei F, Ezema F 2016 *Electrochim. Acta* **198** 220–30
[5] Pendashteh A, Rahmanifar M, Mousavi M F 2014 *Ultrason. Sonochem.* **21** 643–52
[6] Hsu Y-K, Chen Y-C, Lin Y-G 2014 *Electrochim. Acta* **139** 401–7
[7] Liu Y, Qioa Y, Zhang W, Hu P, Chen C, Li Z, Yuan L, Hu Xianluo and Huang Y 2014 *J. Alloys Compd.* **586** 208–15
[8] Ghadge T, Lokhande B J 2016 *J. Mater. Sci.* **51** 9879–88
[9] Lamberti A, Fontana M, Bianco S, Tresso E 2016 *Int. J. Hydrogen Energy* **41** 11700–8
[10] Balela M D L, Pelicano C M O and Lockman Z 2017 *J. Mater. Sci.* **52** 2319–28
[11] Shi F, Li L, Wand X-L, Gu C-D and Tu J-P 2014 *RSC Adv.* **4** 41910–21
[12] Endut Z, Hamdi M, Basirub W J 2013 *Thin Solid Films* **528** 213–16

Carbon monoxide gas sensing using zinc oxide film deposited by spray pyrolysis

J. M. G. Leano[1], J. M. L. A. Villapando[1], A. E. Balaaldia[1], G. Gianan[2], F. K. B. Manalo[1], E. A. Florido[2]

[1]San Pablo City Science High School, San Pablo City, Philippines
[2]Institute of Mathematical Sciences and Physics, University of the Philippines Los Banos 4031, Laguna, Philippines

eaflorido@up.edu.ph

Abstract. This study was aimed to determine the carbon monoxide (CO) gas sensing ability of zinc oxide (ZnO) film fabricated by spray pyrolysis on glass substrate heated at 300^0C using 0.2 M zinc acetate precursor solution. The temperature of the precursor solution was maintained at room temperature. Carbon monoxide gas was synthesized by mixing the required amount of formic acid and excess sulfuric acid in the ratio of 1:6 to produce CO gas concentrations of 100, 200, 300, 400, and 500 parts per million (ppm) v/v. There were five trials for each concentration. The films produced exhibited good sensor characteristics such as high linearity in current voltage relationship and voltage response versus concentration. Electrical characterization using the four-point probe showed a linear relationship between current and voltage with resistivity of 0.49 ohm-cm and R^2 value of 0.994 The zinc oxide film exhibited a sensitivity of 0.19 Volt per 100 ppm of CO gas and linearity R^2 value of 0.993.

1. Introduction

Carbon monoxide (CO) is highly toxic and the danger lies in its being tasteless, odourless, and colourless so that it is not easily detectable. CO poisoning can be classified as acute, chronic or occultic. Single exposure to large doses of CO gas can cause acute CO poisoning. Repeated exposure to low levels of CO for an extended period of time cause chronic poisoning in which the harmful effects are manifested at a later time. Occultic poisoning can either be acute or chronic but the cause is misdiagnosed [1]. Death can occur if the concentration inhaled by a person is 35 ppm for over 8 hours as ascribed by the United States National Institute of Occupational Safety and Health (NIOSH). CO combines with haemoglobin to form carboxyhaemoglobin reducing the total capacity of the blood to carry oxygen [2]. It was reported by [3] that pregnant women, children, infants and individuals lost their lives due to air pollution caused by this deadly gas in Lagos, Nigeria. Since low concentrations are equally dangerous to high levels it is important to have extremely sensitive sensors for monitoring. Human activities account for artificial sources of CO (60%) in the environment and the rest are present naturally (40%) [1]. Common sources of CO include motor vehicles, household fires, and appliances that use carbon based fuels[2,4] Other exogenous sources of CO include as car exhaust fumes, furnaces, gas-powered engines, home water heaters, paint removers containing methylene chloride, pool heaters, smoke from all types of fire, tobacco smoke, and wood stoves [5].

Zinc oxide (ZnO) is an n-type metal oxide semiconductor (MOS) that has important electronic, chemical and physical characteristics making it a well-known commercial sensor because of the sensitivity of its properties to variations in its chemical environment. It has a high chemical sensitivity to adsorbed gases making it useful as a gas sensor [6,7].

When gas molecules interact with the semiconductor surface, properties such as resistance, conductivity and surface potential are affected. Oxygen molecules that are adsorbed on ZnO surface bind to its vacancy sites, extracting or trapping electrons from the conduction band (E_c). The electrons are conducted to the surface and become strongly attached as charged oxygen anions. This results to an increase in the surface resistance and the formation of a space-charge layer. When reducing gases such as CO react with the adsorbed oxygen CO_2 is produced with a consequent release of electrons back to the conduction and increase in conductivity of the semiconductor [8].

There are many different methods of depositing semiconductor films such as metalorganic chemical vapor deposition (MOCVD), molecular beam epitaxy (MBE), liquid phase epitaxy (LPE), sol-gel processing, spray pyrolysis, electrophoretic deposition (EPD) and successive ionic layer adhesion and reaction [9].

Spray pyrolysis is a technique in preparing thin and thick films, ceramic coatings, and powders that is simple and relatively cost-effective. It does not require high-quality substrates or chemicals. Typical spray pyrolysis equipment consists of an atomizer, precursor solution, substrate heater, and temperature controller. Spray parameters that affect the film quality include substrate temperature, precursor solution composition and temperature, and aerosol delivery [10]. Studies showed that zinc oxide with hexagonal wurtzite-type structure can be produced by spray pyrolysis on substrates with temperatures above 500 K [11,12]. Zinc acetate is a good substitute for tin oxide (SnO_2) and is the preferred precursor because of its high vapor pressure [11,13,14,15]. Krunks and Mellikov demonstrated that for solar cell applications it is preferable to use growth temperatures of 625-675 K, indium doping levels of 1-1.5 at.% and precursor concentrations of 0.1-0.2 mol/li to produce ZnO films with good optical and structural qualities [9].

2. Methodology

2.1. Preparation of materials

Glass substrates were prepared with dimensions of 2.0 cm by 0.4 cm and ultrasonically cleaned by ethanol for 15 minutes and dried by air. Commercially available plastic perfume atomizers (60 mL) were cleaned with soap and water and washed with distilled water then ethanol. An improvised enclosure made of card board was made with dimensions of 29 cm x 40.5 cm x 25 cm for the spray pyrolysis set-up.

Zinc acetate dihydrate powder was mixed with distilled water in a concentration of 0.2 M. The solution was stirred using a magnetic stirrer for 60 minutes and then transferred to an atomizer.

2.2. Film fabrication

The spray pyrolysis set up consisted of an atomizer, zinc acetate dihydrate solution, and substrate heater maintained at 300^0C. The solution maintained at room temperature was sprayed intermittently on the heated glass substrate using an atomizer for a total duration of 30 minutes. There was an interval of five seconds between successive sprays. Three replications for each trial was made totalling fifteen films for the five different gas concentrations to be tested. The films were annealed at 300^0C after deposition.

2.3. Film characterization

The I-V characteristics of the films were determined using the four-point probe while the morphology of the film was examined by scanning atomic force microscopy.

2.4. Gas sensing

The sensor was connected as one arm of a Wheatstone bridge supplied with 5 V DC power to test the gas response of the film sensor. The sensor was placed inside an airtight container of known volume filled with varying amounts of carbon monoxide. The concentrations used were 100, 200, 300, 400, and 500 ppm v/v. which were generated by mixing the corresponding amounts of formic acid and sulphuric acid in a beaker inside the gas chamber using a magnetic stirrer. Three trials were conducted per concentration.

3. Data and Results

The atomic force microscopy (AFM) image of the zinc oxide film is shown in Figure 1-A. The histogram of the AFM image showed high variations on vertical dimensions suggesting elongated structure that is consistent with nanorod-like elongated structures. The structure consisted of rod-like structures as well as nanowires. The large surface area offered by the nanorods and nanowires is beneficial to the function of the film sensor since gas sensing is a surface phenomenon. This result is supported by previous studies that showed the formation of nanorods when the substrate temperature is above 500 K [8,9]. The substrate temperature used in this study was 573 K.

Electrical characterization using the four-point probe showed a linear relationship between current and voltage with resistivity of 0.49 ohm-cm and R^2 value of 0.994, as shown in Figure 1-B

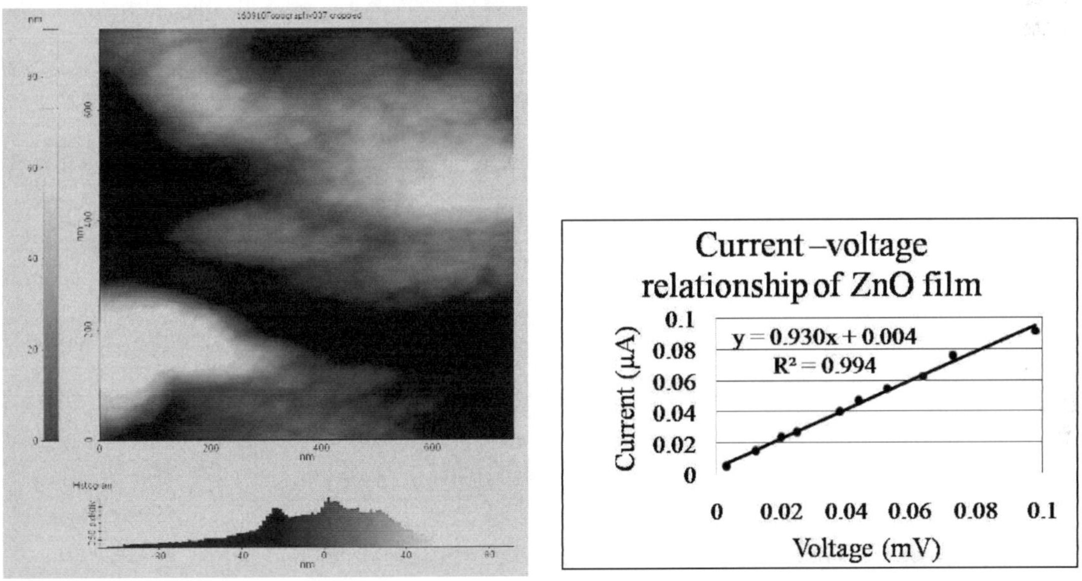

A B

Figure 1. A. Atomic microscopy (AFM) image of the zinc oxide film showing nanorods and nanowires at 10k (a) and 20k magnification. **B.** IV curve of the zinc oxide film showing linear relationship between current and voltage.

Figure 2 shows the graph of voltage response of the ZnO film with time. The maximum change in voltage output of the film was observed at 200 seconds after which the gas was released. The vertical shift after 200 seconds indicates a fast recovery of the film at 90% followed by a slower recovery (90 to 100%) up to 600 seconds. The voltage response was calculated from this graph using the difference between the initial voltage reading before exposure to gas of the film and the lowest final voltage reading

Figure 3 shows the voltage response of the ZnO film sensor to different carbon monoxide concentrations. Each data point is the average of the voltage response from each of the three replicates.

Results showed a sensitivity of 0.19 volts per 100 ppm of CO concentration with high linearity and R^2 value of 0.994.

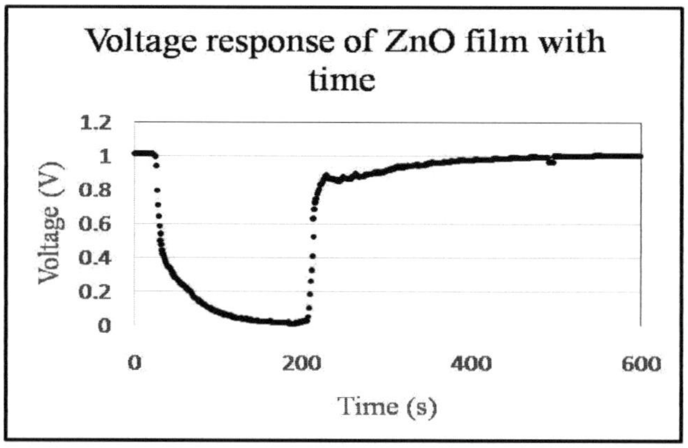

Figure 2. Graph of voltage response of the ZnO film with time with maximum voltage output at 200 seconds after which the gas was released. The vertical shift after 200 seconds indicates a fast 90% recovery of the film followed by a slower recovery (90 to 100%) up to 600 seconds.

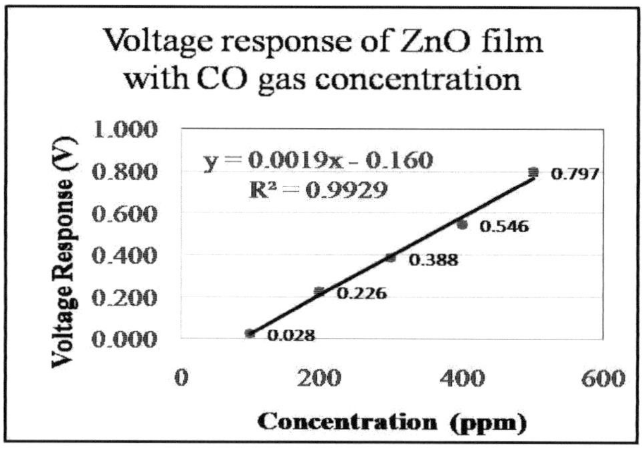

Figure 3. Voltage response of the ZnO film with concentration in ppm (v/v).

4. Conclusion

This study was able to show that good quality zinc oxide thin films can be produced using low cost method and materials. The 0.2 M Zinc acetate precursor solution at room temperature was sprayed intermittently for 30 seconds on a 300^0C glass substrate with an interval of five seconds between sprays. The films produced exhibited good sensor characteristics such as high linearity in current voltage relationship and voltage response versus concentration. Electrical characterization using the four-point probe showed a linear relationship between current and voltage with resistivity of 0.49 ohm-cm and R^2 value of 0.994. The zinc oxide film exhibited a sensitivity of 0.19 Volt per 100 ppm of CO gas and R^2 value of 0.993.This would be beneficial to the production of low-cost sensor instruments with cheap replaceable sensors. It is also recommended to test the reusability of the sensor produced by the pyrolysis method and the repeatability of the results.

Acknowledgements

The authors wish to thank the Philippine Council for Industry, Energy and Emerging Technology Research and Development (PCIEERD) Department of Science and Technology (DOST) for the project funds, and the Institute of Mathematical Sciences and Physics (IMSP) University of the Philippines Los Banos (UPLB) as implementing agency of the research project.

References

[1] Wright J 2002 Chronic and occult carbon monoxide poisoning: we don't know what we're missing *Emerg Med J.* **19** 386–390

[2] Meredith T and Vale A 1988 Carbon monoxide poisoning *British Med. J.* 77-79

[3] John K S and Feyisayo K 2013 Air pollution by carbon monoxide poisonous gas (CO) in Lagos area Southwestern Nigeria *Atmospheric and Climate Sciences* **3** 510-514

[4] Myers R, Linberg S, Cowley R 1979 Carbon monoxide poisoning: The injury and its treatment. *JACEP* **8** 479-484

[5] Varon J and Marik Paul 1997 Carbon monoxide poisoning *The Internal Journal of Emergency and Intensive Care Medicine* **1** **N2:** http://www.ispub.com/journals/IJEICM/Vol 1N2/CO.htm

[6] Weng J ,et al. 2004 *Thin Solid Films* **478** 25

[7] Wellings J, Chaure N, Heavens S, and Dharmadas I 2008 *Thin Solid Films* **516** 3893

[8] Chakraborty S, Sen A and Maiti H S 2006 *Sensors and Actuators B* **115** 610–613

[9] Hsu C, Chen K, Tsai T and Hsueh T 2013 *Sensors and Actuators B* **182** 190

[10] Perednis D and Gauckler L J 2005 Thin film deposition using spray pyrolysis *J. of Electroceramics* **14** 103-111

[11] Krunks M and Enn E 1995 Zinc oxide thin films by the spray pyrolysis method *Thin Solid Films* 270 33-36

[12] Hassan H S, Kashyout A B, Morsi I, and Nasser A A A 2014 Synthesis, characterizaton and fabrication of gas sensor devices using ZnO and ZnO:In *Beni-Suef Uni. J. of Basic and App. Sci.* **3** 216-221

[13] Caillaud F, Smithand A, and Baumard J F 1990 *J. Eur. Ceram. Sm.* **6** 313

[14] Ayouchi R, F. Martin, Leinena D, and Ramos-Barrado J R 2003 Growth of pure ZnO thin films prepared by chemical spray pyrolysis on silicon *J. of Crystal Growth* **247** 497–504

[15] Nunes P, Fernandes B, Fortunato E, Vilarinho P, and Martins R 1999 Performances presented by zinc oxide thin films deposited by spray pyrolysis *Thin Solid Films* **337** 176-179